GRAPHENE, CARBON NANOTUBES, and NANOSTRUCTURES

Techniques and Applications

Devices, Circuits, and Systems

Series Editor
Krzysztof Iniewski
CMOS Emerging Technologies Inc., Vancouver, British Columbia, Canada

GRAPHENE, CARBON NANOTUBES, and NANOSTRUCTURES

Techniques and Applications

Edited by
James E. Morris • Kris Iniewski

CRC Press
Taylor & Francis Group
Boca Raton London New York

CRC Press is an imprint of the
Taylor & Francis Group, an **informa** business

CRC Press
Taylor & Francis Group
6000 Broken Sound Parkway NW, Suite 300
Boca Raton, FL 33487-2742

First issued in paperback 2017

© 2013 by Taylor & Francis Group, LLC
CRC Press is an imprint of Taylor & Francis Group, an Informa business

No claim to original U.S. Government works
Version Date: 2012924

ISBN 13: 978-1-138-07728-7 (pbk)
ISBN 13: 978-1-4665-6056-7 (hbk)

Library of Congress Cataloging-in-Publication Data

Graphene, carbon nanotubes, and nanostuctures : techniques and applications / editors, James E. Morris, Kris Iniewski.
    p. cm. -- (Devices, circuits, and systems)
  Includes bibliographical references and index.
  ISBN 978-1-4665-6056-7 (hardback)
    1. Nanostructured materials. 2. Graphene 3. Nanotubes. I. Morris, James E. II. Iniewski, Krzysztof.

TA418.9.N35G74 2013
621.3815--dc23   2012036917

**Visit the Taylor & Francis Web site at
http://www.taylorandfrancis.com**

**and the CRC Press Web site at
http://www.crcpress.com**

Contents

About the Editors

James (Jim) Morris is an ECE professor at Portland State University, Oregon, and professor emeritus at SUNY–Binghamton; he has served as a department chair at both and is an IEEE fellow. He has BSc and MSc degrees in physics from the University of Auckland, New Zealand, and a PhD in Electrical Engineering from the University of Saskatchewan, Canada, and was the first director of Binghamton's Institute for Research in Electronics Packaging. Jim has served as treasurer of the IEEE Components Packaging and Manufacturing Technology (CPMT) Society (1991–1997), BoG member (1996–1998, 2011–2013), VP for conferences (1998–2003), distinguished lecturer (2000–present), *CPT Transactions* associate editor (1998–present), and IEEE Nanotechnology Council representative (2007–present), and won the 2005 CPMT David Feldman Outstanding Contribution Award. He has edited four books on electronics packaging, including one on nanopackaging, coauthored another, and has two more in preparation. He established the Nanotechnology Council Nanopackaging TC, serves as the NTC awards chair, and contributes to *IEEE Nanotechnology Magazine*. He was general conference chair of Adhesives in Electronics (1998), Advanced Packaging Materials (2001), Polytronic (2004), and IEEE NANO (2011), and serves on multiple conference program committees, e.g., as chair of the nanopackaging program committee for IEEE NANO 2012. His research is currently focused on integrated circuits, nanoelectronics, and nanoelectronics packaging. He is actively involved in international engineering education, and at the local IEEE Oregon level has chaired both the Education Society and CPMT chapters. He spent 2008–2009 in visiting positions with the University of Greenwich, Chalmers University of Technology, Dresden University of Technology, Helsinki University of Technology (with a Nokia-Fulbright Fellowship), and University of Canterbury, New Zealand (on an Erskine Fellowship), and held a Royal Academy of Engineering Distinguished Visiting Fellowship at Loughborough University of Technology (UK) last year.

Krzysztof (Kris) Iniewski is managing R&D at Redlen Technologies, Inc., a startup company in Vancouver, Canada. Redlen's revolutionary production process for advanced semiconductor materials enables a new generation of more accurate, all-digital, radiation-based imaging solutions. Kris is also a president of CMOS Emerging Technologies (www.cmoset.com), an organization of high-tech events covering communications, microsystems, optoelectronics, and sensors. In his career Dr. Iniewski has held numerous faculty and management positions at University of Toronto, University of Alberta, SFU, and PMC-Sierra, Inc. He has published over 100 research papers in international journals and conferences. He holds 18 international patents granted in the United States, Canada, France, Germany, and Japan. He is a frequent invited speaker and has consulted for multiple organizations internationally. He has written and edited several books for IEEE Press, Wiley, CRC

Press, McGraw Hill, Artech House, and Springer. His personal goal is to contribute to healthy living and sustainability through innovative engineering solutions. In his leisurely time Kris can be found hiking, sailing, skiing, or biking in beautiful British Columbia. He can be reached at kris.iniewski@gmail.com.

Contributors

Matteo Altissimo
Melbourne Centre for Nanofabrication
Clayton, Victoria, Australia

Hernán L. Calvo
Institut für Theorie der Statistischen
 Physik
RWTH Aachen University
Aachen, Germany

Yang Chai
Department of Applied Physics
Hong Kong Polytechnic University
Hong Kong

Philip C.H. Chan
Department of Electronic and
 Information Engineering
Hong Kong Polytechnic University
Hong Kong

Chih-hung Chang
Oregon Process Innovation Center
Microproducts Breakthrough Institute
School of Chemical, Biological and
 Environmental Engineering
Oregon State University
Corvallis, Oregon

Antao Chen
Applied Physics Laboratory and
 Electrical Engineering
University of Washington
Seattle, Washington
and
Department of Physics
University of South Florida
Tampa, Florida

Yu-Mo Chien
Department of Electrical and Computer
 Engineering
McGill University
Montreal, Canada, Quebec

Tomasz Fałat
Faculty of Mikrosystem Electronics and
 Photonics
Wrocław University of Technology
Wrocław, Poland

Jan Felba
Faculty of Mikrosystem Electronics and
 Photonics
Wrocław University of Technology
Wrocław, Poland

Luis E.F. Foa Torres
Instituto de Física Enrique Gaviola
 (IFEG)–CONICET
FaMAF
Universidad Nacional de Córdoba
Ciudad Universitaria
Córdoba, Argentina

Ricardo Izquierdo
Génie Microélectronique
Département d'Informatique
Université du Québec à Montréal
Montréal, Quebec

Jason Johnson
Department of Electrical and Computer
 Engineering
University of Florida
Gainesville, Florida

Toru Kato
Department of Advanced Materials
    Science
Graduate School of Frontier Sciences
University of Tokyo
Chiba, Japan

Jeffry Kelber
Department of Chemistry
University of North Texas
Denton, Texas

Neerav Kharche
Computational Center for
    Nanotechnology Innovations
Department of Physics, Applied Physics
    and Astronomy
Rensselaer Polytechnic Institute
Troy, New York

Yuan Li
Department of Electronic and Computer
    Engineering
Hong Kong University of Science and
    Technology
Hong Kong

Aihua Liu
Laboratory for Biosensing
Qingdao Institute of Bioenergy and
    Bioprocess Technology
Chinese Academy of Sciences
Qingdao, China

Haitao Liu
Department of Chemistry
University of Pittsburgh
Pittsburgh, Pennsylvania

Antonio Maffucci
Department of Electrical Engineering
University of Cassino and Southern Lazio
Cassino, Italy

Sergey A. Maksimenko
Institute for Nuclear Problems
Belarus State University
Minsk, Belarus

C. Forestiere Giovanni Miano
Department of Electrical Engineering
University of Naples Federico II
Naples, Italy

Andrzej Mościcki
Amepox Microelectronics, Ltd.
Lodz, Poland

Sho Nakahara
Department of Advanced Materials
    Science
Graduate School of Frontier Sciences
University of Tokyo
Chiba, Japan

Saroj K. Nayak
Department of Physics, Applied Physics
    and Astronomy
Rensselaer Polytechnic Institute
Troy, New York

Thanh C. Nguyen
Centre for Neural Engineering
University of Melbourne
and
National ICT Australia
Victorian Research Laboratory
Parkville, Victoria, Australia

Horacio M. Pastawski
Instituto de Física Enrique Gaviola
    (IFEG)–CONICET
FaMAF
Universidad Nacional de Córdoba
Ciudad Universitaria
Córdoba, Argentina

Brian K. Paul
Oregon Process Innovation Center
Microproducts Breakthrough Institute
School of Mechanical, Industrial and
    Manufacturing Engineering
Oregon State University
Corvallis, Oregon

Huajun Qiu
Laboratory for Biosensing
Qingdao Institute of Bioenergy and
    Bioprocess Technology
and
Key Laboratory for Biofuels
Chinese Academy of Sciences
Qingdao, China

Wanzhi Qiu
Centre for Neural Engineering
University of Melbourne
and
National ICT Australia
Victorian Research Laboratory
Parkville, Victoria, Australia

Stephan Roche
CIN2 (ICN-CSIC)
Catalan Institute of Nanotechnology
Universidad Autónoma de Barcelona
and
Institució Catalana de Recerca i Estudis
    Avançats (ICREA)
Barcelona, Spain

Si-Ok Ryu
School of Chemical Engineering
Yeungnam University
Gyeongbuk, Republic of Korea

Takehiko Sasaki
Department of Complexity Science and
    Engineering
Graduate School of Frontier Sciences
University of Tokyo
Chiba, Japan

Efstratios Skafidas
Centre for Neural Engineering
University of Melbourne
and
National ICT Australia
Victorian Research Laboratory
Parkville, Victoria, Australia

G.Ya. Slepyan
School of Electrical Engineering
    Faculty of Engineering
Department of Physical Electronics,
Tel Aviv University
Tel Aviv, Israel

Anita Smolarek
Amepox Microelectronics, Ltd.
Lodz, Poland

Paul G. Spizzirri
Melbourne Centre for Nanofabrication
Clayton, Victoria, Australia

Sven Stauss
Department of Advanced Materials
    Science
Graduate School of Frontier Sciences
University of Tokyo
Chiba, Japan

Minghui Sun
Department of Electronic and Computer
    Engineering
Hong Kong University of Science and
    Technology
Hong Kong

Sumedh P. Surwade
Department of Chemistry
University of Pittsburgh
Pittsburgh, Pennsylvania

Kazuo Terashima
Department of Advanced Materials
    Science
Graduate School of Frontier Sciences
University of Tokyo
Chiba, Japan

Ant Ural
Department of Electrical and Computer
    Engineering
University of Florida
Gainesville, Florida

Danling Wang
Applied Physics Laboratory and
    Electrical Engineering
University of Washington
Seattle, Washington

Laurens H. Willems van Beveren
School of Physics
University of Melbourne
Parkville, Victoria, Australia

Zhiyong Xiao
Department of Electronic and Computer
    Engineering
Hong Kong University of Science and
    Technology
Hong Kong

Huikai Xie
Department of Electrical and Computer
    Engineering
University of Florida
Gainesville, Florida

Min Zhang
School of Computer & Information
    Engineering
Shenzhen Graduate School
Peking University
China

Ying Zhou
Department of Electrical and Computer
    Engineering
University of Florida
Gainesville, Florida

1 Carbon Nanotubes
From Electrodynamics to Signal Propagation Models

Antonio Maffucci, Sergey A. Maksimenko,
Giovanni Miano, and G.Ya. Slepyan

CONTENTS

1.1 INTRODUCTION

Carbon nanotubes (CNTs) are recently discovered materials [1] made by rolled sheets of graphene of diameters of the order of nanometers and lengths up to millimeters. Because of their outstanding electrical, thermal, and mechanical properties [2,3], CNTs are proposed as *emerging* materials in nanoelectronics [4,5], for fabricating nano-interconnects [6–8], nanopackages [9], nano-transistors [10], nano-passives [11], and nano-antennas [12,13].

Recently, the theoretical predictions have been confirmed by the first real-world CNT-based electronic devices, like the CNT bumps for nanopackaging applications presented in [14] or the CNT wiring of a prototype of a digital integrated circuit, one of the first examples of successful CNT-CMOS integration [15,16].

Given these perspectives, many efforts have been made in literature to derive models able to describe the electrical propagation along carbon nanotubes. The electromagnetic response of carbon nanotubes has been widely examined in frequency ranges from microwave to the visible, taking properly into account the graphene crystalline [17,18]. This requires, in principle, a quantum mechanical approach, as done in [19], where the model is derived by using numerical simulations based on first principles. Alternatively, phenomenological approaches are possible like those proposed in [20], based on the Luttinger liquid theory. Another possible way is given by semiclassical approaches, based on simplified models that yield approximated but analytically tractable results. An example is given by the imposing boundary value problems for Maxwell equations transformed into integral equations [12,18,21], which are numerically solved. A second possibility consists in modeling the transport in the frame of the transmission line (TL) theory: Examples are given by [21], where the CNT is modeled as an electron waveguide, or by [22–26].

Section 1.2 presents the transport equation for carbon nanotubes, first considering isolated CNTs and then considering the case of interacting CNT shells, when the tunneling effect is considered. The transport equations are derived from an electrodynamical model that describes the behavior of the conduction electrons (the so-called π-electrons and by the Maxwell equations). An isolated carbon nanotube shell is described as a quantum wire, that is, a structure with transverse dimensions comparable to the characteristic de-Broglie wavelength of the π-electrons. In the low-frequency regime only intraband transitions of π-electrons with unchanged transverse quasi-momentum are allowed. The intraband transition contributes to the axial conductivity, but not to the transverse conductivity, and excites an azimuthally symmetric electric current density, which leads to an experimentally observed conductivity peak in the terahertz frequency range [27]. In this hypothesis, a simple but physically meaningful transport model may be obtained by describing the electron cloud as if it were a fluid moving under the action of an external field, assuming that the electric fields, due to their collective motion and to the external sources, are smaller than the atomic crystal field and also slowly varying on atomic length and timescales. In these conditions the π-electrons behave as quasi-classical particles and the equations governing their dynamics are the classical equations of motion, provided that the electron mass is replaced by an effective mass, which endows the interaction with the positive ion lattice [21,28].

Under the same conditions, the transport model may be obtained by solving the Boltzmann transport equation in quasi-classical limit, which is the approach reported in Section 1.2 and described in detail in [29]. A similar approach has been adopted, for instance, in [30]. A major feature of this model is the evaluation of the longitudinal dynamic conductivity of the CNT, expressed in terms of the *number of effective conducting channels*, which depends on CNT chirality, size, and temperature [29–31]. This number estimates the subbands passing near or through the nanotube Fermi level, which effectively contribute to the conduction.

The above model is useful to describe single-wall carbon nanotubes (SWCNTs), multiwall carbon nanotubes (MWCNTs), and bundles made by them, in the assumptions that the tunneling currents between CNTs are negligible. This is the case for the MWCNT models and the CNT bundle models presented, for instance, in [11,32,33].

When operating frequencies reach the terahertz range, the tunneling effect may be no longer negligible. The transport model is then modified in Section 1.2 to account for this phenomenon, following the stream of what was done in [34], based on the density matrix formalism and Liouville's equation [35].

In Section 1.3 the transport models are used to derive circuital representation for the propagation of an electrical signal along CNT interconnects, in the frame of the TL theory, in the transverse electromagnetic (TEM) hypothesis of propagation, assuming low-frequency and low-voltage bias conditions.

The TL model for CNT interconnects is formally given by a lossy resistance-inductance-capacitance (RLC) model, where the per unit length (p.u.l.) inductance and capacitance exhibit two new terms, besides the classical electromagnetic ones: a *kinetic inductance* and a *quantum capacitance*. The kinetic inductance takes into account the effects of the π-electron inertia, whereas the quantum capacitance describes the quantum pressure arising from the zero-point energy of the π-electrons (e.g., [23–26]). When the intershell tunneling is considered, the TL equations exhibit spatial dispersion, which may significantly affect the quality of the signals [34].

Finally, Section 1.4 shows some examples of CNT interconnects. Realistic scenarios of on-chip interconnects are considered, and performance analysis is carried out, comparing conventional copper realization with innovative SWCNT or MWCNT ones.

1.2 ELECTRODYNAMICS OF CARBON NANOTUBES

1.2.1 BAND STRUCTURE OF A SINGLE CNT SHELL

A carbon nanotube is made by rolled-up sheets of a monoatomic layer of graphite (graphene), whose lattice is depicted in Figure 1.1a. In the direct space the nanotube unit cell is the cylindrical surface generated by:

1. The *chiral* vector $\mathbf{C} = n\mathbf{a_1} + m\mathbf{a_2}$, where n and m are integers, $\mathbf{a_1}$ and $\mathbf{a_2}$ are the basis vectors of the graphene lattice, of length $|\mathbf{a_1}| = |\mathbf{a_2}| = a_0 = \sqrt{3}b_0$, with $b_0 = 0.142$ nm being the interatomic distance.
2. The *translational* vector $\mathbf{T} = t_1\mathbf{a_1} + t_2\mathbf{a_2}$, of length T, with $t_1 = (2m+n)/d_R$, $t_2 = -(2n+m)/d_R$, and $d_R = \gcd[(2m+n),(2n+m)]$ (gcd is the greatest common divisor). In the Cartesian system (x, y) with the origin at the center of a graphene hexagon and the x axis oriented along the hexagon side, the coordinates of $\mathbf{a_1}$ and $\mathbf{a_2}$ are given by $\mathbf{a_1} = (\sqrt{3}a_0/2, a_0/2)$ and $\mathbf{a_2} = (\sqrt{3}a_0/2, -a_0/2)$.

A CNT shell is obtained by rolling up the graphene sheet in such a way that the circumference of the tube is given by the *chiral* vector $\mathbf{C}$, perpendicular to the axis of the tube. The carbon nanotube radius r_c is therefore given by

$$r_c = \frac{a_0}{2\pi}\sqrt{n^2 + nm + m^2} \qquad (1.1)$$

Nanotubes with $n = 0$ (or $m = 0$) are called *zigzag* CNTs, those with $n = m$ are the *armchair* CNTs, and those with $0 < n \neq m$ are the *chiral* CNTs.

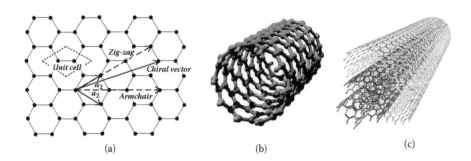

FIGURE 1.1 (a) The unrolled lattice of a carbon nanotube: *lattice basis vectors* of graphene, the *unit cell* of graphene, and the *chiral vector* of the tube graphene lattice. (b) A SWCNT. (c) A MWCNT.

A single-wall carbon nanotube (SWCNT) is made by a single shell (Figure 1.1b), usually with a radius of the order of fractions or of few nanometers. Instead, a multiwall carbon nanotube (MWCNT) is made by several nested shells (Figure 1.1c), with a radius ranging from tens to hundreds of nanometers, separated by the van der Waals distance $\delta = 0.34$ nm.

The graphene layer is a zero-gap semiconductor; thus the number density of the conduction electrons is equal to zero when the absolute temperature is zero. Nevertheless, when a graphene layer is rolled up, it may become either metallic or semiconducting, depending on its geometry (e.g., [2,3]). The general condition to obtain a metallic CNT is $|n - m| = 3q$, where $q = 0, 1, 2, \ldots$; therefore armchair CNTs are always metallic, whereas zigzag CNTs are metallic only if $m = 3q$ with $q = 1, 2,$ .... In all the other cases the CNT behaves as a semiconductor. Statistically, all the chiralities have the same probability, and thus in a population of CNT shells one-third of the total number are metallic and two-thirds are semiconducting. Finally, we have to note that for $m, n \to \infty$ (that is, for $r_c \to \infty$) any carbon nanotube shell tends to the graphene sheet.

To study the band structure of a generic CNT shell, let us analyze the *graphene reciprocal lattice* (see Figure 1.2). In this space we consider the Cartesian coordinate system (k_x, k_y) having the origin at the center of a hexagon, the k_y axis oriented along the hexagon side, and the k_x axis orthogonal to k_y. The basis vectors for the reciprocal lattice are given by $\mathbf{b}_1\left(2\pi/\sqrt{3}a_0, 2\pi/a_0\right)$ and $\mathbf{b}_2\left(2\pi/\sqrt{3}a_0, -2\pi/a_0\right)$.

The first Brillouin zone of a CNT shell is the set $S = \{s_1, s_2, \ldots, s_N\}$ of N parallel segments generated by the orthogonal basis vectors $\mathbf{K}_1 = \left(-t_2\mathbf{b}_1 + t_1\mathbf{b}_2\right)/N$ and $\mathbf{K}_2 = m\mathbf{b}_1 + n\mathbf{b}_2/N$ (Figure 1.4). The $\mathbf{K}_\mu$ points of the segment s_μ are given by

$$\mathbf{K}_\mu(k) = k\frac{\mathbf{K}_2}{|\mathbf{K}_2|} + \mu\mathbf{K}_1 \text{ for } -\frac{\pi}{T} < k \leq \frac{\pi}{T} \text{ and } \mu = 0, 1, \ldots, N-1 \qquad (1.2)$$

The vectors $\mathbf{K}_1$ and $\mathbf{K}_2$ are related to the direct lattice basis vectors $\mathbf{C}$ and $\mathbf{T}$ through: $\mathbf{C}\cdot\mathbf{K}_1 = 2\pi$, $\mathbf{T}\cdot\mathbf{K}_1 = 0$, $\mathbf{C}\cdot\mathbf{K}_2 = 0$, and $\mathbf{T}\cdot\mathbf{K}_2 = 2\pi$; therefore $|\mathbf{K}_1| = 1/r_c$ and $|\mathbf{K}_2| = 2\pi/L$. The first Brillouin zone is equivalent to that of N one-dimensional

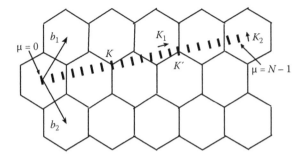

FIGURE 1.2   The reciprocal graphene lattice and the first Brillouin zone referred to a CNT shell.

systems with the same length L. The position of the middle point of s_μ is given by $\mu \mathbf{K}_1$ and the distance between two adjacent segments is $\Delta k_\perp = 1/r_c$. The longitudinal wave vector k is almost continuous assuming the CNT shell length to be large compared with the length of the unit cell; on the contrary, the transverse wave vector $k_\perp$ is quantized: $\mu \Delta k_\perp$ with $\mu = 0,1,\ldots,N-1$.

In the zone-folding approximation the dispersion relation for the SWCNT consists of $2N$ one-dimensional energy subbands given by

$$E_\mu^{(\pm)}(k) = E_g^{(\pm)}\left(k\frac{\mathbf{K}_2}{|\mathbf{K}_2|}+\mu\mathbf{K}_1\right) \text{ for } -\frac{\pi}{L} < k \le \frac{\pi}{L} \text{ and } \mu = 0,1,\ldots,N-1 \qquad (1.3)$$

where the function $E_g^{(\pm)}(\cdot)$ is the dispersion relation of the graphene layer, which is given, in the nearest-neighbor tight-binding approximation, by the following expression:

$$E_g^{(\pm)}(\mathbf{k}) = -\gamma\left[1+4\cos\left(\frac{\sqrt{3}k_x a_0}{2}\right)\cos\left(\frac{k_y a_0}{2}\right)+4\cos^2\left(\frac{k_y a_0}{2}\right)\right]^{1/2} \qquad (1.4)$$

The conduction and the valence energy band of the π-electrons are obtained by putting $+$ and $-$ in (1.4), respectively. Here $\gamma = 2.7$ eV is the carbon-carbon interaction energy. The valence and conduction bands of the graphene touch at the graphene Fermi points. Only the energy subbands that pass through or are close to the Fermi level contribute significantly to the nanotube axial electric current, as will be pointed out in the next section.

1.2.2   TRANSPORT EQUATION FOR A SINGLE CNT SHELL

The dynamics of the π-electrons in the μth subband are described by the distribution function $f_\mu^{(\pm)} = f_\mu^{(\pm)}(z,k,t)$, which satisfies the quasi-classical Boltzmann equation [34]:

$$\frac{\partial f_\mu^{(\pm)}}{\partial t} + v_\mu^{(\pm)} \frac{\partial f_\mu^{(\pm)}}{\partial z} + \frac{e}{\hbar} E_z \frac{\partial f_\mu^{(\pm)}}{\partial k} = -v\left(f_\mu^{(\pm)} - f_{0,\mu}^{(\pm)}\right) \tag{1.5}$$

where e is the electron charge, $\hbar$ is the Planck constant, $E_z = E_z(z,t)$ is the longitudinal component of the electric field at the nanotube surface, $v_\mu^{(\pm)}(k) = dE_\mu^{(\pm)}/d(\hbar k)$ is the longitudinal velocity, v is the relaxation frequency, and $f_{0,\mu}^{(\pm)}(k) = F\left[E_\mu^{(\pm)}(k)\right]/2\pi^2 r_c$ is the equilibrium distribution function, where $F[E]$ is the Dirac-Fermi distribution function with electrochemical potential equal to zero:

$$F[E] = \frac{1}{e^{E/k_B T} + 1} \tag{1.6}$$

in which k_B is the Boltzmann constant and T is the nanotube absolute temperature.

Assuming a time-harmonic regime for the electric field and the surface current density

$$E_z(z,t) = \mathrm{Re}\left\{\hat{E}_z e^{i(\omega t - \beta z)}\right\}, \quad J_z(z,t) = \mathrm{Re}\left\{\hat{J}_z e^{i(\omega t - \beta z)}\right\} \tag{1.7}$$

the constitutive equation for the CNT shell may be written as

$$\hat{\sigma}_{zz}(\beta,\omega)\,\hat{J}_z = \hat{E}_z \tag{1.8}$$

having introduced the CNT longitudinal conductivity $\hat{\sigma}_{zz}(\beta,\omega)$ in the wavenumber and frequency domain. In order to evaluate this parameter, let us assume small perturbations of the distribution functions around the equilibrium values, i.e., $f_\mu^{(\pm)} = f_{0\mu}^{(\pm)} + \mathrm{Re}\left\{\delta f_{1\mu}^{(\pm)} \exp[i(\omega t - \beta z)]\right\}$. From (1.5) we obtain

$$\delta f_\mu^{(\pm)}(k,\omega) = \frac{i}{\hbar} \frac{\partial f_{0\mu}^{(\pm)}}{\partial k} \frac{e\hat{E}_z}{\omega - v_\mu^{(\pm)}\beta - iv} \tag{1.9}$$

hence $\hat{\sigma}_{zz}(\beta,\omega)$ is given by

$$\hat{\sigma}_{zz}(\beta,\omega) = \frac{ie^2}{\hbar} \sum_{\pm} \sum_{\mu=0}^{N-1} \int_{-\pi/L}^{\pi/L} \frac{\partial f_{0\mu}^{(\pm)}}{\partial k} \frac{v_\mu^{(\pm)}}{\omega - v_\mu^{(\pm)}\beta - iv} dk \tag{1.10}$$

For all the subbands that give a meaningful contribution to the conductivity we may put $v_\mu^{(+)} \cong v_F$ in the kernel of (1.10). This assumption is well founded for the Fermi level subbands of metallic shells, whereas for the other subbands it slightly overestimates the effects of the spatial dispersion.

Starting from (1.8) and (1.10), we obtain the *constitutive equation in the spatial and frequency domain*,

$$\left(\frac{i\omega}{v}+1\right)J_z = \frac{1}{v\left(\frac{v}{i\omega}+1\right)}v_F^2\frac{\partial \rho_s}{\partial z}+\sigma_c E_z \tag{1.11}$$

where $\rho_s(z,\omega)$ is the surface charge density and

$$\sigma_c = \frac{v_F}{\pi r_c v R_0}M \tag{1.12}$$

is the long wavelength static limit for the axial conductivity. In (1.12) we have introduced the *quantum resistance* $R_0 = \pi\hbar/e^2 \cong 12.9$ kΩ, the Fermi velocity $v_F \approx 0.87\cdot10^6$ m/s, the relaxation frequency v, and the *equivalent number of conducting channels*, defined as [26]

$$M = \frac{2\hbar}{v_F}\sum_{\mu=0}^{N-1}\int_0^{\pi/L}\left(v_\mu^+\right)^2\frac{dF}{dE}\bigg|_{E=E_\mu^+}dk \tag{1.13}$$

This parameter may be interpreted as the average number of subbands around the Fermi level of the CNT shell. It depends on the number of segments s_μ passing through the two circles of radius k_{eff} and centered at the two inequivalent Fermi points of graphene. Figure 1.3 shows the typical configurations obtained for metallic and semiconducting carbon nanotubes.

This number increases as the CNT radius increases, since $\Delta k_\perp = 1/r_c$. In addition, since the radius k_{eff} is a function of the absolute temperature T, M depends on temperature too; the behavior of M with temperature and CNT diameter is plotted in Figure 1.4 for metallic and semiconducting CNTs.

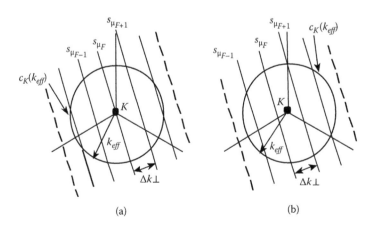

(a)   (b)

FIGURE 1.3 The segments s_μ crossing the circumference $c_k(k_{eff})$ for (a) a metallic and (b) a semiconducting CNT shell.

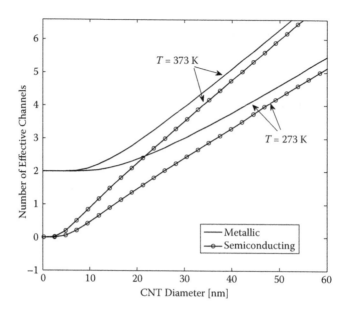

FIGURE 1.4 Equivalent number of conducting channels versus CNT shell diameter, computed at $T = 273$ K and $T = 373$ K.

The constitutive relation (1.11), which can be regarded as a nonlocal Ohm's law, may be rewritten as follows, by applying the charge conservation law and assuming the current to be uniformly distributed along the CNT contour, as shown in [21]:

$$i\omega L_K I(z,\omega) = \frac{1}{1+v/i\omega} \frac{1}{i\omega C_q} \frac{\partial^2 I(z,\omega)}{\partial z^2} + E_z(z,\omega) - RI(z,\omega) \qquad (1.14)$$

Here we have introduced the per unit length (p.u.l.) *kinetic inductance* L_K, the p.u.l. *quantum capacitance* C_q, and the p.u.l. resistance R, defined as

$$L_k = \frac{R_0}{2v_F M}, \quad C_q = \frac{2M}{v_F R_0}\left(1+\frac{v}{i\omega}\right), \quad R = v L_K = \frac{v R_0}{2v_F M} \qquad (1.15)$$

Equation (1.14) may be regarded as a balance of the momentum of the π-electrons and represents their transport equation: The term on the lefthand side represents the electron inertia. The first term on the righthand side represents the quantum pressure arising from the zero-point energy of the π-electrons, the second term describes the action of the collective electric field, and the third one is a relaxation term due to the collisions. Note that the relaxation frequency may be expressed as

$$v = \frac{v_F}{l_{mfp}} \qquad (1.16)$$

where l_{mfp} is the mean free path of the electrons. In conventional conductors, l_{mfp} is of the order of some nanometers, and therefore it is $v \to \infty$; hence (1.11) becomes a local relation. The mean free path of CNTs, instead, may extend up to the order of micrometers; therefore the range of nonlocality is relatively large. An approximated expression for the mean free path is provided in Section 1.3.

The expressions of parameters (1.15) generalize those currently used in literature. We have introduced a correction factor $(1+v/i\omega)$ in C_Q, taking into account the dispersive effect introduced by the losses, and the possibility to account for the proper number of channels. Assuming metallic CNT shells with small radius, it is $M = 2$, and so neglecting the above correction factor, from (1.15) we derive the expressions commonly adopted in literature for metallic SWCNTs (e.g., [23,24]):

$$L_k = \frac{R_0}{4v_F}, \quad C_Q = \frac{4}{v_F R_0} \tag{1.17}$$

1.2.3 Transport Equation for Two CNTs with Intershell Tunneling

When considering a collection of CNT shells (for instance, several SWCNTs in a bundle or the nested shells of MWCNTs) operating at frequencies in the terahertz band, but below interband optical transitions, we must take into account the effect of intershell tunneling conductance. To include this effect, let us analyze the simple case of the double-wall carbon nanotube (DWCNT) depicted in Figure 1.5. The approach may be generalized to any MWCNT.

We assume the worst case, i.e., the case where this effect is maximized. Since the tunnel coupling is fast diminishing quantity with the increase of the difference between the shell circumferences C_1 and C_2, the worst case is given when the combinations of chiral indexes (m_1, n_1) and (m_2, n_2) minimize such a difference mentioned. In view of this, we assume both shells to be conducting and achiral.

In the considered frequency range, we account for two types of the electron motion: intraband motion within a shell and tunneling transitions between shells. It should be noted that tunneling transitions are only of importance in the regions of the degeneracy of energy spectra of different shells, which are in the neighborhood of the Fermi points.

Let us start with the boundary conditions for the electric field on the shell surfaces, expressed in the wavenumber domain as [34,35]

$$\hat{J}_{z1} = \hat{\sigma}_{11}(\beta,\omega)\hat{E}_{z1} + \hat{\sigma}_{12}(\beta,\omega)\hat{E}_{z2} \quad \hat{J}_{z2} = \hat{\sigma}_{21}(\beta,\omega)\hat{E}_{z1} + \hat{\sigma}_{22}(\beta,\omega)\hat{E}_{z2} \tag{1.18}$$

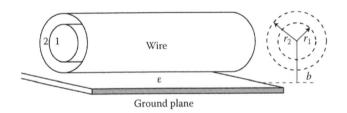

FIGURE 1.5 DWCNT above a ground plane.

where

$$\hat{\sigma}_{11}(\beta,\omega) = \frac{ie^2(\omega - i\nu)}{2\pi\hbar C_1\beta^2} \left[g(\beta,\omega) + \varphi(\beta,\omega) \right] \tag{1.19}$$

$$\hat{\sigma}_{12}(\beta,\omega) = \frac{ie^2(\omega - i\nu)}{2\pi\hbar C_1\beta^2} \left[g(\beta,\omega) - \varphi(\beta,\omega) \right] \tag{1.20}$$

are, respectively, the self and intershell conductivities referred to conductor 1. Expressions for $\hat{\sigma}_{21}(\beta,\omega)$ and $\hat{\sigma}_{22}(\beta,\omega)$ follow from (1.19) and (1.20) by substitutions: $C_1 \to C_2$, $m_1 \to m_2$, $n_1 \to n_2$.

The functions $g(\beta, \omega)$ and $\varphi(\beta, \omega)$ are evaluated from the Liouville equation solution and are given by [35]

$$g = \hbar \sum_{l,l',\mu} \int_{-\pi/T}^{\pi/T} \frac{F[E_\mu^{(l)}(k+\beta) + \hbar\omega_t^{(l')}] - F[E_\mu^{(l)}(k) + \hbar\omega_t^{(l')}]}{E_\mu^{(l)}(k+\beta) - E_\mu^{(l)}(\beta) - \hbar\omega + i\hbar\omega} dk \tag{1.21}$$

$$\varphi = \hbar \sum_{l,l',\mu} \int_{-\pi/T}^{\pi/T} \frac{F[E_\mu^{(l)}(k+\beta) + \hbar\omega_t^{(l')}] - F[E_\mu^{(l)}(k) + \hbar\omega_t^{(l')}]}{E_\mu^{(l)}(k+\beta) - E_\mu^{(l)}(\beta) + 2\hbar\omega_t^{(l')} - \hbar\omega + i\hbar\omega} dk \tag{1.22}$$

In these formulae the relaxation frequency ν is assumed to be the same for both shells, the parameters l, l' belong to the finite sets of symbols $\{-, +\}$, and $\omega_t^{(\pm)} = \pm\omega_t$, where ω_t is the *frequency of the intershell tunneling*. The summation over μ takes into account the contribution of all the subbands passing near or through the Fermi level.

Approximated expressions for (1.19) and (1.20) are obtained by using a two-term expansion of the function $E_\mu^{(\lambda)}(k+\beta)$, and using the Dirac-like dispersion law for the accountable subbands:

$$\hat{\sigma}_{11}(\beta,\omega) = -i\sigma_{C1} \frac{\nu}{\omega'(1-\hat{y})} \left(1 - \frac{\omega_t}{2} \frac{\hat{\gamma}}{\hat{y}} \right), \quad \hat{\sigma}_{12}(\beta,\omega) = -i\sigma_{C1} \frac{\nu}{\omega'(1-\hat{y})} \frac{\omega_t}{2} \frac{\hat{\gamma}}{\hat{y}} \tag{1.23}$$

where

$$\hat{\gamma}(\beta,\omega) = \frac{2\omega_t - \omega'(1+\hat{y})}{(2\omega_t - \omega')^2 - \beta^2 v_F^2} + \frac{2\omega_t + \omega'(1+\hat{y})}{(2\omega_t + \omega')^2 - \beta^2 v_F^2} \tag{1.24}$$

$$\hat{y}(\beta,\omega) = \frac{\beta^2 v_F^2}{\omega'^2}, \quad \omega' = \omega - i\nu \tag{1.25}$$

and the coefficient σ_{C1} is given by (1.12) with $M = M_1$ and $C = C_1$. For $\omega_t \to 0$ it is $\hat{\sigma}_{12} \to 0$, whereas $\hat{\sigma}_{11}$ reduces to (1.11).

To move to spatial domain, let us arrange (1.17) and (1.18) and (1.23)–(1.25), to obtain

$$\left(R+i\omega L_k\right)\mathbf{I}(z,\omega)=\frac{1}{1+\nu/i\omega}\frac{1}{i\omega C_q}\frac{\partial^2\mathbf{I}(z,\omega)}{\partial z^2}+\left(1-\frac{1}{2}\omega_t\Theta y^{-1}\gamma\right)\mathbf{E}_z(z,\omega) \quad (1.26)$$

which generalizes (1.14). Here L_k, R, and C_q are diagonal 2×2 matrices characterizing the shells of the DWCNT, whose nth entry is given by (1.15), using $M = M_n$.

The quantity Θ is the 2×2 matrix whose entries are given by $\Theta = (-1)^{i+j}$, whereas y and γ are the operator functions obtained by the substitution $\beta \rightarrow -i\partial/\partial z$ in (1.24) and (1.25). The transport of the π-electrons in the two shells is governed by Equation (1.26) coupled to the nondiagonal part of the matrix Θ.

To have an idea of the coupling, we may introduce in the wavenumber-frequency domain the ratio ζ between the mutual conductivity $\hat{\sigma}_{12}$ and the conductivity of the inner shell obtained in the absence of the tunneling current: $\varsigma = \omega_t\hat{\gamma}/2\hat{y}$. This dimensionless quantity only depends on the dimensionless groups ω/ω_t, λ/λ_t, and ν/ω_t, where $\lambda = 2\pi/\beta$ and $\lambda_t = (2\pi/\omega_t)v_F$. Typical values of the parameters are $\nu \approx 10^{12}\ s^{-1}$, $v_F \approx 0.87\cdot10^6$ m/s, and $\omega_t \approx 10^{13}$ rad/s [34]; therefore it is $\lambda_t \approx 600$ nm and $\nu/\omega_t \approx 0.1$.

Figure 1.6 shows the module of ς versus ω/ω_t for different values of λ/λ_t and $\nu/\omega_t = 0.1$. For characteristic wavelengths around $\lambda_t/2$ the coupling between the shells due to the tunneling is important at all frequencies. For characteristic wavelengths shorter than $\lambda_t/2$ the values of $|\zeta|$ drop at low frequencies and the graphs show a peak that moves toward high frequencies in the short wavelength. Also, for characteristic wavelengths higher than $\lambda_t/2$ this value drops at low frequencies, but the curves increase with a dip at $\omega = \omega_t$. In the frequency range up to terahertz the effects of the intershell tunneling currents are important when the characteristic wavelength of the electric signals is comparable to $\lambda_t/2$ or greater than λ_t. The sharp peak at $\lambda_t/2$ has an amplitude equal to 0.25 and a full width at half maximum of roughly $0.04\lambda_t$.

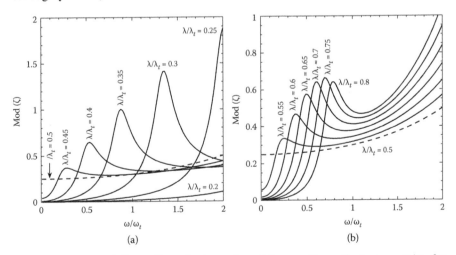

FIGURE 1.6 Intershell coupling: absolute value of the coupling ratio ς versus ω/ω_t for $\nu/\omega_t = 0.1$, with λ/λ_t ranging from (a) 0.2 to 0.5 and (b) 0.5 to 0.8.

1.3 TRANSMISSION LINE MODELS FOR CARBON NANOTUBE INTERCONNECTS

1.3.1 CNTs as Future Material for Nano-Interconnects

The state of the art of the literature in terms of simulations and experimental results coming from the first measurements led to convergent conclusions:

1. Bundles of SWCNTs or MWCNTs may be effectively used to replace copper in nano-interconnect materials, as recently shown in the first examples of CNT-CMOS integration [14–16].
2. Good quality CNT bundle on-chip interconnects made by this material outperform copper in terms of electrical, thermal, and mechanical performance, at least at the intermediate and global levels, whereas at the local level the behavior is comparable [11,32,33,36].
3. For vertical vias or interconnects for packaging applications, very high-density CNT bundles must be obtained [37].

The use of CNT interconnects is therefore strongly related to the possibility of achieving a high-quality fabrication process, which must provide low-contact resistance, good direction control, and compatibility with CMOS technology.

Parallel to this effort in improving the fabrication techniques, attention has been paid in literature to deriving more and more refined models of such interconnects, able to account in a circuital environment for their peculiar behavior.

A simple model for an interconnect made by a bundle of CNTs may be obtained by coupling Maxwell equations to the CNT constitutive relation and modeling the propagation in the frame of the multiconductor transmission line (MTL) theory, as in the conceptual scheme in Figure 1.7.

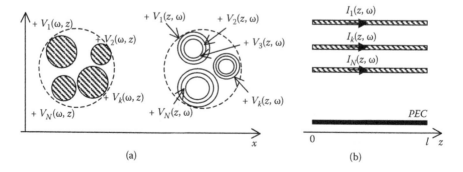

FIGURE 1.7 A bundle of CNTs modeled as a multiconductor interconnect, where a perfect electric conductor (PEC) ground is the reference for the voltages: (a) longitudinal view; transverse section of (b) a SWCNTs' bundle and (c) a MWCNTs' bundle.

1.3.2 MULTICONDUCTOR TRANSMISSION LINE MODEL FOR CNT INTERCONNECTS WITHOUT TUNNELING EFFECT

Let us refer to Figure 1.7 and introduce the vectors of the N voltages, currents, p.u.l. electrical charge, and magnetic flux at given z, in the frequency domain: $\mathbf{V}(\omega,z) = [V_1(\omega,z)....V_N(\omega,z)]^T$, $\mathbf{I}(\omega,z) = [I_1(\omega,z)....I_N(\omega,z)]^T$, $\mathbf{Q}(\omega,z) = [Q_1(\omega,z)....$ $Q_N(\omega,z)]^T$, and $\mathbf{\Phi}(\omega,z) = [\Phi_1(\omega,z)....\Phi_N(\omega,z)]^T$.

In the quasi-TEM assumption, the propagation is governed by the TL equations:

$$i\omega\mathbf{\Phi}(\omega,z) + \frac{\partial\mathbf{V}(\omega,z)}{\partial z} = -\mathbf{E}(\omega,z), \quad i\omega\mathbf{Q}(\omega,z) + \frac{\partial\mathbf{I}(\omega,z)}{\partial z} = 0 \tag{1.27}$$

which may be rewritten in terms of currents and voltages by imposing

$$\mathbf{\Phi}(\omega,z) = L_m\mathbf{I}(\omega,z), \quad \mathbf{Q}(\omega,z) = C_e\mathbf{V}(\omega,z) \tag{1.28}$$

where C_e and L_m are, respectively, the classical p.u.l. electric capacitance and magnetic inductance matrices.

Let us assume that the tunneling between the shells is negligible: This means that the low-energy band structure of each shell is the same as if it were isolated, and thus it is possible to write (1.14) for the generic nth shell:

$$i\omega L_{K,n}I_n(z,\omega) = \frac{1}{1+\nu_n/i\omega}\frac{1}{i\omega C_{Q,n}}\frac{\partial^2 I_n(z,\omega)}{\partial z^2} + E_{z,n}(z,\omega) - R_nI_n(z,\omega) \tag{1.29}$$

Using (1.29) to derive the longitudinal field $E_{z,n}(z,\omega)$ and replacing it in (1.27) and (1.28), we get the MTL equations:

$$-\frac{d\mathbf{V}(z,\omega)}{dz} = (R+i\omega L)\mathbf{I}(z,\omega), \quad -\frac{d\mathbf{I}(z,\omega)}{dz} = i\omega C\mathbf{V}(z,\omega) \tag{1.30}$$

with the p.u.l. parameter matrices given by

$$L = \alpha_C^{-1}(L_m + L_k), \quad C = C_e, \quad R = \nu L_k\alpha_C^{-1}, \quad \alpha_C = (I + C_eC_q^{-1}) \tag{1.31}$$

where I is the identity matrix, and the other matrices are given by

$$\nu = diag(\nu_k), \quad L_k = diag(L_{kk}), \quad C_q = diag(C_{qk}) \tag{1.32}$$

In (1.32) each entry is computed by using (1.15).

Equation (1.30) describes a lossy TL where the quantum effects are combined to the classical electrical and magnetic ones in the definition of the p.u.l. parameters.

Note that up to hundreds of gigahertz, and for typical values of the quantum capacitance for interconnects, it is usually $\alpha_C \approx 1$.

Although the CNT bundle is modeled as a MTL, in practical applications any CNT bundle is used to carry a single signal; i.e., all the CNTs are fed in parallel. Therefore a CNT bundle above a ground may be described by an equivalent single TL, as in Figure 1.8.

The parameters of this equivalent single TL may be rigorously derived from the MTL model in (1.30)–(1.32), imposing the parallel condition. Alternatively, assuming the typical arrangements proposed for practical applications, we may derive an approximated expression for these parameters (e.g., [17]). First, we assume in (1.31) that $a_C \approx 1$. Next, since the CNT bundles intended for practical use are very dense, assuming all the CNTs in parallel, the p.u.l. capacitance of a CNT bundle C_b of external diameter D with respect to a ground plane located at a distance h from the bundle center may be approximated by the p.u.l. capacitance to the ground of a solid wire of diameter D:

$$C_b = \frac{2\pi\varepsilon}{\ln(2h/D)} \tag{1.33}$$

As for the p.u.l. inductance of the equivalent single TL, the magnetic component may be computed from the vacuum space electrostatic capacitance, since it is $L_m = \mu_0\varepsilon_0 C_{e0}^{-1}$, where C_{e0} may be computed exactly or may be approximated by (1.33). The kinetic inductance L_{kb} of a bundle of N CNTs may be simply given by the parallel of the N kinetic inductance L_{kn} associated to any single CNT. Recalling (1.15), it is

$$L_{kb} = \left[\sum_{n=1}^{N} \frac{1}{L_{kn}}\right]^{-1} = \left[\sum_{n=1}^{N} \frac{2v_F M_n}{R_0}\right]^{-1} \tag{1.34}$$

where M_n is the number of equivalent channels of the nth CNT (see Figure 1.4).

Assuming, for instance, the bundle to be made by N SWCNTs, with one-third metallic, it is

$$L_{kb} = \frac{1}{N}\frac{3R_0}{4v_F} \tag{1.35}$$

Finally, the p.u.l. resistance of a single CNT may be found using (1.15) and (1.16):

$$R_n = vL_{Kn} = \frac{vR_0}{2v_F M_n} = \frac{R_0}{2l_{mfp}M_n} \tag{1.36}$$

A complete model for l_{mfp} must include all the scattering mechanisms (defect, acoustic, optical, and zone boundary phonons). However, for temperatures 300 K $< T < 600$ K and longitudinal electric field $E_z < 0.54$ V/μm, an accurate fitting of experimental data provides [30]

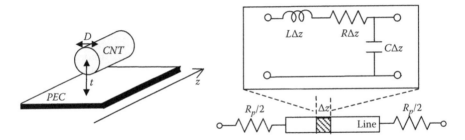

FIGURE 1.8 (a) A CNT interconnect above a PEC ground. (b) Equivalent circuit: elementary cell (inset) and lumped contact resistances.

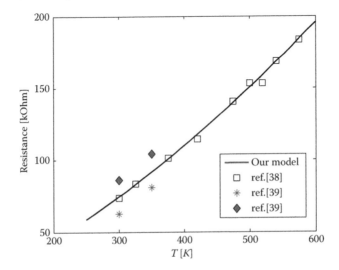

FIGURE 1.9 Resistance of a CNT shell versus temperature: fitting model compared to experimental data in [38] and [39].

$$\frac{1}{l_{mfp}} \approx \frac{2}{D}\Big[k_1 + k_2 T + k_3 T^2\Big] \tag{1.37}$$

where T and D are the CNT temperature and diameter, respectively, and the fitting coefficients $k_1 = 3.005 \cdot 10^{-3}$, $k_2 = -2.122 \cdot 10^{-5}\,\mathrm{K}^{-1}$, and $k_3 = 4.701 \cdot 10^{-8}\,\mathrm{K}^{-2}$.

The resistance of a single CNT shell for varying temperature values derived from the above model is compared in Figure 1.9 to available experimental data.

The resistance of N carbon nanotube shells is simply given by the parallel of N resistances (36).

Note that beside the p.u.l. resistance, any CNT shell introduces a lumped contact resistance R_p (see Figure 1.8b). This resistance is given by the parallel of the contact resistance of each single conducting channel $R_0 + R_{par}$, where R_{par} depends on the quality of contacts and tends to zero for ideal contacts. Even in this case, the contact resistance would not vanish, because of the presence of the quantistic term R_0.

1.3.3 TRANSMISSION LINE MODEL FOR CNT INTERCONNECTS WITH TUNNELING EFFECT

The transmission line model (30) is modified if we have to take into account the intershell tunneling currents. Let us refer to the double-wall CNT in Figure 1.5 and the results presented in Section 1.2 for this structure. By using the modified transport equation (1.26) instead of (1.14), we obtain a MTL model, which is formally described by the equations in (1.27), provided that a new definition for the p.u.l. parameters is introduced [34]:

$$L = L_m + L_k + L_{tun}, \quad C^{-1} = C_e^{-1} + C_{tun}^{-1}, \quad R = v(L_k + L_{tun}) \tag{1.38}$$

having assumed again $a_C \approx I$. The inductance and capacitance matrices are corrected by two new terms, which are the following operators:

$$L = L_m + L_k + L_{tun}$$

$$L_{tun} = \frac{\omega_t \omega'}{2 v_F^3 \omega} \Theta C_e^{-1} \int_{-\infty}^{+\infty} K_1(z - z'; \omega') \{.\} dz' \tag{1.39}$$

$$C_{tun}^{-1} = -\frac{\omega_t \omega'}{2 v_F} \Theta C_e^{-1} \int_{-\infty}^{+\infty} K_2(z - z'; \omega') \{.\} dz' \tag{1.40}$$

The kernels of these operators are given by

$$K_1(z; \omega') = \sin\left(\frac{2\omega_t}{v_F}|z|\right) \exp\left(-i\frac{\omega'}{v_F}|z|\right) \tag{1.41}$$

$$K_2(z; \omega') = \frac{1}{2i} \exp\left(-i\frac{\omega'}{v_F}|z|\right)$$

$$\times \left[\frac{1}{2\omega_t - \omega'} \exp\left(i\frac{2\omega_t}{v_F}|z|\right) + \frac{1}{2\omega_t + \omega'} \exp\left(-i\frac{2\omega_t}{v_F}|z|\right)\right] \tag{1.42}$$

　　The presence of these operators changes the mathematical nature of the model, which moves from a differential model to an integro-differential one. From a physical point of view, the operators L_{tun} and C_{tun} introduce a spatial dispersion. It is possible to show that in the terahertz range the norm of L_{tun} may rise up to 60% of the total norm of L for CNT lengths greater than 0.5 μm [34]. It is also possible to show that in the same conditions the effect of C_{tun} may be neglected, and hence we can again assume $C = C_e$ in (1.38).

1.4   STUDY OF SIGNAL PROPAGATION PROPERTIES ALONG CNT INTERCONNECTS

A realistic application for CNT interconnects is given by the so-called on-chip inter-connects, i.e., the electrical routes inside the different layers that compose a chip. These interconnects route the signal from the inner layers at contact with the transis-tor (local level) to the package pins that are connected to the outside (global level). In the next year the typical dimensions of the transistor gate will decrease to few tens of nanometers [4]: For such dimensions it is not possible to scale down the conventional interconnects, since materials like copper will exhibit dramatic loss of performance, mainly due to the steep increase of resistivity. This is the reason why carbon nanotubes are considered "the ideal interconnect technology for next-generation ICs" [36]. In this section we analyze typical arrangements for on-chip interconnects, modeled with the TL model presented in Section 1.3.

1.4.1   PERFORMANCE OF A CNT ON-CHIP INTERCONNECT

Let us consider the on-chip interconnect in Figure 1.10, which shows a stripline configuration. The signal trace is assumed to be constituted by a solid Cu conductor, by a SWCNT, or by a MWCNT bundle. The driver is modeled by a voltage source with a series resistance R_{dr} and a parallel capacitance C_{out}. The terminal receiver is a load capacitor C_L.

The electrical and geometrical parameters are given in Table 1.1 and refer to a global level interconnect for the 22 nm technology node [4].

The copper realization has been studied by assuming for copper resistivity the value reported in Table 1.1, which is slightly higher than the usual value for bulk copper, due to electromigration phenomena [40]. As for the SWCNT realization, we assume that the line section is filled by SWCNTs of diameters of 1 and 2 nm, with a filling factor of 80%. To simulate realistic conditions, only one-third of the CNT population is metallic. In addition, a parasitic lumped resistance of $R_{par} = 50$ kΩ for each CNT channel.

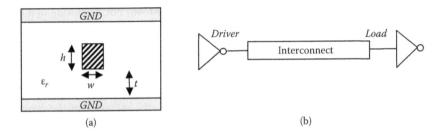

(a)   (b)

FIGURE 1.10   On-chip interconnect: (a) cross and (b) circuit.

TABLE 1.1

**ITRS Values of the Parameters for a Global Level
On-Chip Interconnect at the 22 nm Technology Node**

w [nm]	t [nm]	h [nm]	R_{dr} [kΩ]	C_{out} [fF]	C_L [fF]	ε_r [SiO$_2$]	ρ_{Cu} [$\mu\Omega\cdot$cm]
160	96	76.8	0.16	4.9	14	2.2	2.94

Source: ITRS, *International Technology Roadmap for Semiconductors*, 2009, http://public.itrs.net.

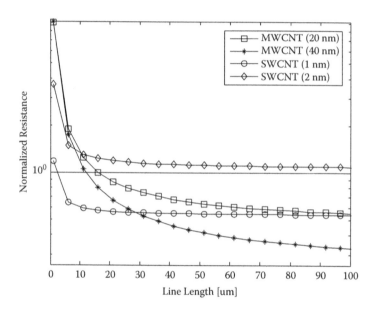

FIGURE 1.11 Per unit length resistance of a global level interconnect, normalized to copper value, versus line length: Bundles of MWCNTs and SWCNTs with different diameters are compared.

Finally, for the MWCNT realization we analyze two possible arrangements, corresponding to the values of external diameter $D_{out} = [20; 40]$ nm, assuming that the inner diameter is $D_{in} = 0.5D_{out}$ and that the shells are separated by the van der Waals distance of 0.34 nm.

Let us first assume the line temperature to be constant at a fixed room temperature $T = T_0 = 293$ K.

Figure 1.11 shows the computed line per unit length resistance, normalized to the value obtained with the copper realization. Due to the influence of the contact resistance, all the CNT realizations show worse performance with respect to the Cu one for shorter lengths. For lengths greater than 10 μm, the MWCNT solutions are always better than the Cu ones, and the performance is even better as length

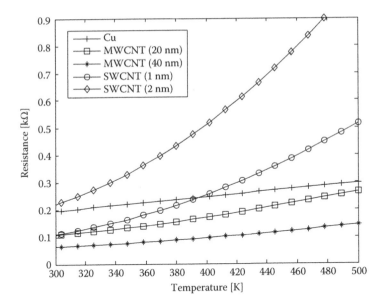

FIGURE 1.12 Resistance of a global level interconnect of length 0.1 mm versus temperature: Bundles of MWCNTs and SWCNTs with different diameters are compared.

increases. The SWCNT solution may be better or worse than Cu, depending on the chosen CNT diameter (which affects, of course, the total number of the CNTs in the bundle). In this case, an optimal solution would be the use of 1 nm diameter SWCNTs. Note that typical lengths for global level interconnects are of the order of 0.1 mm.

Let us now analyze the sensibility of the resistance to the temperature variation, which is an important issue for on-chip interconnects, given their major problem of Joule heating. A simple temperature-dependent model for copper resistivity is given by

$$\rho(T) = \rho_0[1 + \alpha_0(T - T_0)] \tag{1.43}$$

where ρ_0 is the value given in Table 1.1 and the coefficient $\alpha_0 = 2.65 \cdot 10^{-3}$.

As for the CNT realization, the increase of temperature affects the number of conducting channels and the mean free path, as pointed out in Section 1.2. Figure 1.12 shows the behavior of the resistance versus temperature, assuming a line length of 0.1 mm. The SWCNT solutions are the most sensitive to temperature increase, whereas the MWCNT ones are the more stable, the larger the CNT diameter used.

Let us now investigate the so-called signal integrity (SI) performance of the considered channel (Figure 1.10b), assuming for the chip interconnect an operating temperature of 400 K and a length of 0.1 mm. Let us consider the MWCNT case with a 40 nm diameter and the SWCNT one with a 1 nm diameter. The computed values of the line parameters are reported in Table 1.2.

Figures 1.13 and 1.14 show the eye diagram obtained by considering a transmission of a digital signal at a rate of 50 and 100 Gbaud/s. The three channels show

TABLE 1.2

Computed RLC Parameters for the Global Level On-Chip Interconnect with Length 0.1 mm, for Cu, SWCNT, and MWCNT Realizations

	R [kΩ]	C [fF]	L [pH]
Cu	0.24	0.32	0.76
SWCNT	0.25	0.32	0.83
MWCNT	0.09	0.32	1.47

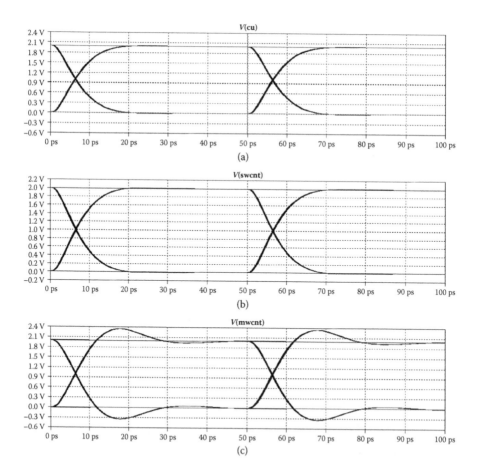

FIGURE 1.13 Eye diagram for the transmission of a digital signal at 20 Gbaud/s rate: (a) Cu, (b) SWCNT, and (c) MWCNT realization.

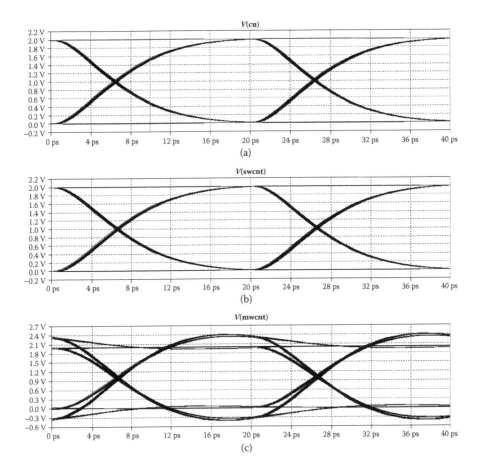

FIGURE 1.14 Eye diagram for the transmission of a digital signal at 50 Gbaud/s rate: (a) Cu, (b) SWCNT, and (c) MWCNT realization.

similar performances at 20 Gbaud/s: Each channel shows an excellent response to this high-frequency digital signal. The MWCNT realization exhibits a sort of ringing due to the presence of a higher inductance (mainly related to the kinetic term). The three channels may also be used at 50 Gbaud/s, without significant degradation of the performance, unless there is an increase of the jitter for the MWCNT case.

1.5 CONCLUSIONS

In this chapter the electrodynamics of the conduction electrons of carbon nanotubes (CNTs) have been described, and models are derived to study the propagation of signals along CNT interconnects. The conduction electrons are seen as a fluid moving under the influence of the collective field and of the interaction with the fixed ion lattice. The transport equation is derived in the quasi-classical limit, allowing the formulation of a frequency domain constitutive equation for the CNTs, in terms of a nonlocal Ohm's law. The CNT conductivity is strongly influenced by two parameters, which account for the electron inertia and for the quantum pressure. The conductivity may be expressed in terms of the effective number of conducting channels, a parameter that counts the number of subbands significantly involved in the conduction.

The model is then extended to the case of interacting CNT shells, where the tunneling effect is considered. Self and mutual conductivities are then defined to account for this phenomenon.

By coupling the above CNT constitutive equations to Maxwell equations, a transmission line model is derived, able to describe the propagation of signals along CNT bundles, in terms of simple RLC distributed lines. The intershell tunneling occurring at terahertz is then accounted for, with a suitable recasting of the transmission line equations, which now include spatial dispersion terms.

REFERENCES

1. S. Ijima. Helical microtubules of graphitic carbon. *Nature*, 354(6348), 56–58, 1991.
2. R. Saito, G. Dresselhaus, and S. Dresselhaus. *Physical Properties of Carbon Nanotubes*. London: Imperial College Press, 1998.
3. M.P. Anantram and F. Léonard. Physics of carbon nanotube electronic devices. *Report Progress in Physics*, 69, 507, 2006.
4. ITRS. *International Technology Roadmap for Semiconductors*. 2009. http://public.itrs.net.
5. P. Avouris, Z. Chen, and V. Perebeinos. Carbon-based electronics. *Nature Nanotechnology*, 2(10), 605–613, 2007.
6. M.J. Hagmann. Isolated carbon nanotubes as high-impedance transmission lines for microwave through terahertz frequencies. *IEEE Transactions on Nanotechnology*, 4(2), 289–296, 2005.
7. F. Kreupl, A.P. Graham, G.S. Duesberg, W. Steinhogl, M. Liebau, E. Unger, and W. Honlein. Carbon nanotubes in interconnect applications. *Microelectronic Engineering*, 64(1–4), 399–408, 2002.
8. J. Li, Q. Ye, A. Cassell, H.T. Ng, R. Stevens, J. Han, and M. Meyyappan. Bottom-up approach for carbon nanotube interconnects. *Applied Physics Letters*, 82(15), 2491–2493, 2003.
9. J.E. Morris. *Nanopackaging: Nanotechnologies and Electronics Packaging*. New York: Springer, 2008.
10. H.W.C. Postma, T. Teepen, Z. Yao, M. Grifoni, and C. Dekker. Carbon nanotube single-electron transistors at room temperature. *Science*, 293(5527), 76–79, 2001.
11. H. Li, C. Xu, N. Srivastava, and K. Banerjee. Carbon nanomaterials for next-generation interconnects and passives: Physics, status, and prospects. *IEEE Transactions on Electron Devices*, 56(9), 1799–1821, 2009.
12. G.W. Hanson. Fundamental transmitting properties of carbon nanotube antennas. *IEEE Transactions on Antennas Propagation*, 53, 3426, 2005.
13. S.A. Maksimenko, G.Ya. Slepyan, A.M. Nemilentsau, and M.V. Shuba. Carbon nanotube antenna: Far-field, near-field and thermal-noise properties. *Physica E*, 40, 2360, 2008.
14. I. Soga, D. Kondo, Y. Yamaguchi, T. Iwai, M. Mizukoshi, Y. Awano, K. Yube, and T. Fujii. Carbon nanotube bumps for LSI interconnect. In *Proceedings of Electronic Components and Technology Conference*, May 2008, pp. 1390–1394.
15. G.F. Close, S. Yasuda, B. Paul, S. Fujita and H.-S. Philip Wong. A 1 GHz integrated circuit with carbon nanotube interconnects and silicon transistors. *Nano Letters*, 8(2), 706–709, 2009.

16. X. Chen et al. Fully integrated graphene and carbon nanotube interconnects for gigahertz high-speed CMOS electronics. *IEEE Transactions on Electronic Devices*, 57, 3137–3143, 2012.

17. G.Ya. Slepyan, S.A. Maksimenko, A. Lakhtakia, O. Yevtushenko, and A.V. Gusakov. Electrodynamics of carbon nanotubes: Dynamics conductivity, impedance boundary conditions, and surface wave propagation. *Physical Review B*, 60, 17136, 1999.

18. S.A. Maksimenko, A.A. Khrushchinsky, G.Y. Slepyan, and O.V. Kibis. Electrodynamics of chiral carbon nanotubes in the helical parametrization scheme. *Journal of Nanophotonics*, 1, 013505, 2007.

19. Y. Miyamoto, S.G. Louie, and M.L. Cohen. Chiral conductivities of nanotubes. *Physical Review Letters*, 76, 2121–2124, 1996.

20. P.J. Burke. Luttinger liquid theory as a model of the gigahertz electrical properties of carbon nanotubes. *IEEE Transactions on Nanotechnology*, 1, 129–144, 2002.

21. G. Miano and F. Villone. An integral formulation for the electrodynamics of metallic carbon nanotubes based on a fluid model. *IEEE Transactions on Antennas and Propagation*, 54, 2713, 2006.

22. J.J. Wesström. Signal propagation in electron waveguides: Transmission-line analogies. *Physical Review B*, 54, 11484–11491, 1996.

23. P.J. Burke. An RF circuit model for carbon nanotubes. *IEEE Transactions on Nanotechnology*, 2, 55, 2003.

24. S. Salahuddin, M. Lundstrom, and S. Datta. Transport effects on signal propagation in quantum wires. *IEEE Transactions on Electron Devices*, 52, 1734–1742, 2005.

25. A. Maffucci, G. Miano, and F. Villone. A transmission line model for metallic carbon nanotube interconnects. *International Journal of Circuit Theory and Applications*, 36(1), 31, 2008.

26. A. Maffucci, G. Miano, and F. Villone. A new circuit model for carbon nanotube interconnects with diameter-dependent parameters. *IEEE Transactions on Nanotechnology*, 8, 345–354, 2009.

27. G.Ya. Slepyan, M.V. Shuba, S.A. Maksimenko, C. Thomsen, and A. Lakhtakia. Terahertz conductivity peak in composite materials containing carbon nanotubes: Theory and interpretation of experiment. *Physical Review B*, 81, 205423, 2010.

28. C. Forestiere, A. Maffucci, and G. Miano. Hydrodynamic model for the signal propagation along carbon nanotubes. *IEEE Transactions on Nanophotonics*, 4, 041695, 2010.

29. G. Miano, C. Forestiere, A. Maffucci, S.A. Maksimenko, and G.Y. Slepyan. Signal propagation in carbon nanotubes of arbitrary chirality. *IEEE Transactions on Nanotechnology*, 10, 135–149, 2011.

30. S. Salahuddin, M. Lundstrom, and S. Datta. Transport effects on signal propagation in quantum wires. *IEEE Transactions on Electron Devices*, 52, 1734–1742, 2005.

31. C. Forestiere, A. Maffucci, and G. Miano. On the evaluation of the number of conducting channels in multiwall carbon nanotubes. *IEEE Transactions on Nanotechnology*, 10(6), 1221–1223, 2011.

32. A. Naeemi and J.D. Meindl, Performance modeling for single- and multiwall carbon nanotubes as signal and power interconnects in gigascale systems. *IEEE Transactions on Electron Devices*, 55(10), 2574–2582, 2008.

33. A. Maffucci, G. Miano, and F. Villone. Performance comparison between metallic carbon nanotube and copper nano-interconnects. *IEEE Transactions on Advanced Packaging*, 31(4), 692–699, 2008.

34. C. Forestiere, C. Maffucci, A. Maksimenko, S.A. Miano, and G. Slepyan. Transmission line model for multiwall carbon nanotubes with intershell tunneling. *IEEE Transactions on Nanotechnology*, 11(3), 554–564, 2012.

35. M.V. Shuba, G.Ya. Slepyan, S.A. Maksimenko, C. Thomsen, and A. Lakhtakia. Theory of multiwall carbon nanotubes as waveguides and antennas in the infrared and the visible regimes. *Physical Review B*, 79, 155403, 2009.

36. H. Li, C. Xu, and K. Banerjee. Carbon nanomaterials: The ideal interconnect technology for next-generation ICs. *IEEE Design and Test of Computers*, 4, 20–31, 2010.

37. A. Maffucci. Carbon nanotubes in nanopackaging applications. *IEEE Nanotechnology Magazine*, 3(3), 22–25, 2009.

38. C.T. Avedisian, R.E. Cavicchi, P.M. McEuen, X. Zhou, W.S. Hurst, and J.T. Hodges. High temperature electrical resistance of substrate-supported single walled carbon nanotubes. *Applied Physics Letters*, 93, 252108, 2008.

39. A. Pop, D.A. Mann, K.E. Goodson, and H. Dai. Electrical and thermal transport in metallic single-wall carbon nanotubes on insulating substrates. *Journal of Applied Physics*, 101, 093710, 2007.

40. W. Steinhogl, G. Schindler, G. Steinlesberger, M. Traving, and M. Engelhardt. Comprehensive study of the resistivity of copper wires with lateral dimensions of 100 nm and smaller. *Journal of Applied Physics*, 97, 023706/1-7, 2005.

2 Quasi-Particle Electronic Structure of Pristine and Hydrogenated Graphene on Weakly Interacting Hexagonal Boron Nitride Substrates

Neerav Kharche and Saroj K. Nayak

CONTENTS

2.1 INTRODUCTION

Owing to its extraordinary electronic, optical, thermal, and mechanical properties and immense potential for nanoelectronic applications, graphene has attracted enormous attention of both academic and industrial research groups [1–9]. Graphene is a promising material for several technological applications, such as field effect transistors, photodetectors, electrodes in solar cells and batteries, interconnects, etc. [3,9–12]. When integrated into functional devices the electronic properties of graphene are heavily influenced by the surrounding materials in these devices. For example, the surrounding dielectrics screen the scattering potential of charged impurities, which limit the conductivity of graphene. As a result, the conductivity of graphene can be modulated over three orders of magnitude by changing its dielectric environment [13]. Likewise, the plasmon excitations in graphene are strongly affected due to the nonlocal screening of electron-electron interactions by the substrate, even in the limit of a weak van der Waals (vdW) interaction [14].

Typically graphene devices are fabricated on SiO_2 and SiC substrates; however, carrier mobility in graphene supported on these substrates reduces significantly due to charged surface states, surface roughness, and surface optical phonons [15,16]. Recently, graphene supported on a hexagonal boron nitride (hBN) substrate was found to exhibit much higher mobility than any other substrate [17,18]. High mobility in graphene on hBN is attributed to an extremely flat surface of hBN, which practically eliminates the scattering due to substrate surface roughness and much weaker coupling of electrons in graphene with surface optical phonons of hBN [17,19]. The hBN layer can also be used as a gate dielectric separating the graphene channel and the gate electrode in field effect devices [20]. Moreover, novel field effect tunneling transistors based on the graphene-hBN tunnel junctions have been recently demonstrated [21]. Here we present the electronic structure calculations of a graphene-hBN interface to aid the understanding of experimental devices based on these heterostructures. The sublattice asymmetry induced on the graphene lattice by the highly polar hBN surface is found to modulate the band gap at the Dirac point of graphene [22].

The zero band gap in graphene is a major impediment for its applications in logic devices [3]. A band gap can be opened through quantum confinement by patterning graphene into the so-called graphene nanoribbons (GNRs) [23–25]. However, it is difficult to control the band gap in GNRs due to its strong dependence on the width and edge geometry. Alternatively, the band gap can be opened by chemical functionalization of graphene with a variety of species, such as H, F, OH, etc. [26,27]. The band gap opening by H-functionalization/hydrogenation of graphene has especially been a subject of several recent experimental and theoretical studies, which show that the band gap of graphene can be tuned by controlling the degree of hydrogenation [28–31]. Here we show that the screening of electron-electron interactions by the polarization of surrounding hBN layers can significantly reduce the band gaps in hydrogenated graphene [22]. Such polarization effects are expected to also play an important role in other emerging monolayer materials, such as MoS_2, WS_2, hBN, etc.

2.2   APPROACH

An effective single particle theory such as the density functional theory (DFT) cannot fully capture the effect of dielectric screening, and a more accurate many-body approach that includes the modification of self-energy due to the dielectric screening is required. Here we use many-body perturbation theory based on Green's function (G) and the screened Coulomb potential (W) approach, i.e., the GW approximation.

Heterostructures of graphene (or graphone) and hBN are modeled using the repeated-slab approach, where the slabs periodic in the xy plane are separated by a large enough vacuum region along the z direction so that their interaction with the periodic images is negligible. The electronic structure calculations are performed in the framework of first-principles DFT at the level of the local density approximation (LDA) as implemented in the ABINIT code [32]. The Troullier-Martins norm-conserving pseudopotentials and the Teter-Pade parametrization for the exchange correlation functional are used [33,34]. A large enough vacuum of 10 Å in the z direction is used to ensure negligible interaction between periodic images. The Brillouin zone is sampled using Monkhorst-Pack meshes [35] of different size, depending on the size of the unit cell: $18 \times 18 \times 1$ for Bernal stacked graphene on hBN, $6 \times 6 \times 1$ for misaligned graphene on hBN, and $18 \times 18 \times 1$ for graphone. Wavefunctions are expanded in plane waves with an energy cutoff of 30 Ha. The quasi-particle corrections to the LDA band structure are calculated within the G_0W_0 approximation, and the screening is calculated using the plasmon-pole model [36]. The Coulomb cutoff technique proposed by Ismail-Beigi et al. is used to minimize the spurious interactions with periodic replicas of the system [37]. The convergence of GW band gap is carefully tested.

Graphene and hBN are bonded by van der Waals interactions. The equilibrium distance between graphene (or graphone) and hBN is determined using the Vienna ab initio simulation package (VASP), which provides a well-tested implementation of vdW interactions [38,39]. The projected augmented wave (PAW) pseudopotentials [40], the Perdew-Burke-Ernzerhof (PBE) exchange correlation functional in the GGA approximation [41], and the DFT-D2 method of Grimme [42] are used. The same values of energy cutoff, vacuum region, and k-point grids as in the ABINIT calculations are used in VASP calculations. The optimized geometries and the electronic structures calculated using VASP and ABINIT without including vdW interactions are found to agree well with each other.

2.3   RESULTS AND DISCUSSION

2.3.1   GRAPHENE-HBN INTERFACE

There have been several recent proposals on using hBN as a substrate [17], a dielectric [20], and a tunneling barrier [21] in graphene-based field effect devices. The electronic transport in such devices is controlled by the band offsets at the graphene-hBN interface. The band offsets can be extracted from the band structure of the supercell of a composite system where graphene is adsorbed on a hBN surface. A schematic of graphene adsorbed on a hBN surface is shown in the inset of Figure 2.1(a), and an ideal Bernal stacking arrangement of graphene with respect to the top hBN layer is shown in Figure 2.1(a). The stacking arrangement of Figure 2.1(a), where half of the C atoms in graphene are positioned above the B atoms and the rest of the C atoms are positioned above the center of hBN hexagons, has been found to be a lowest-energy configuration when both graphene and hBN have the same lattice constant [43,44].

The distance between graphene and the hBN surface is determined using the DFT-D2 method of Grimme as implemented in the VASP code [39,42]. The equilibrium distance is found to be 3.14 Å, indicating that graphene is weakly bonded to the hBN surface. A surface-averaged charge density of $|\psi(z)|^2 = \iint dxdy |\psi(x,y,z)|^2$, a typical wavefunction near the Dirac point of graphene supported on hBN is depicted by grey circles in Figure 2.1(b). The probability density $|\psi(z)|^2$ peaks at ≈ 0.4 Å from the graphene layer and decays rapidly beyond this distance. The wavefunction near the Dirac point can be well approximated by a p_z-like orbital given by $\psi(z) = N|z|e^{-\zeta|z|}$. The probability density of the parameterized wavefunction with $N = 1.87$ Å^{-1} and $\zeta = 2.39$ Å^{-1} compares well with that obtained from the DFT simulation, as shown in Figure 2.1(b). The same parameterization works well for the freestanding graphene, indicating that the graphene orbitals do not hybridize with hBN orbitals, and the electronic structure of graphene is only weakly influenced by the hBN substrate.

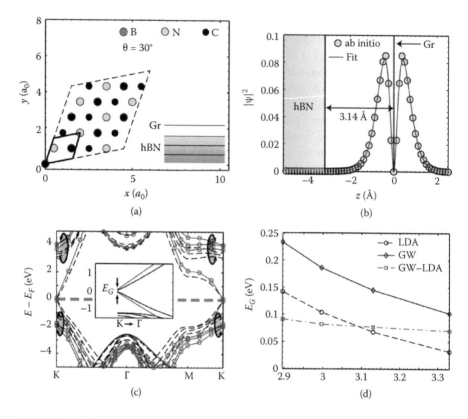

FIGURE 2.1 (a) Atomistic schematic of a Bernal stacked graphene on a hBN substrate. A Bernal stacked graphene lattice is rotated by 30° with respect to the adjacent hBN layer. The solid black lines depict the unit cell. (b) Surface-averaged density $|\psi(z)|^2$ for states near the Dirac point of graphene supported on a hBN substrate. (c) DFT and GW band structures of a composite graphene-hBN system shown in the inset of (a). The bands contributed by hBN are identified separately. The inset shows a small band gap opening in graphene at the k-point. (d) DFT and GW band gaps of Bernal stacked graphene on a hBN substrate as a function of the distance between graphene and the hBN surface.

The DFT and GW band structures of the graphene-hBN system are depicted in Figure 2.1(a) and shown in Figure 2.1(c). Three monolayers of the hBN substrate are included in the electronic structure calculation to ensure the convergence of the GW band gap. Due to the weak interaction between graphene and hBN, the linear band structure of graphene near the Dirac point is preserved and a small band gap is opened up at the Dirac point, as shown in the inset of Figure 2.1(c). The band gap opening in graphene is attributed to the sublattice asymmetry induced on the graphene lattice by a weak electrostatic interaction between graphene and Bernal stacked hBN, which unlike graphene has highly polar B-N bonds [45]. As shown in Figure 2.1(d), the graphene band gap can be modulated by changing the distance of the graphene layer from the hBN substrate. The sublattice asymmetry induced by the polar hBN lattice increases as the distance of the graphene layer from the hBN

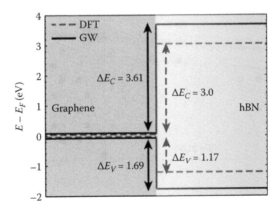

FIGURE 2.2   Band edges at the graphene-hBN interface obtained from the band structure of Figure 2.1(c). DFT underestimates the hBN band gap and the resulting band offsets.

surface decreases, which results in a higher band gap. The GW-corrected band gap of the Bernal graphene on hBN is found to be higher by ≈0.1 eV than the DFT band gap (Figure 2.1(d)).

The bands contributed by hBN are identified in Figure 2.1(c). The band offsets at the graphene-hBN interface extracted from the composite band structure of Figure 2.1(c) are shown in Figure 2.2. There are two important points to be noted from Figure 2.2—first, the asymmetric conduction and valence band offsets. The signatures of such asymmetry have been observed in recent experiments on tunneling across the graphene-hBN interface [21,46]. Second, the DFT band gap of hBN is smaller by ≈1 eV than the GW-corrected band gap, and as a result, DFT underestimates band offsets by ≈0.5 eV. The graphene-hBN band offsets extracted from the recent experimental data on the graphene-hBN interface are in better agreement with the GW band offsets. This clearly indicates the importance of using advanced methods such as the GW approximation in accurate prediction of band offsets at the interfaces of novel materials.

The ideal Bernal stacking considered above is unlikely to be achieved in realistic devices because of the 1.7% lattice mismatch between graphene and hBN. Indeed, recent experiments report the formation of Moiré patterns at the graphene-hBN interface [18,47]. Moreover, contrary to the above calculations, the experimental data do not show any evidence of the band gap opening at the Dirac point of graphene supported on hBN [17]. To address the discrepancy between the above calculations and the experimental data, we compute band gaps in the graphene-hBN system where the stacking arrangement deviates from the ideal Bernal stacking.

The Bernal stacked graphene lattice is rotated by 30° with respect to the underlying hBN layer. This stacking arrangement can be simulated using a small unit cell depicted by the solid black lines in Figure 2.1(a). The unit cell size is, however, much larger when the stacking arrangement deviates from the ideal Bernal stacking. Here we consider three different misalignments where graphene is rotated by 21.8°, 32.2°, and 13.2° with respect to the adjacent hBN monolayer. The supercells for these rotations are shown in Figure 2.3(a–c), and they contain 28 (14 C, 7 B, 7 N), 52 (26 C, 13

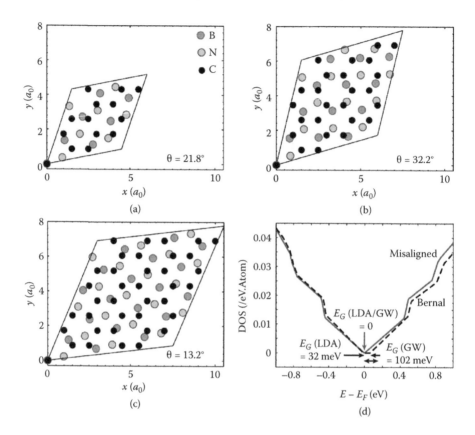

FIGURE 2.3 Atomistic schematics of commensurate unit cells of a misaligned graphene-hBN system, where graphene is rotated by (a) 21.8°, (b) 32.2°, and (c) 13.2° with respect to the underlying hBN layer. (c) Density of states (DOS) of Bernal stacked (Figure 2.1(a)) and misaligned (Figure 2.3(a)) graphene supported on a hBN substrate illustrating the band gap closing due to misalignment. Due to the high computational requirements, only one hBN layer is included in the DOS calculation.

B, 13 N), and 76 (38 C, 19 B, 19 N) atoms, respectively. The supercells are constructed using the commensuration conditions presented in [48]. Only one layer of hBN is included to reduce the computational requirement associated with the GW calculation.

As pointed out earlier, the band gap opening in the Bernal stacked graphene on hBN results from the sublattice asymmetry induced on the graphene lattice by the underlying highly polar hBN lattice. When the stacking arrangement deviates from the Bernal stacking, the sublattice asymmetry induced by the hBN is significantly reduced. Figure 2.3(d) shows the DFT density of states (DOS) and the DFT and GW band gaps at the Dirac point in the Bernal stacked (Figure 2.1(a)) and mis-aligned (Figure 2.3(a)) graphene on hBN. As expected, DFT underestimates the band gap of Bernal stacked graphene on hBN compared to the more accurate GW calculation. The DFT band gap of the misaligned graphene on hBN is 0. The gap-less nature of misaligned graphene on hBN could be associated with the fact that

DFT underestimates the band gaps. To rule out this possibility, we carried out GW calculations on the misaligned supercells shown in Figure 2.3. The GW-corrected band gap of graphene calculated using all three misaligned supercells remains 0. The calculations presented here explain the experimentally observed zero band gap nature of graphene on hBN. As pointed out earlier, only one hBN layer is included in the simulations of misaligned supercells. Including more layers of hBN is not expected to open up a band gap because the induced sublattice asymmetry on the graphene lattice due to hBN layers decreases rapidly as the distance between graphene and hBN layers increases.

2.3.2   GRAPHONE-HBN HETEROSTRUCTURES

Several recent experimental and theoretical studies have shown that the electronic structure of graphene can be significantly altered by chemical functionalization with H, F, Cl, and transition metal atoms [26,31,49,50]. In these graphene-derived materials, the hybridization of the functionalized C atom changes from sp^2 to sp^3, which results in band gap opening. Hydrogenation of graphene is of particular interest because several experimental and theoretical studies have demonstrated that the band gap of hydrogenated graphene can be tuned by varying the degree of hydrogenation, and this process can even be reversed to recover pristine graphene [31]. Compared to the graphene-hBN system, the hBN substrate has a very different effect on the band gap of hydrogenated graphene [22]. In the following, we investigate the effect of hBN substrate on the band gap of hydrogenated graphene. A single-sided semihydrogenated graphene, also called graphone, is used as an example; however, the results are also applicable to graphene with different hydrogen coverage.

Figure 2.4(a) shows the atomistic schematics of graphone where C atoms belonging to one sublattice are hydrogenated [29]. Graphone can be synthesized by selectively desorbing hydrogen from one side of graphane. The change of hybridization of C atoms from sp^2 to sp^3 upon hydrogenation results in a nonplanar atomic structure.

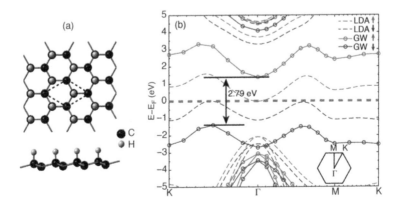

FIGURE 2.4   Atomistic schematics of single-sided semihydrogenated graphene (graphone). The unit cell is depicted by the dotted lines. (b) DFT and GW band structures of graphone. The inset shows the Brillouin zone and high-symmetry directions.

The optimized atomic structure of graphone in our calculations is virtually identical to the earlier studies [29]. Graphone has a ferromagnetic ground state with a magnetic moment of ≈ 1 μ_B located on each nonfunctionalized C atom.

The DFT and GW band structures of graphone are shown in Figure 2.4(b). The DFT (LDA) band structure of graphone shows a large (≈ 1.5 eV) spin splitting near the Fermi level but no band gap. The GW-corrected band structure shows a large band gap of 2.79 eV and a spin splitting of ≈ 4 eV throughout the whole Brillouin zone. A large spin splitting makes graphone an attractive material for spintronic applications [51].

When graphone is deposited on the hBN substrate, it binds weakly to the hBN surface via the van der Waals interaction. The equilibrium distance between graphone and hBN obtained from VASP simulations that include the van der Waals interaction using the DFT-D2 method of Grimme is found to be 3.13 Å, as shown in the atomistic schematic of Figure 2.5(a). The surface-averaged density distribution $|\psi(z)|^2$ of states at the valence band maximum and the conduction band minimum are shown in Figure 2.5(b). Similar to graphene, $|\psi(z)|^2$ decays to an almost negligible value before it reaches the hBN surface, indicating that graphone orbitals do not hybridize with hBN orbitals. Unlike graphene, where the wavefunctions near the band extrema are composed mainly of p_z orbitals, in graphone, the wavefunctions near the band extrema are composed of both s and p orbitals. Both spin-split bands near the Fermi level (Figure 2.4(b)) have appreciable s and p contributions, but the p contribution dominates over the s contribution [29].

Now we consider the effect of substrate, hBN in this case, on the band structure of graphone. The DFT and GW band structures of a composite bilayer of graphone and a hBN monolayer are shown in Figure 2.6(a). If the additional bands in Figure 2.6(a)

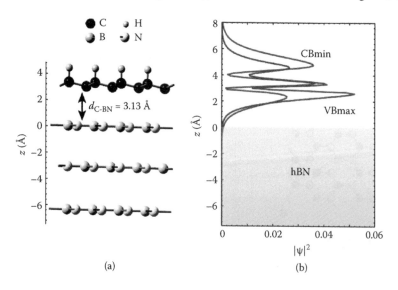

(a)   (b)

FIGURE 2.5 (a) Atomistic schematic of graphone supported on a hBN substrate. (b) Surface-averaged density $|\psi(z)|^2$ for states of graphone at the valence band maximum and the conduction band minimum.

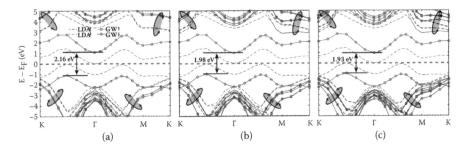

FIGURE 2.6 DFT and GW band structures of graphone-hBN supercell including (a) one, (b) two, and (c) three layers of hBN. The band gap is converged within 0.05 eV for three or more layers of hBN. The bands contributed by hBN are identified separately.

contributed by the hBN monolayer are excluded, the DFT band structure of graphone deposited on the hBN monolayer is virtually identical to that of the freestanding graphone shown in Figure 2.4(b). This is because of the fact that graphone and hBN layers are weakly bound and are essentially electronically isolated systems within the DFT approximation, which does not include the long-range Coulomb interactions. The GW-corrected bands of graphone, however, are significantly altered when it is deposited on the hBN monolayer.

The GW band gap of freestanding graphone (Figure 2.4(b)) decreases from 2.79 eV to 2.16 eV when deposited on the hBN monolayer. The reduction in GW band gap is related to the polarization effects at the graphone-hBN interface. The dielectric properties, and hence the polarization, depend on the thickness of the dielectric material [52]. To accurately estimate the band gap reduction in the adsorbed graphone layer due to the polarization response of the hBN surface, we carry out a convergence study with respect to the thickness of the hBN layer. The DFT and GW band structures of graphone adsorbed on one, two, and three monolayers of hBN are shown in Figure 2.6(a–c). Including more layers of hBN does not affect the DFT bands contributed by the graphone layer. As expected, the polarizability of the hBN layer increases with the number of monolayers, and as a result, the GW band gap reduction in graphone also increases. The comparison of the band structures of graphone deposited on the four-monolayer-thick hBN layer (not shown) and the band structures in Figure 2.6(a–c) indicates that the band gap of graphone is converged within 0.05 eV when three monolayers of hBN are included to mimic the hBN substrate.

The quasi-particle GW band gaps of freestanding graphone and the hBN substrate-supported graphone are 2.79 eV (Figure 2.4(b)) and 1.93 eV (Figure 2.6(c)), respectively, which corresponds to a 0.86 eV band gap reduction. The surrounding dielectric environment strongly influences the electron-electron interactions in nanostructures, which in turn affect their band gaps. Here the band gap reduction in hBN-supported graphone is a result of more effective screening of electron-electron interaction by hBN compared to the screening by the vacuum in freestanding graphone. The screened Coulomb interaction (W) in graphone is reduced by the polarization of the hBN surface, which consequently lowers the band gap. Such band gap reductions have been previously reported in GW calculations on zero-dimensional systems such as molecules adsorbed on metallic or insulating substrates and

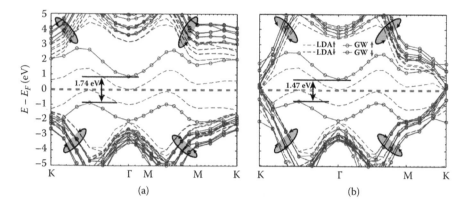

FIGURE 2.7 DFT and GW band structures of (a) graphone sandwiched between hBN layers on both sides and (b) graphone supported on a graphite substrate. The bands contributed by hBN and graphite are identified separately.

quantum dots embedded in a dielectric medium, and one-dimensional systems such as GNRs adsorbed on a hBN substrate and CNTs embedded in a dielectric medium [22,53–59].

Monolayer materials such as graphene and its derivatives, when used in realistic devices, are often covered with another material such as the gate dielectric or the metal contact. In graphene-based field effect devices, hBN can be used as a substrate as well as a dielectric separating the channel and the metal contact [17,20,21]. The DFT and GW band structures of graphone surrounded by a bottom hBN substrate andaz top hBN dielectric are shown in Figure 2.7(a). Three layers of hBN on each side of graphone are included in the simulation domain to attain convergence of the GW band gap. The band gap of hBN substrate-supported graphone reduces from 1.93 eV (Figure 2.6(c)) to 1.74 eV (Figure 2.7(a)) when covered with a top hBN layer. The reduction in the screened Coulomb potential, W in graphone due to the additional screening from the top hBN dielectric layer, results in further lowering of the graphone band gap. However, the further band gap reduction due to the additional screening from the top hBN dielectric layer is much smaller.

Hydrogenated graphene can be synthesized by hydrogenation of the (0001) graphite surface [60]. Although this process might result in a different configuration of hydrogenated graphene compared to graphone, the polarization of the underlying graphite surface will have a similar effect on the band gap of the hydrogenated graphene on the surface. As an example of such a system, we show the DFT and GW band structures of graphone supported on the graphite substrate. Compared to the band gap reduction of 0.86 eV in hBN-supported graphone, graphite-supported graphone shows a larger band gap reduction of 1.32 eV. This is because of higher polarizability of graphite than of hBN. Thus the band gap reduction is dependent on the electronic structure and dielectric properties of the substrate [55]. The band gap reduction in hydrogenated graphene due to the surrounding dielectric materials can be estimated within a reasonable accuracy using a semiclassical image charge model presented in [22].

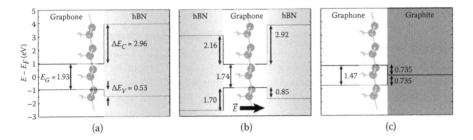

FIGURE 2.8   Band offsets extracted from GW band structures in (a) Figure 2.6(c), (b) Figure 2.7(a), and (c) Figure 2.7(b).

The operation of devices employing two-dimensional materials is strongly influenced by the band offsets at their interfaces with the surrounding materials [61]. The conduction and valence band offsets (denoted by ΔE_C and ΔE_V, respectively) of graphone when integrated into graphone-hBN and graphone-graphite heterostructures are shown in Figure 2.8. The band offsets at the graphone-graphite interface are symmetric, while the band offsets at the graphone-hBN interface are asymmetric. The valence band offset at the interface between the nonhydrogenated side of graphone and hBN is different in Figure 2.8(a) and (b). This is because of the smaller band gap of graphone in the hBN-graphone-hBN heterostructure compared to the graphone-hBN heterostructure. The band offsets at the two graphone-hBN interfaces in the hBN-graphone-hBN heterostructure are also asymmetric. The bands of the hBN layer on the hydrogenated side of graphone are lower in energy than the bands of the hBN layer on the nonhydrogenated side. This asymmetry is a consequence of the electric field depicted by a thick arrow in Figure 2.8(b), which is generated by a dipole layer due to a slight net positive charge on the hydrogenated C atom and a slight net negative charge on the nonhydrogenated C atom [22].

2.4   CONCLUSION

In summary, the electronic structures of atomically thin materials such as graphene and hydrogenated graphene are found to be strongly influenced by the surrounding materials, hBN in this case. The highly polar hBN surface induces a sublattice asymmetry in the perfect Bernal stacked graphene, resulting in a band gap opening on the order of 0.1 eV at the Dirac point of graphene. The induced asymmetry is significantly reduced when the stacking arrangement deviates from the Bernal stacking, which consequently results in closing of the graphene band gap. Hydrogenation of graphene can open up large band gaps; specifically, the band gap of graphone is larger than 2.5 eV. When deposited on the hBN substrate, the band gap of graphone is lowered by ≈1 eV due to the polarization of the hBN substrate surface. Effects such as the band gap modulation by interaction with the surrounding materials can have a profound impact on the devices employing two-dimensional monolayer materials. The band offsets suggest that hBN can also be used as a gate dielectric in the devices based on graphene and hydrogenated graphene.

ACKNOWLEDGMENTS

This work is supported partly by the Interconnect Focus Center funded by the MARCO program of SRC and the State of New York, NSF PetaApps grant number 0749140, and an anonymous gift from Rensselaer. Computing resources of the Computational Center for Nanotechnology Innovations at Rensselaer partly funded by the State of New York and of nanoHUB.org funded by the National Science Foundation have been used for this work. We thank Timothy Boykin, Mathieu Luisier, and Gerhard Klimeck for helpful discussions.

REFERENCES

1. ITRS. *International Technology Roadmap for Semiconductors.* 2009. Available from http://www.itrs.net/Links/2009ITRS/Home2009.htm.
2. A.K. Geim and K.S. Novoselov. The rise of graphene. *Nature Materials*, 6, 183–191, 2007.
3. F. Schwierz. Graphene transistors. *Nature Nanotechnology*, 5, 487–496, 2010.
4. K.S. Novoselov, A.K. Geim, S.V. Morozov, D. Jiang, Y. Zhang, S.V. Dubonos, I.V. Grigorieva, and A.A. Firsov. Electric field effect in atomically thin carbon films. *Science*, 306, 666–669, 2004.
5. K.S. Novoselov, A.K. Geim, S.V. Morozov, D. Jiang, M.I. Katsnelson, I.V. Grigorieva, S.V. Dubonos, and A.A. Firsov. Two-dimensional gas of massless Dirac fermions in graphene. *Nature*, 438, 197–200, 2005.
6. R.R. Nair, P. Blake, A.N. Grigorenko, K.S. Novoselov, T.J. Booth, T. Stauber, N.M.R. Peres, and A.K. Geim. Fine structure constant defines visual transparency of graphene. *Science*, 320, 1308–1308, 2008.
7. A.A. Balandin, S. Ghosh, W.Z. Bao, I. Calizo, D. Teweldebrhan, F. Miao, and C.N. Lau. Superior thermal conductivity of single-layer graphene. *Nano Letters*, 8, 902–907, 2008.
8. Y.M. Lin, A. Valdes-Garcia, S.J. Han, D.B. Farmer, I. Meric, Y.N. Sun, Y.Q. Wu, C. Dimitrakopoulos, A. Grill, P. Avouris, and K.A. Jenkins. Wafer-scale graphene integrated circuit. *Science*, 332, 1294–1297, 2011.
9. T. Mueller, F.N.A. Xia, and P. Avouris. Graphene photodetectors for high-speed optical communications. *Nature Photonics*, 4, 297–301, 2010.
10. H. Park, J.A. Rowehl, K.K. Kim, V. Bulovic, and J. Kong. Doped graphene electrodes for organic solar cells. *Nanotechnology*, 21, 505204, 2010.
11. M.H. Liang and L.J. Zhi. Graphene-based electrode materials for rechargeable lithium batteries. *Journal of Materials Chemistry*, 19, 5871–5878, 2009.
12. R. Murali, Y.X. Yang, K. Brenner, T. Beck, and J.D. Meindl. Breakdown current density of graphene nanoribbons. *Applied Physics Letters*, 94, 243114, 2009.
13. F. Chen, J.L. Xia, D.K. Ferry, and N.J. Tao. Dielectric screening enhanced performance in graphene FET. *Nano Letters*, 9, 2571–2574, 2009.
14. J. Yan, K.S. Thygesen, and K.W. Jacobsen. Nonlocal screening of plasmons in graphene by semiconducting and metallic substrates: First-principles calculations. *Physical Review Letters*, 106, 146803, 2011.
15. J.H. Chen, C. Jang, S.D. Xiao, M. Ishigami, and M.S. Fuhrer. Intrinsic and extrinsic performance limits of graphene devices on SiO2. *Nature Nanotechnology*, 3, 206–209, 2008.
16. L.A. Ponomarenko, R. Yang, T.M. Mohiuddin, M.I. Katsnelson, K.S. Novoselov, S.V. Morozov, A.A. Zhukov, F. Schedin, E.W. Hill, and A.K. Geim. Effect of a high-kappa environment on charge carrier mobility in graphene. *Physical Review Letters*, 102, 206603, 2009.

17. C.R. Dean, A.F. Young, I. Meric, C. Lee, L. Wang, S. Sorgenfrei, K. Watanabe, T. Taniguchi, P. Kim, K.L. Shepard, and J. Hone. Boron nitride substrates for high-quality graphene electronics. *Nature Nanotechnology*, 5, 722–726, 2010.

18. J.M. Xue, J. Sanchez-Yamagishi, D. Bulmash, P. Jacquod, A. Deshpande, K. Watanabe, T. Taniguchi, P. Jarillo-Herrero, and B.J. Leroy. Scanning tunnelling microscopy and spectroscopy of ultra-flat graphene on hexagonal boron nitride. *Nature Materials*, 10, 282–285, 2011.

19. J. Schiefele, F. Sols, and F. Guinea. Temperature dependence of the conductivity of graphene on boron nitride. arXiv:1202.2440v1 [cond-mat.mtrl-sci], 2012.

20. I. Meric, C. Dean, A. Young, J. Hone, P. Kim, K.L. Shepard, and IEEE. Graphene field-effect transistors based on boron nitride gate dielectrics. In *2010 International Electron Devices Meeting—Technical Digest*, 2010, pp. 556–559.

21. L. Britnell, R.V. Gorbachev, R. Jalil, B.D. Belle, F. Schedin, A. Mishchenko, T. Georgiou, M.I. Katsnelson, L. Eaves, S.V. Morozov, N.M.R. Peres, J. Leist, A.K. Geim, K.S. Novoselov, and L.A. Ponomarenko. Field-effect tunneling transistor based on vertical graphene heterostructures. *Science*, 335, 947–950, 2012.

22. N. Kharche and S.K. Nayak. Quasiparticle band gap engineering of graphene and graphone on hexagonal boron nitride substrate. *Nano Letters*, 11, 5274–5278, 2011.

23. A.D. Hernandez-Nieves, B. Partoens, and F.M. Peeters. Electronic and magnetic properties of superlattices of graphene/graphane nanoribbons with different edge hydrogenation. *Physical Review B*, 82, 165412, 2010.

24. M.Y. Han, B. Ozyilmaz, Y.B. Zhang, and P. Kim. Energy band-gap engineering of graphene nanoribbons. *Physical Review Letters*, 98, 206805, 2007.

25. Y.W. Son, M.L. Cohen, and S.G. Louie. Energy gaps in graphene nanoribbons. *Physical Review Letters*, 97, 216803, 2006.

26. L.Z. Li, R. Qin, H. Li, L.L. Yu, Q.H. Liu, G.F. Luo, Z.X. Gao, and J. Lu. Functionalized graphene for high-performance two-dimensional spintronics devices. *ACS Nano*, 5, 2601–2610, 2011.

27. D.A. Abanin, A.V. Shytov, and L.S. Levitov. Peierls-type instability and tunable band gap in functionalized graphene. *Physical Review Letters*, 105, 086802, 2010.

28. D. Haberer, D.V. Vyalikh, S. Taioli, B. Dora, M. Farjam, J. Fink, D. Marchenko, T. Pichler, K. Ziegler, S. Simonucci, M.S. Dresselhaus, M. Knupfer, B. Buchner, and A. Gruneis. Tunable band gap in hydrogenated quasi-free-standing graphene. *Nano Letters*, 10, 3360–3366, 2010.

29. J. Zhou, Q. Wang, Q. Sun, X.S. Chen, Y. Kawazoe, and P. Jena. Ferromagnetism in semihydrogenated graphene sheet. *Nano Letters*, 9, 3867–3870, 2009.

30. R. Balog, B. Jorgensen, L. Nilsson, M. Andersen, E. Rienks, M. Bianchi, M. Fanetti, E. Laegsgaard, A. Baraldi, S. Lizzit, Z. Sljivancanin, F. Besenbacher, B. Hammer, T.G. Pedersen, P. Hofmann, and L. Hornekaer. Bandgap opening in graphene induced by patterned hydrogen adsorption. *Nature Materials*, 9, 315–319, 2010.

31. D.C. Elias, R.R. Nair, T.M.G. Mohiuddin, S.V. Morozov, P. Blake, M.P. Halsall, A.C. Ferrari, D.W. Boukhvalov, M.I. Katsnelson, A.K. Geim, and K.S. Novoselov. Control of graphenes properties by reversible hydrogenation: Evidence for graphane. *Science*, 323, 610–613, 2009.

32. X. Gonze, B. Amadon, P.M. Anglade, J.M. Beuken, F. Bottin, P. Boulanger, F. Bruneval, D. Caliste, R. Caracas, M. Cote, T. Deutsch, L. Genovese, P. Ghosez, M. Giantomassi, S. Goedecker, D.R. Hamann, P. Hermet, F. Jollet, G. Jomard, S. Leroux, M. Mancini, S. Mazevet, M.J.T. Oliveira, G. Onida, Y. Pouillon, T. Rangel, G.M. Rignanese, D. Sangalli, R. Shaltaf, M. Torrent, M.J. Verstraete, G. Zerah, and J.W. Zwanziger. ABINIT: First-principles approach to material and nanosystem properties. *Computer Physics Communications*, 180, 2582–2615, 2009.

33. N. Troullier and J.L. Martins. Efficient pseudopotentials for plane-wave calculations. *Physical Review B*, 43, 1993–2006, 1991.

34. S. Goedecker, M. Teter, and J. Hutter. Separable dual-space Gaussian pseudopotentials. *Physical Review B*, 54, 1703–1710, 1996.

35. H.J. Monkhorst and J.D. Pack. Special points for Brillouin-zone integrations. *Physical Review B*, 13, 5188–5192, 1976.

36. M.S. Hybertsen and S.G. Louie. Electron correlation in semiconductors and insulators: Band gaps and quasiparticle energies. *Physical Review B*, 34, 5390–5413, 1986.

37. S. Ismail-Beigi. Truncation of periodic image interactions for confined systems. *Physical Review B*, 73, 233103, 2006.

38. G. Kresse and J. Furthmuller. Efficiency of ab-initio total energy calculations for metals and semiconductors using a plane-wave basis set. *Computational Materials Science*, 6, 15–50, 1996.

39. T. Bucko, J. Hafner, S. Lebegue, and J.G. Angyan. Improved description of the structure of molecular and layered crystals: Ab initio DFT calculations with van der Waals corrections. *Journal of Physical Chemistry A*, 114, 11814–11824, 2010.

40. G. Kresse and D. Joubert. From ultrasoft pseudopotentials to the projector augmented-wave method. *Physical Review B*, 59, 1758–1775, 1999.

41. J.P. Perdew, K. Burke, and M. Ernzerhof. Generalized gradient approximation made simple. *Physical Review Letters*, 77, 3865–3868, 1996.

42. S. Grimme. Semiempirical GGA-type density functional constructed with a long-range dispersion correction. *Journal of Computational Chemistry*, 27, 1787–1799, 2006.

43. B. Sachs, T.O. Wehling, M.I. Katsnelson, and A.I. Lichtenstein. Adhesion and electronic structure of graphene on hexagonal boron nitride substrates. *Physical Review B*, 84, 195414, 2012.

44. G. Giovannetti, P.A. Khomyakov, G. Brocks, P.J. Kelly, and J. van den Brink. Substrate-induced band gap in graphene on hexagonal boron nitride: Ab initio density functional calculations. *Physical Review B*, 76, 073103, 2007.

45. E.J. Kan, H. Ren, F. Wu, Z.Y. Li, R.F. Lu, C.Y. Xiao, K.M. Deng, and J.L. Yang. Why the band gap of graphene is tunable on hexagonal boron nitride. *Journal of Physical Chemistry C*, 116, 3142–3146, 2012.

46. L. Britnell, R.V. Gorbachev, R. Jalil, B.D. Belle, F. Schedin, M.I. Katsnelson, L. Eaves, S.V. Morozov, A.S. Mayorov, N.M.R. Peres, A.H. Castro Neto, J. Leist, A.K. Geim, L.A. Ponomarenko, and K.S. Novoselov. Electron tunneling through ultrathin boron nitride crystalline barriers. *Nano Letters*, 12, 1707–1710, 2012.

47. R. Decker, Y. Wang, V.W. Brar, W. Regan, H.Z. Tsai, Q. Wu, W. Gannett, A. Zettl, and M.F. Crommie. Local electronic properties of graphene on a BN substrate via scanning tunneling microscopy. *Nano Letters*, 11, 2291–2295, 2011.

48. S. Shallcross, S. Sharma, and O.A. Pankratov. Quantum interference at the twist boundary in graphene. *Physical Review Letters*, 101, 056803, 2008.

49. K.J. Jeon, Z. Lee, E. Pollak, L. Moreschini, A. Bostwick, C.M. Park, R. Mendelsberg, V. Radmilovic, R. Kostecki, T.J. Richardson, and E. Rotenberg. Fluorographene: A wide bandgap semiconductor with ultraviolet luminescence. *ACS Nano*, 5, 1042–1046, 2011.

50. J.T. Robinson, J.S. Burgess, C.E. Junkermeier, S.C. Badescu, T.L. Reinecke, F.K. Perkins, M.K. Zalalutdniov, J.W. Baldwin, J.C. Culbertson, P.E. Sheehan, and E.S. Snow. Properties of fluorinated graphene films. *Nano Letters*, 10, 3001–3005, 2010.

51. I. Zutic, J. Fabian, and S. Das Sarma. Spintronics: Fundamentals and applications. *Reviews of Modern Physics*, 76, 323–410, 2004.

52. K. Natori, D. Otani, and N. Sano. Thickness dependence of the effective dielectric constant in a thin film capacitor. *Applied Physics Letters*, 73, 632–634, 1998.

53. J.B. Neaton, M.S. Hybertsen, and S.G. Louie. Renormalization of molecular electronic levels at metal-molecule interfaces. *Physical Review Letters*, 97, 216405, 2006.

54. K.S. Thygesen and A. Rubio. Renormalization of molecular quasiparticle levels at metal-molecule interfaces: Trends across binding regimes. *Physical Review Letters*, 102, 046802, 2009.
55. J.M. Garcia-Lastra, C. Rostgaard, A. Rubio, and K.S. Thygesen. Polarization-induced renormalization of molecular levels at metallic and semiconducting surfaces. *Physical Review B*, 80, 245427, 2009.
56. J.M. Garcia-Lastra and K.S. Thygesen. Renormalization of optical excitations in molecules near a metal surface. *Physical Review Letters*, 106, 187402, 2011.
57. A. Franceschetti and A. Zunger. Addition energies and quasiparticle gap of CdSe nanocrystals. *Applied Physics Letters*, 76, 1731–1733, 2000.
58. X. Jiang, N. Kharche, P. Kohl, T.B. Boykin, G. Klimeck, M. Luisier, P.M. Ajayan, and S.K. Nayak. Giant quasiparticle band gap modulation in graphene nanoribbons supported on weakly interacting surfaces. Submitted for publication, 2012.
59. M. Rohlfing. Redshift of excitons in carbon nanotubes caused by the environment polarizability. *Physical Review Letters*, 108, 087402, 2012.
60. K.S. Subrahmanyam, P. Kumar, U. Maitra, A. Govindaraj, K. Hembram, U.V. Waghmare, and C.N.R. Rao. Chemical storage of hydrogen in few-layer graphene. *Proceedings of the National Academy of Sciences of the United States of America*, 108, 2674–2677, 2011.
61. J. Robertson. Band offsets of wide-band-gap oxides and implications for future electronic devices. *Journal of Vacuum Science and Technology B*, 18, 1785–1791, 2000.

3 On the Possibility of Observing Tunable Laser-Induced Bandgaps in Graphene

Hernán L. Calvo, Horacio M. Pastawski,
Stephan Roche, and Luis E.F. Foa Torres

CONTENTS

3.1 INTRODUCTION

Graphene offers a wealth of fascinating opportunities for the study of truly two-dimensional physics [1]. The last years have seen an unprecedented pace of development regarding the study of its electronic, mechanical, and optical properties, among many others. More recently, the interplay between these properties is making its way to the forefront of graphene research [2–5], opening up many promising paths for technology. Here, we focus on the interplay between optical and electronic properties, and we address the issue of tunability of the latter using a laser field. In the following we give a brief overview to this topic, with emphasis on our recent work [11], which we extend in the forthcoming sections.

In 2008 Syzranov and collaborators [6] predicted the emergence of dynamical gaps in graphene when illuminated by a laser of frequency Ω. The authors [6,7] studied the effects of these gaps, located at $\pm\hbar\Omega/2$ above/below the charge neutrality point on the electronic properties of n-p junctions illuminated by a linearly polarized laser. Later on, Oka and Aoki [8] predicted the emergence of another gap at the charge neutrality point when changing the polarization from linear to circular; this was confirmed in [9]. Similar results were also presented for bilayer graphene by Abergel and Chakraborty [10].

In what follows, we will show that these strong renormalizations of the electronic structure of graphene can be rationalized in a transparent way using the Floquet theory, which allows us to explore electronic and transport properties of materials in the presence of oscillating electromagnetic fields. The origin of the dynamical gaps will be shown to be related to an inelastic Bragg's scattering mechanism occurring in a higher-dimensional space (Floquet space). In [11] we followed this line of research trying to answer questions such as: What is the role of the laser polarization on these dynamical gaps, and which experimental setup would be needed to observe them? Our results showed that the polarization could be used to strongly modulate the associated electrical response. Furthermore, the first atomistic simulations of the electrical response (dc conductance) of a large graphene ribbon (of about 1 micron lateral size) were presented. Our results hinted that a transport experiment carried out while illuminating with a laser in the mid-infrared could unveil these unconventional phenomena. There are also additional ingredients that add even more interest to this proposal: the current interest in finding novel ways to open bandgaps in graphene and the promising prospects for optoelectronics applications.

Other recent studies have also contributed to different aspects of this field, including the analogy between the electronic spectra of graphene in laser fields with that of static graphene superlattices [12], the possible manifestation of additional Dirac points induced by a superimposed lattice potential [13], a study of the effects of radiation on graphene ribbons showing that for long ribbons there is a ballistic regime where edge states transport dominates [14], a proposal to unveil the dynamical gaps using photon-echo experiments [15], and the effects of radiation on tunneling [16] or tunneling times [17]. The influence of mentioned laser-induced bandgaps on the polarizability was also recently discussed in [18]. On the experimental side, the only available experiments are in the low-intensity regime where renormalizations such as those mentioned before are not important [19]. Further theoretical calculations within a master equation approach in that regime were also presented in [20].

Another captivating possibility is the generation of a topological insulator through ac fields as proposed in [21,22]. The idea is that ac fields could be used to generate topologically protected states that conduct electricity along the sample edges. We expect for this to become one of the most promising research lines in the field of ac transport [23–25].

In the following sections we discuss the interplay of a laser field in both the electronic structure and transport properties of graphene. In Section 3.2 we introduce the so-called **k.p** model and describe in detail the underlying mechanisms for the opening of energy gaps both in the vicinity of the low-energy Dirac point and at the energies $\varepsilon = \pm\hbar\Omega/2$, where new symmetries are superimposed by the laser field. We derive approximate formulas for the new dynamical gaps and contrast them with numerical data. In Section 3.3 we turn to a π-orbitals tight-binding model where the explicit structure of the lattice is taken into account. Under this scenario, we first incorporate the time-dependent field by using the Floquet theory, and then we calculate the local density of states for a linearly polarized laser. When studying the electronic properties of a graphene layer weakly interacting with a boron-nitride substrate, new Dirac points are found to occur at higher energies related with the wavelength of the Moiré pattern induced by the graphene/boron-nitride mismatch as reported in [13]. In our case, laser illumination induces a superimposed spatial pattern [12], which may lead to a similar phenomenon when the laser polarization is linear. Departure from linear polarization leads to a suppression of these new Dirac points and dynamical gaps develop. Furthermore, we analyze the transport properties of the model by calculating the dc conductance for laser wavelengths in the mid-infrared domain and explore the laser-induced dips for different choices of the parameters. Finally, in Section 3.4 we summarize the obtained results and discuss their impact on possible future applications.

3.2 DIRAC FERMIONS: K.P APPROACH

Because of the honeycomb symmetry dictating the arrangement of carbon atoms in graphene, the electronic states that are relevant to the transport properties are located close to the two independent Dirac points (valleys) K and K' alternately arranged at the edges of the Brillouin zone. For a clean sample (i.e., without any impurity or distortions), and by considering that the proposed external ac field does not introduce any intervalley coupling, we can treat both degeneracy points separately.

For each one of the two valleys, the electronic states can be accurately described by the **k.p** method where the energy spectrum is assumed to depend linearly on the momentum **k**. The electronic states are then computed by an envelope wave function $\Psi = (\Psi_A, \Psi_B)^T$. These two components refer to the two interpenetrating sublattices A and B [26,27], and are usually called the sublattice pseudospin degrees of freedom, owing to their analogy with the actual spin of the carriers. The energies and wave functions are thus solutions of the Dirac equation:

$$\mathbf{H}\Psi = \varepsilon\Psi \qquad (3.1)$$

with **H** the Hamiltonian operator of the system and ε the eigenenergies. The time periodic electromagnetic field, with period $T = 2\pi/\Omega$, is assumed to be created by a monochromatic plane wave traveling along the z axis, perpendicular to the plane defined by the graphene monolayer. The vector potential associated with the laser field is thus written as

$$\mathbf{A}(t) = \Re\left[\mathbf{A}_0 e^{-i\Omega t}\right] \qquad (3.2)$$

where $\mathbf{A}_0 = A_0(1, e^{i\varphi})$ refers to the intensity $A_0 = E/\Omega$ and polarization φ of the field. For this choice of the parameters, $\varphi = 0$ yields a linearly polarized field $\mathbf{A}(t) = A_0 \cos \Omega t\ (\mathbf{x} + \mathbf{y})$, while $\varphi = \pi/2$ results in a circularly polarized field $\mathbf{A}(t) = A_0\ (\cos \Omega t\ \mathbf{x} + \sin \Omega t\ \mathbf{y})$. Notice here that the **k.p** approach preserves circular symmetry around the degeneracy point, and in consequence, there is no preferential choice for the in-plane axes **x** and **y**. As we shall see in Section 3.3, this is not the case in the tight-binding model for which an explicit orientation of the lattice has to be fixed. Therefore the graphene electronic states in the presence of the ac field are encoded in the Hamiltonian:

$$\mathbf{H}(t) = v_F \boldsymbol{\sigma} \cdot \left[\mathbf{p} - e\mathbf{A}(t) \right] \tag{3.3}$$

where $v_F \approx 10^6$ m/s is the typical value for the Fermi velocity in graphene and $\boldsymbol{\sigma} = (\hat{\sigma}_x, \hat{\sigma}_y)$ is the vector of Pauli matrices describing the pseudospin degree of freedom.

We operate in the nonadiabatic regime in which the electronic dwell time (i.e., the traverse time along the laser spot) is larger than the period T of the laser, such that the electron experiences several oscillations of the laser field. Hence the Floquet theory represents a suitable approach to describe such electron-photon scattering processes. The Hamiltonian presented in Equation (3.3) is thus expanded into a composite basis between the usual Hilbert space and the space of time periodic functions. This new basis defines the so-called Floquet pseudostates $|\mathbf{k}, n\rangle_{\pm}$, where $\mathbf{k}$ is the electronic wave vector, characterized by its momentum, and n is the Floquet index associated with the n mode of a Fourier decomposition of the modulation. The subindex refers to the alignment of the pseudospin with respect to the momentum. Considering the K valley, a $+$ $(-)$ labeled state has its pseudospin oriented parallel (antiparallel) to the electronic momentum. In this representation of the Hamiltonian, the time dependence introduced by the external potential is thus replaced by a series of system's replicas arising from the Fourier decomposition. In this so-called Floquet space, one has to solve a time-independent Schrödinger equation with Floquet Hamiltonian $\mathbf{H}_F = \mathbf{H} - i\hbar\partial_t$. Recursive Green's function techniques [28] can be exploited to obtain both the dc component of the conductance and the density of states (DoS) from the so-called Floquet Green's functions [29].

The matrix elements of the resulting Floquet Hamiltonian are then computed as [30]

$$H_{i,j}^{(m,n)} = \frac{1}{T} \int_0^T dt\ H_{i,j}(t) e^{i(m-n)\Omega t} + m\hbar\Omega\delta_{mn}\delta_{ij} \tag{3.4}$$

where δ stands for the Kronecker symbol, i and j indicate the pseudospin orientation with respect to the momentum, and m and n are Fourier indices related with the field modulation. Therefore for a given vector $\mathbf{k} = k\,(\cos\alpha, \sin\alpha)$, the m diagonal block matrix contains both the electronic kinetic terms $\pm\,\hbar v_F\,k$ and the m component of the modulation of the field $m\hbar\Omega$, i.e.,

$$\mathbf{H}^{m,m} = \hbar v_F \begin{pmatrix} k + m\Omega/v_F & 0 \\ 0 & -k + m\Omega/v_F \end{pmatrix} \tag{3.5}$$

where we have transformed the standard basis (the one related with the nonequivalent sublattices) into a diagonal basis for the diagonal block. Due to the particular sinusoidal time dependence of the field, only inelastic transitions involving the absorption or emission of a single photon are allowed, i.e., $\Delta m = \pm 1$, and the corresponding off-diagonal block matrix connecting Floquet states with a different number of photons reads

$$\mathbf{H}^{(m,m+1)} = \begin{pmatrix} \gamma_1 & \gamma_2 \\ -\gamma_2 & -\gamma_1 \end{pmatrix} \tag{3.6}$$

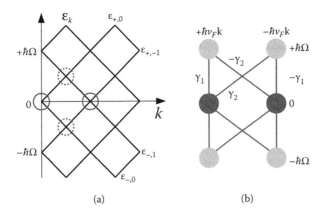

FIGURE 3.1   (a) Scheme of the quasi-energies as a function of the electronic momentum k. The relevant crossing regions, marked by circles with dotted and solid lines, yield the dynamical and central gap, respectively. (b) Representation of the Floquet Hamiltonian for the **k.p** approach: Circles correspond to Floquet states and lines denote inelastic hopping elements. (From Calvo, H.-L., et al., *Applied Physics Letters*, 98, 232103, 2011. Copyright © 2011 American Institute of Physics. With permission.)

and it is responsible for the photon-assisted tunneling processes. Here, the direct hopping term $\gamma_1 = eA_0 v_F/2$ $(\cos\alpha + e^{-i\varphi}\sin\alpha)$, with $\alpha = \tan^{-1}(k_y/k_x)$, sets the transition amplitude between Floquet states with same pseudospin character. On the contrary, the off-diagonal term $\gamma_2 = ieA_0 v_F/2$ $(e^{-i\varphi}\cos\alpha - \sin\alpha)$ introduces an inelastic back-scattering process that enables simultaneously both m and pseudospin transitions. This leads to the analog of an inelastic Bragg reflection as found in other contexts for inelastic scattering in carbon nanotubes [31,32].

In Figure 3.1 we provide a sketch of the Floquet spectrum for a given orientation of the momentum. The complete spectrum involves an average over all possible orientations, and it looks like a series of superimposed Dirac cones touching at energies $m\hbar\Omega$. The effect that the ac field induces on the electronic structure manifests whenever states with different pseudospin cross. Because of the hopping elements γ_1 and γ_2, a new family of gaps will open in the vicinity of the crossings shown by circles in Figure 3.1a. This defines a region where no states are available and the transport becomes drastically suppressed. As we will discuss in the next lines, the width of these gaps strongly depends on the intensity, frequency, and polarization of the field. For $eA_0 v_F < \hbar\Omega$ these effects are clear for energies close to a half-integer of $\hbar\Omega$, where a *dynamical gap* [6] opens as φ increases. On the other hand, as pointed out by Oka and Aoki [8], around the Dirac point another gap emerges and becomes pronounced in the circularly polarized case. As we will make clear later, the crucial difference between these two is the involved number of (photon-assisted) tunneling processes it takes to backscatter the conduction electrons. For the considered frequencies and intensities of the field, processes beyond the second order can be ignored.

In the mentioned studies [6,8], the authors have considered lasers in two ranges: the far infrared with $\hbar\Omega = 29$ meV or the visible range [8]. In the former, the gaps were predicted to be of about 6 meV for a photocurrent generated in a p-n junction;

in the latter, the photon energy of about 2 eV is much larger than the typical optical phonon energy 170 meV, and severe corrections to the transport properties due to dissipation of the excess energy via electron–(optical) phonon interactions can be expected. Moreover, as we pointed out in [11], appreciable effects in this last case required a power above 1 W/μm^2, which could compromise the material stability. To overcome both limitations we quantitatively explore the interaction with a laser in the mid-infrared range $\lambda = 2\text{–}10\ \mu$m, where photon energies can be made smaller than the typical optical phonon energy while keeping a much lower laser power.

3.2.1 DYNAMICAL GAPS UNVEILED THROUGH THE K.P MODEL

The leading order process in the interaction with the electromagnetic field occurs at $\varepsilon \sim \pm\hbar\Omega/2$. In this region, we can consider an effective two-level system in which the relevant states are $\left|\mathbf{k},0\right\rangle_+$ and $\left|\mathbf{k},1\right\rangle_-$. The difference between the eigenenergies is estimated as twice the hopping between the mentioned states, and the resulting gap is

$$\Delta_{k=\Omega/2v_F} \approx eA_0 v_F \sqrt{1-\cos\varphi\sin\alpha} \tag{3.7}$$

The same analysis can be done for the crossing between the symmetrically disposed states $\left|k,0\right\rangle_-$ and $\left|k,1\right\rangle_+$ retrieving the same value for the energy gap. This clearly shows a linear dependence with the field strength and, for a fixed value of A_0, is independent of the photon frequency. Furthermore, it is interesting to note that when averaging over all orientations of the wave vector, no net gap opens in the linearly polarized case ($\varphi = 0$) since for the particular value $\alpha = \pi/4$ the hopping vanishes and the backscattering mechanism is suppressed. For the calculation of the DoS, we use the Floquet Green's function technique, in which we define the following Green's function in the product space as

$$\mathbf{G}(\varepsilon,\mathbf{k}) = \left(\varepsilon\mathbf{I} - \mathbf{H}_F(\mathbf{k})\right)^{-1} \tag{3.8}$$

where the $\mathbf{k}$-vector dependence of the Floquet Hamiltonian enters in both the diagonal blocks (amplitudes) and off-diagonal ones (directions). The DoS related with the $m = 0$ level includes the sum over all possible values of $\mathbf{k}$ and writes

$$v_0(\varepsilon) = -\frac{1}{\pi}\int_0^\infty k\,dk\int_0^{2\pi}\frac{d\alpha}{2\pi}\Im\left[G_{+,+}^{0,0}(\varepsilon,\mathbf{k})+G_{-,-}^{0,0}(\varepsilon,\mathbf{k})\right] \tag{3.9}$$

In Figure 3.2a we show an example of the calculated DoS around the dynamical gap region for different values of the polarization. As can be seen, even in the linearly polarized case there is a strong modification in the DoS that would resemble the usual Dirac point for a Fermi energy around $\varepsilon \sim \pm\hbar\Omega/2$.

The decreasing number of allowed states for different orientations of the electronic momentum becomes evident through the depletion region. By increasing the polarization one finds that immediately for finite values of φ a gap is opened and

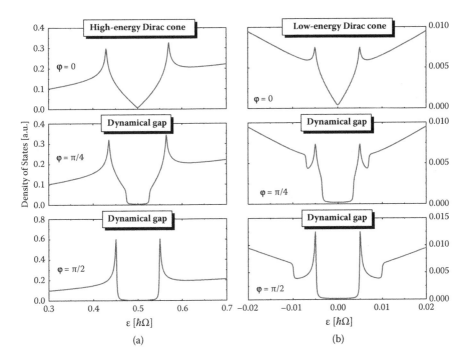

FIGURE 3.2   Density of states in graphene for several values of the laser polarization. (a) Lowest-order effect around $\varepsilon \sim \hbar\Omega/2$, $k \sim \Omega/2v_F$. (b) Second-order effects around ($\varepsilon \sim 0$, $k \sim 0$). The chosen parameters are given by $eA_0v_F/\hbar\Omega = 0.1$.

reaches its maximum for the circularly polarized case. Under this situation, it is clear from Equation (3.7) that there is no dependence with the orientation of the wave vector, which can be interpreted in terms of the recovered axial symmetry, characterized by the standard one-dimensional van Hove singularities emerging at the edges of the gap.

3.2.2   ENERGY GAP TUNING AROUND ENGINEERED LOW-ENERGY DIRAC CONES

To describe this mechanism, we refer to the schematic representation of the Floquet Hamiltonian depicted in Figure 3.1b. Here, the leading contribution around the Dirac point ($\varepsilon \approx 0$ and $k \approx 0$) comes from the four paths connecting the crossing states $|\mathbf{k},0\rangle_+$ and $|\mathbf{k},0\rangle_-$ through the first neighboring states. For the calculation, we observe that the mixing between these states is produced by the presence of the neighboring states $|\mathbf{k},\pm1\rangle_\pm$. The effective Hamiltonian can be reduced by a decimation procedure [28] in which we keep the relevant states at $m = 0$. The resulting correction from the diagonal terms of the self-energy cancels out exactly, such that the energy difference only arises from the effective hopping according to

$$\Delta_{k=0} \approx 2\frac{\left(eA_0v_f\right)^2}{\hbar\Omega}\sin\varphi \qquad (3.10)$$

The quadratic dependence with the field strength is due to the fact that the mixing between these two states involves both the absorption and emission of a single photon. Additionally, the inverse dependence with the frequency quantifies the amount of energy absorbed and emitted during the tunneling event. For the constraint $eA_0v_F < \hbar\Omega$ considered here, this gap is much smaller than the dynamical gap discussed before. Note also that there is no dependence of the gap with the orientation of the **k**-vector, since we are assuming $k = 0$. By inspecting Equation (3.10), it is easy to observe that the maximum value for the gap is reached in the circularly polarized case, while no net gap opens for the linearly polarized case. According to the present approximation, we notice that in this last case no ac effects should be observed in the DoS. However, as can be seen in Figure 3.2b, there is a strong modification around the Dirac point region reflected by an increased slope in comparison with the usual DoS without any laser field. To explain this, it is convenient to include the effect induced by the crossing between states $|\mathbf{k},-1\rangle_+$ and $|\mathbf{k},1\rangle_-$ around $\varepsilon = 0$ and $k = \Omega/v_F$ (right circle with solid line in Figure 3.1a). Although in this case the total energy difference going from $|k,0\rangle_+$ to $|k,0\rangle_-$ is $2\hbar\Omega$, we are in the resonant condition where transitions via auxiliary states $|\mathbf{k},\pm1\rangle_+$ involve the same energy difference $\hbar\Omega$. The presence of states $|\mathbf{k},2\rangle_+$ also contributes to the self-energy correction, and the effective hopping term connecting the $m = 0$ states originates a gap:

$$\Delta_{k=\Omega/v_F} \approx \frac{(eA_0v_F)^2}{2\hbar\Omega}\left[5 - 2\cos\varphi\sin 2\alpha - 3\cos^2\varphi\sin^2 2a\right]^{1/2} \quad (3.11)$$

that drops to zero for $\varphi = 0$, $\alpha = \pi/4$ (as in the dynamical gap) and reaches its maximum for $\varphi = \pi/2$. This contribution is of the same order as $\Delta_{k=0}$ and responsible for the modification of the DoS close to the Dirac point even for $\varphi = 0$. It is interesting to observe that, in addition to the gap, this transition yields the emergence of two surrounding peaks that persist even in the linear case.

In addition, we observe that this effect reduces by approximately a half the predicted value of Oka and Aoki [8]. Here we neglect higher-order contributions since we consider that the strength of the laser is small compared to the driving frequency. However, as recently discussed in [33], in the opposite limit where the number of involved photons is large, additional scattering processes enable the existence of states around the studied regions and the gaps are effectively closed.

3.3   TIGHT-BINDING MODEL

In the previous sections, we have shown how the laser field modifies the electronic structure of two-dimensional graphene, at least in any experiment carried out over a time much larger than the period T. A natural question is if these effects would be observable in a transport experiment and how. To such end we turn now to the calculation of the transport response at zero temperature using Floquet theory applied to a tight-binding (TB) π-orbitals Hamiltonian. As we shall see, the correspondence between this model and the **k.p** approach becomes evident in the *bulk limit*, where the width of the ribbon is of the order of the laser's spot.

In the real space defined by the sites of the two interpenetrating sublattices, the electromagnetic field can be accounted for through an additional phase in the hopping γ_{ij} connecting two adjacent sites $\mathbf{r}_i$ and $\mathbf{r}_j$ through the Peierls substitution:

$$\gamma_{ij} = \gamma_0 \exp\left(i \frac{2\pi}{\Phi_0} \int_{r_i}^{r_j} \mathbf{A}(t) \cdot d\mathbf{r} \right) \tag{3.12}$$

where $\gamma_0 \approx 2.7$ eV is the hopping amplitude at zero field and Φ_0 is the quantum of magnetic flux. The intensity of the magnetic vector potential is assumed to be constant along the whole sample where the irradiation takes place. Therefore the computation of such a phase is simply the scalar product between the vector potential and the vector connecting the two sites.

For numerical convenience, we consider an armchair edge structure for the lattice, and the vector potential is defined as $\mathbf{A}(t) = A_x \cos \Omega t \, \mathbf{x} + A_y \sin \Omega t \, \mathbf{y}$, where x and y coordinates are in the plane of the nanoribbon. It is important to observe that the particular choice of the edges does not restrict the outcoming results once the bulk limit is reached. For the armchair structure we can distinguish three principal orientations according to the angle between $\mathbf{r}_j - \mathbf{r}_i$ and $\mathbf{A}(t)$. By considering the scheme in Figure 3.3b, we define the hopping elements $\gamma_{+,\pm}(t)$ and $\gamma_{2+,0}(t)$, where the subindex denotes the x and y components of two adjacent sites. In this representation, we have

$$\gamma_{+,\pm}(t) = \gamma_0 \exp\left[i \frac{a\pi}{\Phi_0} \left(A_x \cos \Omega t \pm \sqrt{3} A_y \sin \Omega t \right) \right] \tag{3.13}$$

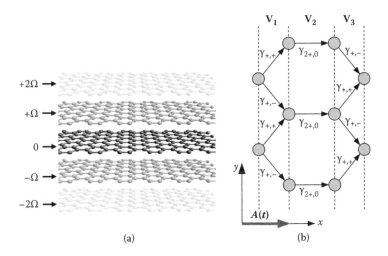

(a)   (b)

FIGURE 3.3   (a) Schematic view of the Floquet space for the tight-binding model in the real space. (b) Opening, longitudinal, and closing hopping matrices according to the relative direction between the adjacent sites and the vector potential in an armchair edge structure.

for those terms with simultaneous x and y components and

$$\gamma_{2+,0}(t) = \gamma_0 \exp\left[i\frac{2a\pi}{\Phi_0}A_x\cos\Omega t\right] \tag{3.14}$$

for the hopping element with only an x component, respectively (see Figure 3.3b). Here $a \approx 0.142$ nm is the nearest carbon-carbon distance. The Fourier decomposition of these expressions is based on the Anger-Jacobi expansion that yields a m transition amplitude in terms of Bessel functions:

$$\gamma_{+,\pm}^m = \gamma_0 \sum_{k=-\infty}^{\infty} i^k J_k(z_x) J_{m-k}(\pm z_y) \tag{3.15}$$

$$\gamma_{2+,0}^m = \gamma_0 i^m J_m(2z_x) \tag{3.16}$$

where $z_y = \sqrt{3}\,\pi a A_y/\Phi_0$ and $z_y = \sqrt{3}\pi a A_y/\Phi_0$. In contrast to the **k.p** approximation, in this situation we may have processes involving the absorption or emission of several photons at once. This, however, decays rapidly with the number of photons, and the main contribution still comes from the renormalization at $m = 0$ and the leading inelastic order $m = \pm 1$.

We compute the hopping matrices (opening, longitudinal, and closing) as those connecting two adjacent transverse layers such that the difference in the Floquet indexes is m (see Figure 3.3b). If N denotes the number of transverse states, i.e., the number of electronic sites in each vertical layer, the hopping matrices are $N \times N$ blocks expressed as

$$\mathbf{V}_1^m = \begin{pmatrix} \gamma_{+,+}^m & \gamma_{+,-}^m & \\ 0 & \gamma_{+,+}^m & \\ & & \ddots \end{pmatrix}, \mathbf{V}_2^m = \begin{pmatrix} \gamma_{2+0}^m & 0 & \\ 0 & \gamma_{2+,0}^m & \\ & & \ddots \end{pmatrix}, \mathbf{V}_3^m = \begin{pmatrix} \gamma_{+,-}^m & 0 & \\ \gamma_{+,+}^m & \gamma_{+,-}^m & \\ & & \ddots \end{pmatrix} \tag{3.17}$$

for the opening, longitudinal, and closing geometries, respectively. Therefore in the natural basis defined on the real space, the total Floquet Hamiltonian is composed by a periodic block tridiagonal structure where the off-diagonal block matrices contain the different $\mathbf{V}_i^m$ terms and the diagonal block only accounts for the site energies $E_m = m\hbar\Omega$ since no gate voltages are assumed. For the proposed values of the field and a typical regularization energy $\eta \approx 30$ μeV, the bulk limit is reached when $N > 10^4$. Hence it is necessary to decompose the Floquet Hamiltonian in terms of transverse modes, since otherwise one should deal with $O(N^3)$ operations, increasing enormously the computation time. For this reason, we limit ourselves to the linearly polarized case in which $A_y = 0$, which is already enough to give a hint on the transport effects. Other choices of the laser orientation or polarization would be subjected to the knowledge of more tricking bases or the employment of parallel computing techniques that are beyond the scope of this work.

The total Hamiltonian is transformed according to the normal modes of the sublattices A and B. The rotation matrix results in a trivial expansion of that in the appendix of [34] in the composite Floquet basis. The spatial part of the Hilbert space is thus reduced, and the hopping matrices now have the dimension $2N + 1$ of the (truncated) Fourier space. The resulting single-mode layers preserve the same Hamiltonian structure with $V_i^m(q) = 2V_i^m \cos [\pi q/(2N + 1)]$ for $i = 1, 3$ and $V_2^m(q) = V_2^m$, where q is the mode number.

The DoS is therefore obtained through the Floquet Green's functions defined as

$$\mathbf{G}_F^q(\varepsilon) = \left(\varepsilon \mathbf{I} - \mathbf{H}_F^q - \mathbf{\Sigma}_F^q(\varepsilon)\right)^{-1} \qquad (3.18)$$

where $\mathbf{H}_F^q$ is the q mode Hamiltonian containing the diagonal block of the Fourier components and is written as the following matrix:

$$\mathbf{H}_F^q(\varepsilon) = \begin{pmatrix} \ddots & & & \\ & -\hbar\Omega & 0 & 0 & \\ & 0 & 0 & 0 & \\ & 0 & 0 & +\hbar\Omega & \\ & & & & \ddots \end{pmatrix} \qquad (3.19)$$

The self-energy $\mathbf{\Sigma}_F^q$ is a correction term arising from the presence of the neighbor sites at both sides of the chain. The effective Hamiltonian $\mathbf{H}_{eff}^q(\varepsilon) = \mathbf{H}_F^q + \mathbf{\Sigma}_F^q(\varepsilon)$ for a transverse layer can be calculated recursively by using a decimation procedure, and the DoS is thus obtained from the resulting Green's function as

$$\nu_0(\varepsilon) = -\frac{1}{\pi}\sum_q \Im\left[\mathbf{G}_F^q(\varepsilon)\right]_{0,0} \qquad (3.20)$$

where the subindex 0, 0 corresponds to the zero photon state. Note the similarity of this expression with Equation (3.9), where the amplitude and direction of the electronic momentum are discretized in the q modes.

By employing this numerical strategy, we compute the DoS for an armchair edge structure and compare it with the one obtained from the $\mathbf{k.p}$ approach. As we can observe in Figure 3.4, there is in general a good agreement between both results. In particular, the depletion region around the resonances $\varepsilon \sim \pm\hbar\Omega/2$ is also reproduced by the TB model. However, a closer look at the peaks reveals small differences between both curves. For the TB method, we see two peaks that surround the peak obtained in

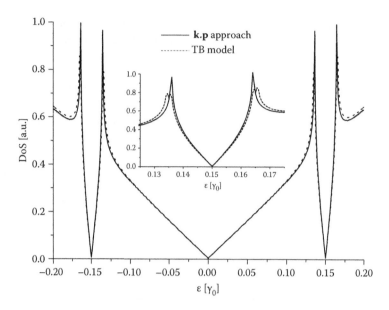

FIGURE 3.4 Density of states for linear polarization of the laser field calculated by both the **k.p** approach (solid line) and tight-binding (TB) model (dotted line). The chosen parameters of the laser are $\hbar\Omega = 0.3\ \gamma_0$ and $A_0 = 0.01\ \Phi_0/\pi a$. Inset: DoS in the depletion region.

the **k.p** model (see inset of Figure 3.4). The origin of the splitting of the peaks in the DoS can be attributed to a trigonal warping effect that is inherently included in the TB Hamiltonian, since the explicit structure of the lattice is taken into account.

This effect is also present in the region close to the Dirac point, where both approaches reproduce an increased slope, compared with the bare result without a laser. This, however, is not visible in Figure 3.4 since the intensity of the laser is too small compared with the frequency. To estimate the relevance of this effect, we explore the DoS around the Dirac point for several values of the intensity (Figure 3.5a) and frequency (Figure 3.5b) of the laser field. In the left panel we fix the frequency at $\hbar\Omega = 0.3\ \gamma_0$ and increase the intensity ($A_0 = E/\Omega$) from 0 to 0.05 $\Phi_0/\pi a$. As we can observe, the effect the intensity induces on the distance between the peaks is approximately linear, as well as the width of the depletion region. In Figure 3.5b we fix the intensity at this final value and decrease the frequency from 0.5 γ_0 to 0.2 γ_0. In this situation the distance between the two peaks remains approximately the same while changing the frequency, thus revealing a small (even negligible) dependence.

The numerical analysis of the DoS can be complemented by considering the mode decomposition shown in Figure 3.6. This representation allows an alternative inspection of the different effects we have been discussing throughout this section. Here, we plot the DoS in grey scale as a function of the energy for each transverse mode.

The sum of all these contributions gives the standard DoS. In the figure, each panel represents a fixed value of the laser intensity, starting from $A_0 = 0$ to 0.05 $\Phi_0/\pi a$. In the absence of a laser field, we can observe (see upper left panel) the linear behavior

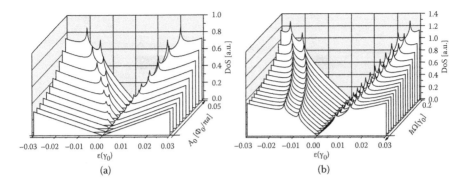

FIGURE 3.5 Density of states around the Dirac point for a linearly polarized field as a function of the laser intensity ($\hbar\Omega = 0.3\,\gamma_0$, left panel) and the frequency ($A_0 = 0.05\,\Phi_0/\pi a$, right panel). The white curve in both figures corresponds to the same choice of the parameters ($\hbar\Omega = 0.3\,\gamma_0$ and $A_0 = 0.01\,\Phi_0/\pi a$).

of the single-mode bandgaps as we move away from the central mode where no gaps occur. This central mode is defined via the identity $V_1^m(N_C) = V_2^m$ in which the system would be homogeneous; thereby no gap should be opened for this mode. In the considered example this results in $N_C \sim 6600$. This point constitutes the symmetry axis where both contributions at left and right sides are the same. As we turn on the laser, a gap is suddenly opened at $\hbar\Omega/2 = 0.15\,\gamma_0$ in each mode (except the central one), and its width increases linearly with the intensity of the laser.

In addition, we observe that the emergence of the splitting of the peaks is manifested by a slight asymmetry in their position with respect to the central mode. For $A_0 > 0.03\,\Phi_0/\pi a$ the contribution from side modes originates the increased slope around the Dirac point region. The DoS in this region also shows an asymmetry in the size of the lobes that originates the splitting of the peaks.

3.3.1 ELECTRONIC TRANSPORT THROUGH IRRADIATED GRAPHENE

In order to estimate the influence of the laser field in a transport experiment, we calculate the two-terminal dc conductance through a graphene stripe of $1 \times 1\ \mu m$ in the presence of a linearly polarized laser. The sample is connected to two leads at both sides, considered a prolongation of the stripe where no laser is applied.

By following the same strategy as we used in the calculation of the DoS, we obtain the dc conductance from the recursive calculation of the Floquet Green's functions. In this sense, the mode decomposition is again a key tool in the numerical implementation of the recursive formula for the self-energy correction. In particular, the semi-infinite leads are incorporated through an initial self-energy where the opening, longitudinal, and closing hopping matrices result in a diagonal. Since in this region no ac fields are applied, the elements of the above hopping matrices are simply $\gamma_{+,\pm}^m = \gamma_{2+,0}^m = \gamma_0\delta_{m,0}$. Therefore as shown in Figure 3.7, the resulting lattice structure for a single mode in the leads is a dimer with alternating hoppings γ_0 and $2\gamma_0\cos[\pi q/(2N+1)]$. In the sample region, however, the effect of the time-dependent

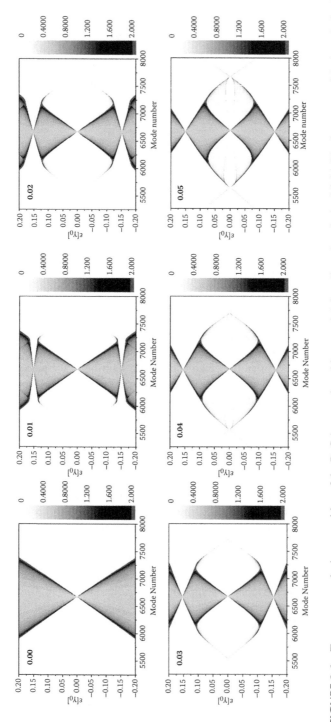

FIGURE 3.6 Transverse mode decomposition of the DoS (grey scale, arbitrary units) for a linearly polarized field. The frequency of the laser is fixed at $\hbar\Omega = 0.3\,\gamma_0$, while the intensity varies from 0 to 0.05 in units of $\Phi_0/\pi a$. The central mode in this example is $N_C \sim 6600$.

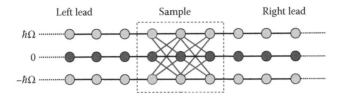

FIGURE 3.7   Scheme of the transport setup of the transverse mode decomposition of the Floquet Hamiltonian. The central region (sample) contains additional hopping elements connecting different Fourier sheets because of the presence of the time-dependent field.

field is described by the hopping elements connecting nearest-transverse layers at different Fourier levels. According to Equations (3.15) and (3.16), these decay fast with the difference between the involved Fourier replicas. For the considered examples in which $eA_0v_F/\hbar\Omega \sim 0.1$ it usually converges for a small (~3) number of photons. The dc conductance is thus calculated through the transmittance via the Floquet Green's functions (see appendix in [29]):

$$T_{RL}(\varepsilon) = \sum_{q,n} 2\Gamma^q_{(R,n)}(\varepsilon)\left|G^q_{(R,n)\leftarrow(L,0)}(\varepsilon)\right|^2 2\Gamma^q_{(L,0)}(\varepsilon) \tag{3.21}$$

where $\Gamma^q_{(\alpha,n)}(\varepsilon) = -\Im\Sigma^q_\alpha(\varepsilon + n\hbar\Omega)$. The dc conductance is then obtained from the Landauer formula [35], and we consider that the system preserves space inversion symmetry, i.e., $T_{RL}(\varepsilon) = T_{LR}(\varepsilon) = T(\varepsilon)$. Therefore, the conductance reads:

$$G(\varepsilon_F) = -\frac{2e^2}{h}\int_{-\infty}^{\varepsilon_F} d\varepsilon T(\varepsilon)\frac{d}{d\varepsilon}f(\varepsilon) \tag{3.22}$$

with $f(\varepsilon) = 1/\left(1+e^{(\varepsilon-\varepsilon_F)/kT}\right)$ the Fermi function. In the zero temperature limit the derivative of $f(\varepsilon)$ results in a Dirac delta, and we recover a linear relation between these two functions:

$$G(\varepsilon_F) = \frac{2e^2}{h}T(\varepsilon_F) \tag{3.23}$$

In the numerical calculation of the conductance, we consider a laser field whose wavelength lies within the mid-infrared region. We show in Figure 3.8 the resulting conductance at zero temperature for different values of the laser power. We consider $\lambda = 10$ μm ($\hbar\Omega \sim 140$ meV) in Figure 3.8a and $\lambda = 2$ μm ($\hbar\Omega \sim 620$ meV) in Figure 3.8b, respectively. For the case where no external laser is applied (grey dashed lines in Figure 3.8a), the conductance shows a linear dependence with the Fermi energy. In this situation there is a perfect transmission and each channel contributes a unit of the conductance quantum $G_0 = 2e^2/h$.

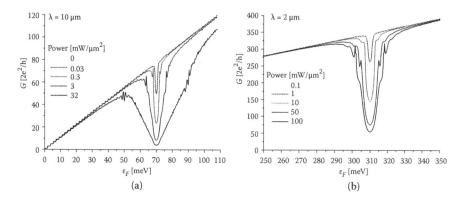

FIGURE 3.8   dc conductance through a graphene stripe of 1×1 μm in the presence of a linearly polarized laser as a function of the Fermi energy and laser power for a laser wavelength of (a) $\lambda = 10$ μm and (b) $\lambda = 2$ μm.

This effect becomes evident in the small plateaus along the whole curve, whose widths depend on the total number of transverse modes, i.e., the width of the sample. When increasing the energy of the carriers, more channels participate in the transport and the resulting conductance increases. However, when we turn on the laser, the interaction with the field induces a significant depletion around the region $\varepsilon_F \sim \hbar\Omega/2$. In the depletion region we can observe that the minimum value at $\varepsilon_F = \hbar\Omega/2$ decreases with the laser power. Additionally, this minimum value approaches zero by increasing the size of the sample (i.e., the irradiated area) since the larger the time spent by the carriers, the higher the backscattering probability. Notice that there are no visible effects close to the Dirac point $\varepsilon_F = 0$. This is due to the fact that the laser polarization is linear, and the position of the channel bands remains unaffected. By comparing the two plots, one can observe that the effect of the laser is relatively more significant in the left panel. For instance, we observe that even for $P \sim 0.03$ mW/μm^2 the depletion is still visible, whereas in the right panel for $P \sim 0.1$ mW/μm^2 this one could be hard to distinguish. This is related to the slope of the gapped modes in the DoS of Figure 3.6 (see, e.g., upper central panel), which approximately goes like

$$\left(\frac{d\Delta}{dN}\right)_{\varepsilon=\hbar\Omega} \propto \frac{\sqrt{P}}{\Omega} \tag{3.24}$$

and therefore for smaller frequencies the number of modes affected by the field is increased, and thus the stronger is the suppression in the conductance. We emphasize that in the above equation the dependence of the slope (and hence the width and depth of the depletion) with the laser power is given by a square root, and in consequence there are still visible effects when decreasing this value even by four orders of magnitude.

Regarding the effect of static disorder, it is usually found that the presence of local impurities or topological defects in the lattice tend to broaden energy gaps produced

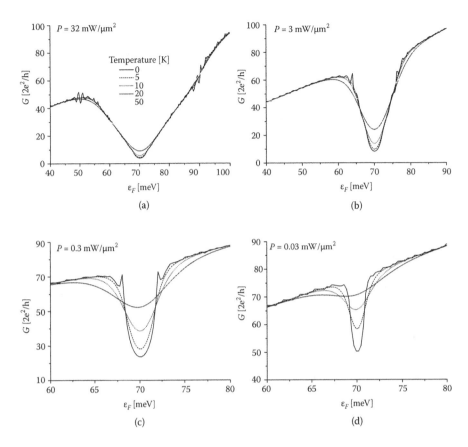

FIGURE 3.9   dc conductance as a function of the Fermi energy and temperature for different values of the laser power. The chosen wavelength of the laser is $\lambda = 10 \ \mu m$.

by confinement (as in a graphene nanoribbon) [36]. *A priori*, a similar effect is to be expected here, and further studies are needed. In order to estimate the robustness of the depletion regions we calculate the conductance for $\lambda = 10 \ \mu m$ and explore its behavior for different values of the temperature. This is shown on Figure 3.9, where each panel corresponds to a fixed value of the laser power. As we take a small finite temperature (~5 K), the plateaus structure is no longer visible and the curve is smoothed. The depletion regions are almost the same for the upper panels (large power) but slightly reduced in the lower ones. Nevertheless, this is still visible for the case $P = 0.03 \ mW/\mu m^2$. By increasing the temperature, the width of the depletion region also increases, whereas its height decreases to the bare value for zero field.

We emphasize that the dip in the conductance persists for small values of the laser power. In particular, even for temperatures around 20 K, the depletion region in the case of a laser power $P = 0.03 \ mW/\mu m^2$ is of the order of several units of G_0; thereby this should be observable in a clean sample. Such features in the conductance could be also resolved in its derivative, where a peak emerges in the region of the depletion.

3.4 CONCLUSIONS

In summary, in this chapter we have shown how a laser field would modify the electronic structure and, more importantly, the transport response (dc conductance) of graphene, where thanks to its low dimensionality and peculiar electronic structure, bandgaps can be generated. For these features to be observable, we conclude that experiments with lasers in the mid-infrared would be particularly timely. The modifications are predicted to arise both around the Dirac point and at $\hbar\Omega/2$, leading to results that are strongly dependent on the laser polarization, thereby enabling its use as a control parameter.

As pointed out already in the introduction, the potential technological impact, and the high level of recent activity in this field, makes this an outstanding area for further research and much-needed experiments.

ACKNOWLEDGMENTS

We acknowledge discussions and correspondence from Junichiro Kono, Gonzalo Usaj, and Frank Koppens, as well as support from CONICET, ANPCyT, and SeCyT-UNC. L.E.F.F.T. acknowledges support from ICTP–Trieste and the Alexander von Humboldt Foundation.

REFERENCES

1. A.H. Castro Neto, F. Guinea, N.M.R. Peres, K.S. Novoselov, and A.K. Geim. The electronic properties of graphene. *Reviews of Modern Physics*, 81, 109–162, 2009.
2. F. Bonaccorso, Z. Sun, T. Hasan, and A.C. Ferrari. Graphene photonics and optoelectronics. *Nature Photonics*, 4, 611–622, 2010.
3. F. Xia, T. Mueller, Y.-M. Lin, A. Valdes-Garcia, and P. Avouris. Ultrafast graphene photodetector. *Nature Nanotechnology*, 4, 839–843, 2009.
4. N.M. Gabor, J.C.W. Song, Q. Ma, N.L. Nair, T. Taychatanapat, K. Watanabe, T. Taniguchi, L.S. Levitov, and P. Jarillo-Herrero. Hot carrier-assisted intrinsic photoresponse in graphene. *Science*, 334(6056), 648–652, 2011.
5. D. Sun, G. Aivazian, A.M. Jones, J.S. Ross, W. Yao, D. Cobden, and X. Xu. Ultrafast hot-carrier-dominated photocurrent in graphene. *Nature Nanotechnology*, 7, 114–118, 2012.
6. S.V. Syzranov, M.V. Fistul, and K.B. Efetov. Effect of radiation on transport in graphene. *Physical Review B*, 78, 045407, 2008.
7. S. Mai, S.V. Syzranov, and K.B. Efetov. Photocurrent in a visible-light graphene photodiode. *Physical Review B*, 83, 033402, 2011.
8. T. Oka and H. Aoki. Photovoltaic hall effect in graphene. *Physical Review B*, 79(R), 081406, 2009.
9. O.V. Kibis. Metal-insulator transition in graphene induced by circularly polarized photons. *Physical Review B*, 81, 165433, 2010.
10. D.S.L. Abergel and T. Chakraborty. Generation of valley polarized current in bilayer graphene. *Applied Physics Letters*, 95, 062107, 2009.
11. H.L. Calvo, H.M. Pastawski, S. Roche, and L.E.F. Foa Torres. Tuning laser-induced band gaps in graphene. *Applied Physics Letters*, 98, 232103, 2011.
12. S.E. Savel'ev and A.S. Alexandrov. Massless Dirac fermions in a laser field as a counterpart of graphene superlattices. *Physical Review B*, 84, 035428, 2011.
13. M. Yankowitz, J. Xue, D. Cormode, J.D. Sanchez-Yamagishi, K. Watanabe, T. Taniguchi, P. Jarillo-Herrero, Ph. Jacquod, and B.J. LeRoy. Emergence of superlattice Dirac points in graphene on hexagonal boron nitride. arxiv: 1202.2870, 2012.

14. Z. Gu, H.A. Fertig, D.P. Arovas, and A. Auerbach. Floquet spectrum and transport through an irradiated graphene ribbon. *Physical Review Letters*, 107, 216601, 2011.

15. O. Roslyak, G. Gumbs, and S. Mukamel. Trapping photon-dressed Dirac electrons in a quantum dot studied by coherent two dimensional photon echo spectroscopy. arxiv: 1112.4233, 2011.

16. A. Iurov, G. Gumbs, O. Roslyak, and D. Huang. Anomalous photon-assisted tunneling in graphene. *Journal of Physics: Condensed Matter*, 24, 015303, 2012.

17. J.-T. Liu, F.-H. Su, H. Wang, and X.-H. Deng. Optical field modulation on the group delay of chiral tunneling in graphene. *New Journal of Physics*, 14, 013012, 2012.

18. M. Busl, G. Platero, and A.-P. Jauho. Dynamical polarizability of graphene irradiated by circularly polarized ac electric fields. arxiv: 1202.3293, 2012.

19. J. Karch, C. Drexler, P. Olbrich, M. Fehrenbacher, M. Hirmer, M.M. Glazov, S.A. Tarasenko, E.L. Ivchenko, B. Birkner, J. Eroms, D. Weiss, R. Yakimova, S. Lara-Avila, S. Kubatkin, M. Ostler, T. Seyller, and S.D. Ganichev. Terahertz radiation driven chiral edge currents in graphene. *Physical Review Letters*, 107, 276601, 2011.

20. L.E. Golub, S.A. Tarasenko, M.V. Entin, and L.I. Magarill. Valley separation in graphene by polarized light. *Physical Review B*, 84, 195408, 2011.

21. T. Kitagawa, E. Berg, M. Rudner, and E. Demler. Topological characterization of periodically driven quantum systems. *Physical Review B*, 82, 235114, 2010.

22. N.H. Lindner, G. Refael, and V. Galitski. Floquet topological insulator in semiconductor quantum wells. *Nature Physics*, 7, 490–495, 2011.

23. J.I. Inoue and A. Tanaka. Photoinduced transition between conventional and topological insulators in two-dimensional electronic systems. *Physical Review Letters*, 105, 017401, 2010.

24. T. Kitagawa, T. Oka, A. Brataas, L. Fu, and E. Demler. Transport properties of nonequilibrium systems under the application of light: Photoinduced quantum Hall insulators without Landau levels. *Physical Review B*, 84, 235108, 2011.

25. B. Dóra, J. Cayssol, F. Simon, and R. Moessner. Optically engineering the topological properties of a spin hall insulator. *Physical Review Letters*, 108, 056602, 2012.

26. P.R. Wallace. The band theory of graphite. *Physical Review*, 71, 622–634, 1947.

27. T. Ando. Theory of transport in carbon nanotubes. *Semiconductor Science and Technology*, 15, R13–R27, 2000.

28. H.M. Pastawski and E. Medina. Tight binding methods in quantum transport through molecules and small devices: From the coherent to the decoherent description. *Revista Mexicana de Fisica*, 47, 1–23, 2001.

29. L.E.F. Foa Torres. Mono-parametric quantum charge pumping: Interplay between spatial interference and photon-assisted tunneling. *Physical Review B*, 72, 245339, 2005.

30. J.H. Shirley. Solution of the Schrodinger equation with a Hamiltonian periodic in time. *Physical Review*, 138, B979–B987, 1965.

31. L.E.F. Foa Torres and S. Roche. Inelastic quantum transport and Peierls-like mechanism in carbon nanotubes. *Physical Review Letters*, 97, 076804, 2006.

32. L.E.F. Foa Torres, R. Avriller, and S. Roche. Nonequilibrium energy gaps in carbon nanotubes: Role of phonon symmetries. *Physical Review B*, 78, 035412, 2008.

33. Y. Zhou and M.W. Wu. Optical response of graphene under intense terahertz fields. *Physical Review B*, 83, 245436, 2011.

34. C.G. Rocha, L.E.F. Foa Torres, and G. Cuniberti. ac transport in graphene-based Fabry-Perot devices. *Physical Review B*, 81, 115435, 2010.

35. S. Kohler, J. Lehmann, and P. Hänggi. Driven quantum transport on the nanoscale. *Physics Reports*, 406, 379–443, 2005.

36. A. Cresti, N. Nemec, B. Biel, G. Niebler, F. Triozon, G. Cuniberti, and S. Roche. Charge transport in disordered graphene-based low dimensional materials. *Nano Research*, 1, 361–394, 2008.

4 Transparent and Flexible Carbon Nanotube Electrodes for Organic Light-Emitting Diodes

Yu-Mo Chien and Ricardo Izquierdo

CONTENTS

In recent years, carbon nanotubes (CNTs) have gained popularity in various fields of applications. This is due to the fact that CNTs possess remarkable properties. CNTs excel in practically all qualities that a material could have, mechanical, electronic, thermal, chemical, etc. CNTs are simply rolled-up sheets of graphene that form cylindrical tubes referred to as single-wall carbon nanotubes (SWCNTs), or multiwall carbon nanotubes (MWCNTs) when more than one tube is concentrically nested together. Their chirality (roll-up vector) and diameter will determine the electronic properties of the CNTs, which could be of either metallic or semiconductor type, and with small or large band gaps [1]. Unlike other conducting materials, the transport of charges within a metallic SWCNT occurs ballistically, with no scattering, thus allowing for very high-current capacity coupled with no heat generation. CNTs are also known to have extraordinary mechanical properties. Comparing CNTs to high-strength steel, they can have one order of magnitude higher in tensile strength (~63 GPa) and Young's modulus (~1 TPa). CNTs are also less dense than Al by about half and are twice as thermally conductive as natural diamond. Thus one can only begin to imagine why CNTs are so attractive in a very large number of applications. In this chapter, we pay particular attention to what CNTs can bring to the field of organic semiconductor devices and flexible devices. More specifically, we see that CNTs can actually be processed into thin films to form large-area electrodes. Thus when doing so, CNTs form a network that is highly transparent and flexible within the visible spectrum, and yet remains highly conductive, which is ideal for organic semiconductor device applications. In the following sections we start with a short introduction to organic semiconductor technology, followed by a review of different types of electrodes used for organic light-emitting diodes (OLEDs) on flexible substrates, and finally, a discussion of CNT transparent electrode fabrication and the process of OLEDs using those CNT electrodes by different research groups.

4.1 ORGANIC SEMICONDUCTOR TECHNOLOGY

Organic semiconductor technology has opened the doors to a whole new spectrum of novel electronics applications that were simply inconceivable with traditional inorganic semiconductors, such as devices on flexible substrates and disposable electronics. The ability to easily build devices that conform to any shape of surface is no longer just a dream. Furthermore, organic semiconductor processing techniques enable the possibility to fabricate on very large areas at very low cost. Two of the main applications of organic semiconductors are organic light-emitting diodes (OLEDs) and photovoltaic (PV) cells. Both types of applications require essentially the same fabrication process and similar device structures; only the operation mechanisms and materials of choice are different. Therefore here we concentrate our discussion on OLEDs only. Nevertheless, the technical aspects developed here could also be adapted to PV cells.

4.2 ORGANIC LIGHT-EMITTING DIODES

OLEDs have been the focus of many research centers across the world in the past two decades, ever since their invention [2]. The result of such intensive global research efforts led to the first commercial OLED-based electronic monochrome displays for car audio equipment before the beginning of the new millennium. More recently, higher-resolution full-color OLED displays have also emerged for cellular phone and television set applications. Furthermore, there has been a lot of work directed toward implementing OLEDs on flexible substrates; however, we have yet to see a

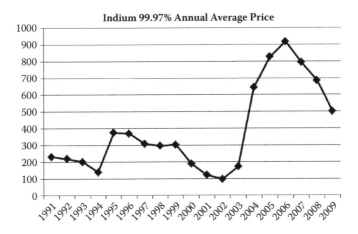

FIGURE 4.1 Annual average price of indium 99.97%. (From Tolcin, A.C., *Mineral Commodity Summaries*, U.S. Geological Survey, January 2010. With permission.)

commercial product available. The main reason behind this comes from the fact that OLED technology heavily relies on transparent conductive oxides (TCOs), more specifically indium tin oxide (ITO) electrodes. Although ITO possesses a combination of desirable qualities such as low resistivity (~2 × 10^{-4} Ω-cm), high visible transparency (~85%) [3], and good commercial accessibility, it also holds properties that are detrimental to organic semiconductor devices. It has often been proven that the nonstoichiometric nature of the ITO surface results in a strong dependence on surface treatments in order to obtain the preferred surface properties (work function, roughness, surface energy, and morphology) needed to optimize device performance [4,5]. ITO has also been plagued with oxygen and indium diffusion into the organic active layer, thus destroying the device as the usage progresses [6–8].

Moreover, indium, one of the main elements in the composition of ITO, is also a rare element that can be expensive to come by. Indium prices fluctuate heavily, governed by the supply and demand. Figure 4.1 shows the annual average price of indium 99.97% over the span of the last 19 years in the United States [9]. When demand rises and supply cannot keep up, prices rise tremendously, as has been witnessed recently with worldwide increased production of flat-panel displays.

Despite the above shortcomings, ITO has been quite successful in a wide range of applications, mainly due to its attractive electronic and optical properties. However, in order to fully take advantage of organic semiconductor technology, the mechanical properties of the electrodes used are also important. The electrode should be able to be as flexible and stress resistant as organic materials. In the case of ITO, it is obvious that it is not suitable for flexible substrate applications, as it is a brittle material that cracks and delaminates when subjected to bending [10]. Furthermore, in order to achieve low-resistivity values, ITO requires relatively high-temperature processing. This means that ITO is not suitable in certain applications, such as in the fabrication of organic top emission devices where one needs to deposit a transparent electrode on top of the organic materials.

4.3 ELECTRODES FOR OLEDS

In order to fully take advantage of organic material properties and be able to build flexible displays, transparent anode alternatives to ITO with properties that are more compatible with organic semiconductor technology and potentially more abundant and lower cost are being sought by researchers. CNT electrodes present themselves as one of the most promising materials for replacing ITO in organic semiconductor devices, which are especially compatible with the mechanical requirements of flexible and robust applications However, as will be discussed, first introductions of CNTs into OLEDs were accomplished by making composites with conductive polymers as either transparent electrodes or hole injector layers, and in some cases as electron injectors. A short review of conductive polymers that are also candidates for transparent flexible electrodes will also be presented.

4.3.1 CONDUCTIVE POLYMER ELECTRODES

First conducting polymers were discovered by Shirakawa et al. [11] in the form of doped polyacetylene in 1977. Since then, many other conducting polymers have emerged. Only a handful of them have been reportedly used as anodes in OLEDs.

4.3.1.1 PANI

The first report of such devices was in 1992 by Gustafsson et al., where a thin film of polyaniline (PANI) was spin-coated on a poly(ethylene terephthalate) (PET) flexible substrate to form the transparent anode [12]. The active emitting material was poly(2-methoxy, 5-(2′-ethyl-hexoxy)-1,4-phenylene-vinylene) (MEH-PPV). The device was completed by evaporating a layer of Ca and Al as cathode. The PANI film was reported to have very low sheet resistance (~100 $\Omega/\square$ at 70% transmittance). The turn-on voltage was measured to be around 1.8 V, which was reported to be very similar to ITO devices, and its work function was 4.3 eV compared to 4.1 eV for ITO at the time. No actual luminescence data value was actually supplied. However, PANI devices seemed to perform similar to the ITO devices. The major drawback of PANI films is that wavelengths under 475 nm were cut off; therefore blue emission devices could not be built with this type of electrode. Figure 4.2 shows the structure and photograph of the PANI OLED devices that were fabricated.

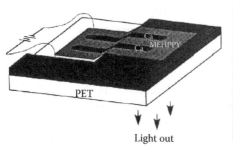

FIGURE 4.2 Structure and photograph of PANI anode MEH-PPV device on PET substrate. (From Gustafsson, G., et al., *Nature*, 1992; 357(6378): 477–479. With permission.)

Another report published in 1994 by Yang et al. also regarding the use of PANI as a hole injection electrode in MEH-PPV devices is of particular interest [13]. It was the first time where a conducting polymer was used as a hole injection layer between ITO and the active organic layer. It was proven that by inserting that conducting polymer in between, it effectively reduced the hole injection barrier height at the ITO/MEH-PPV interface. In this publication it was found that the barrier height at the PANI/MEH-PPV interface was 0.8–0.12 eV, compared to 0.2–0.24 eV for ITO/ MEH-PPV. Given that device performance also depends on the sheet resistance at higher current densities, and that in order to obtain very low sheet resistance of PANI films, one needs a thicker layer of the material, which also reduces dramatically the transmittance, Yang et al. have opted to use a very thin layer of PANI (~600 Å) with 90% transmittance in the visible wavelength backed with an ITO anode, thus resulting in a best of both worlds situation. The combination ITO/PANI anode provided a low sheet resistance and a lower barrier height to the MEH-PPV. It was found that ITO/PANI anodes resulted in a reduced operating voltage of ~30–50%. Furthermore, another group [14] that investigated ITO/PANI anodes concluded that the introduction of a conducting polymer layer not only improved hole injection efficiency, but also had a smoother interface with the organic active layer, resulting in fewer shorts, improved reliability, and lifetime. However, this approach still needs the use of ITO and is not fully compatible with flexible displays.

4.3.1.2   PEDOT:PSS

Another conductive polymer, which is the most commonly used in OLEDs and PV cells today, is poly(3,4-ethylenedioxythiophene):poly(4-styrenesulfonate) (PEDOT:PSS). Commercial availability of PEDOT:PSS in aqueous solutions can be currently found at H.C. Starck and Agfa under the trade names Clevios and Orgacon, respectively. PEDOT:PSS has been mainly used as hole injection layer or buffer layer between ITO and the organic active layer [15]. Both commercial trades have also higher conductivity formulations of their PEDOT:PSS for conductive coating applications that are achieved by altering the ratio of PEDOT and PSS. For typical 1 μm thick higher-conductivity PEDOT:PSS films, the sheet resistance (R_s) is about 3 kΩ/□, which translates to a conductivity of ~30 S/cm. Compared to typical ITO for OLED applications with a conductivity of ~20,000 S/cm (R_s = 15 Ω/□, thickness = 130 nm, 90% transmittance), there seems to be a huge gap in conductivity in order to make this polymer conductive enough to replace ITO. Moreover, for a typical OLED application, PEDOT:PSS thickness is usually less than 1 μm in order to keep transparency high; thus the resulting conductivity at practical levels of transparency is even lower. Fortunately, there have been reports suggesting methods for increasing PEDOT:PSS conductivity without increasing the thickness, thus preserving the optical transparency, by the addition of controlled amounts of a polyalcohol such as glycerol [16,17] and sorbitol [18], or by mixing with various solvents such as dimethyl sulfoxide (DMSO), N,N-dimethyl formamide (DMF), and tetrahydrofuran (THF) [19]. Improvements of as much as two orders of magnitude were observed. Adding a polyalcohol has been said to decrease PEDOT:PSS resistivity by around five to six times, whereas mixing with DMSO could reduce this further by another 10-fold. Because of these reports, it is now possible to obtain much higher-conductivity films without sacrificing transparency.

Many research groups have published results of their OLEDs based on different combinations of PEDOT:PSS preparations. In 2002, Kim et al. [20] compared devices with high-conductivity PEDOT:PSS anodes with and without added glycerol. It was reported that these conducting polymer anodes compared well with ITO devices, but no actual comparison data were presented. It was also shown that devices with PEDOT:PSS with glycerol had a maximum luminescence of 1490 cd/m^2, which was one order of magnitude higher than regular PEDOT:PSS electrode devices. Kim et al. [21] also published another set of results based on a highly conductive Orgacon PEDOT:PSS, which is capable of delivering 150 $\Omega/\square$ films at more than 70% visible transparency. It was reported that these devices showed lower turn-on voltages and better performance than the ITO anode devices at operating voltages lower than 8 V. This behavior is explained by a limited current injection at higher voltages due to the higher sheet resistance of the PEDOT:PSS. However, at low voltages, the better work function match of the conducting polymer to the emission layer wins over the ITO.

Further, similar results have been obtained by other groups with anodes based on high-conductivity PEDOT:PSS (Baytron P HC V4 and PH 500 from H.C. Starck) with added ethylene glycol and DMSO [22,23]. One report showed the results of a comparison between PH 500 + DMSO anodes and ITO transparent anodes in green, red, and blue OLEDs. For all the devices, the polymeric anodes showed better performance than ITO counterparts at low-drive voltages and luminescence. This performance was also attributed to a better work function match and lower hole injection barrier for PEDOT:PSS anodes, and also to its superior optical properties (lower refractive index). In another study, PH 500 and P HC V4 electrodes with conductivities of 300 and 450 S/cm, respectively, were compared with ITO. It was found that the surface morphology differences between P HC V4 and PH 500 gave the upper hand to PH 500 devices with higher luminescence. Further reducing the surface roughness with the addition of the methoxy-substituted 1,3,5-tris[4-(diphenylamino)phenyl]-benzene (TDAPB) hole transport (TDAPB HT) layer on the polymeric electrode not only reduced the surface roughness (0.3 nm) but also allowed for even higher luminescence (10,000 cd/m^2) and better current efficiency up to 3.5 cd/A. Although these performance figures are quite respectable, they are still far from matching those of ITO-based electrode devices.

4.4 CNTs IN OLEDS

In this section, our discussion regarding CNTs for OLED applications will begin. CNTs were first introduced into OLEDs as CNT/polymer composites. Thus we will discuss these composites briefly as they were used in emission layers, hole injection layers, and conducting polymer electrodes. We will then give a more extensive insight into the current status of CNT technology as transparent conducting electrodes.

4.4.1 CNT/POLYMER COMPOSITES

4.4.1.1 Composite Emission Layer

First use of CNTs in OLEDs was reported by Curran et al. in 1998 [24] in the form of "doping" agents for the emitting conjugated polymer material, poly(*m*-phenylenevinylene-*co*-2,5-dioctoxy-*p*-phenylenevinylene) (PmPV). General consensus indicates that emitting polymers in OLEDs need to remain undoped in order to maintain efficient luminescence. Traditional dopants act as trapping sites, quenching the radiative decay of excitons, thus reducing the overall luminescence. However, polymers have low conductivity, which in turn requires the device to operate at very high voltages, and therefore induce large thermal strains on the device. To minimize thermal effects, polymers would require dopants to increase charge carrier mobility and electrical conductivity. Curran et al. have taken advantage of the properties of carbon nanotubes by dispersing small quantities of MWCNTs in PmPV to act as nanometric heat sinks and improve conductivity at the same time. It was claimed that MWCNTs do not chemically dope the polymer; thus the electronic processes of radiative exciton decay are not altered in the emissive polymer. It was found that incorporating nanotubes into PmPV led to an increase in conductivity up to eight orders of magnitude, which reduced the operating voltage considerably and, at the same time, had very low impact on the electroluminescent characteristics. It was also found that MWCNT/PmPV composite devices lasted up to five times longer in air, possibly an indication of the effective nanometric head sink function of the nanotubes.

4.4.1.2 Composite Buffer Layer

Further studies on CNT/polymer composites employing CNTs as composites in hole-blocking layers, electron transport layers, hole transport layers, and hole injection layers in OLED structures have been carried out. We will summarize some of those results here.

Early experiments [25,26] have found that SWCNTs in the hole transport layer (HTL) PmPV/SWCNT composites placed between the hole injection layer (HIL) and the emissive layer (EL) acted has a hole trapping material that block holes from crossing the HT layer. The same phenomenon was also found when SWCNTs were incorporated into PEDOT:PSS [27]. Devices showed poor performance with SWCNT/PEDOT:PSS composites as HIL. Such use of CNTs did not yield improved OLED performance.

However, later studies by Wang et al. [28,29] had different results with MWCNT/composite HILs, which showed lower turn-on voltages and higher luminance. Another recent study has also showed improved performance, in a different manner, for HIL PEDOT:PSS/MWCNT composite devices. Different MWCNT loadings in the composite [30] were studied. The loading of MWCNT was varied as follows: 0, 0.7, 1.5, and 2.5 wt%. It was seen that as the MWCNT ratio was increased in the HIL composite, the turn-on voltage also increased, which suggests that MWCNTs worsen the hole injection and transport efficiencies of the PEDOT:PSS layer. On the other hand, the current efficiency is increased as the MWCNT concentration is increased. This could be explained by charge carriers being trapped by the MWCNT, causing better balance of both carrier types, and thus an increased luminescence current efficiency.

Given that the different studies did not yield consistent results with improved performance in CNT/polymer composite HILs or HTLs, it is safe to conclude that such use of CNTs is not the best.

4.4.1.3 CNT Composite Electrodes

So far, all the devices mentioned containing CNT/polymer composites have been used in conjunction with an ITO electrode. However, there have also been studies that

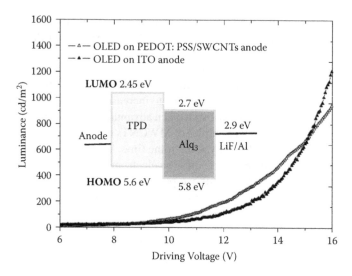

FIGURE 4.3 I-V-L curve of the OLEDs based on PEDOT:PSS/SWCNT and ITO anodes. (From Wang, G.-F., X.-M. Tao, and R.-X. Wang. Flexible organic light-emitting diodes with a polymeric nanocomposite anode. *Nanotechnology,* 2008; 19(14): 145201–1. With permission.)

tried to use CNT composites as ITO replacement, the same way as it was for conducting polymers. An example of composite electrode devices will be shown here.

Wang et al. [31] have fabricated SWCNT/PEDOT:PSS transparent electrode OLEDs and compared them to ITO devices. The devices were fabricated on PET substrates and bending tests were performed. With no surprise, bending test results showed minimal increase in resistance for SWCNT/PEDOT:PSS devices (10% after 1600 cycles). For ITO devices, however, the resistance increased by two orders of magnitude only after 20 cycles. Figure 4.3 shows the luminance curves for two anodes. The performance seems to be similar with lower turn-on voltage for SWCNT/PEDOT:PSS devices, which is not surprising given PEDOT:PSS has a higher work function than ITO leading to a better charge injection and low voltage drive. However, at higher voltage drive, ITO devices will have greater luminance due to a lower resistivity of the anode compared to composite anode devices.

4.4.2 CNT ELECTRODES

In order to minimize complexity and keep device fabrication simple, transparent and conductive electrodes can also be fabricated solely using CNTs, and many research laboratories have been geared toward its achievement. In the upcoming sections, the fabrication of CNT thin films and OLED devices based on such electrodes will be discussed.

4.4.2.1 CNT Thin-Film Fabrication

There are several methods that can be used for fabricating CNT thin films such as drop-drying from solvent, spin coating, airbrushing, and Langmuir-Blodgett deposition [32–35]. Regardless of the method used for forming the actual CNT thin film, as a first step, carbon nanotubes, in their initial powder-like form, need to be dispersed

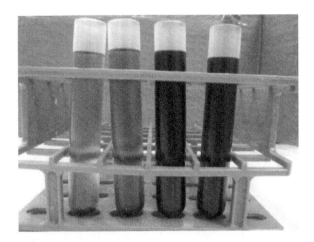

FIGURE 4.4 SWCNTs in aqueous solution after centrifugation.

in some kind of solution, either aqueous or in a solvent. Dispersing CNTs in solution often involves putting the solution with carbon nanotubes through a sonication process in order to unbundle the CNTs as much as possible. The sonication process is usually carried out using an ultrasonic bath or an ultrasonic probe. Although very effective at dispersing CNTs in a solution, sonication is also known to break CNTs to shorter lengths [36]; therefore the power and duration of the sonication need to be controlled carefully so that CNT lengths are preserved as much as possible in order to obtain better conductivities on the resulting film.

After the sonication process, the solution of CNTs is then transferred to an ultracentrifuge to separate larger bundles of CNT from smaller bundles or individual CNTs. Through ultracentrifugation, large undispersed bundles of nanotubes will then be trapped at the bottom of the solution, leaving perfectly homogeneously dispersed CNTs at the top portion. Thus only the top portion of the solution is collected and used in the subsequent steps of the CNT thin-film fabrication. Figure 4.4 shows solutions of SWCNTs dispersed in deionized (DI) water after ultracentrifugation. The different concentrations of SWCNTs in water were achieved by varying the sonication time. A lighter color of solution represents a lower concentration of CNTs with longer nanotubes, whereas a darker color indicates a high concentration but with shorter nanotubes.

4.4.2.1.1 Vacuum Filtration

The most commonly used method for creating thin films of CNT from a dispersed solution is vacuum filtration. This method was originally developed by Wu et al. [37]. A membrane filter is used to filter a small quantity of a well-dispersed CNT in solution. Figure 4.5 shows a typical vacuum filtration system setup that can be used for making CNT thin films. The final thin film will be deposited on the membrane filter, followed by the transfer of the CNT onto the substrate or actual device. The type of filter used will be determined by the type of CNT solution used and the desired method of transfer of CNT thin film onto substrate, as will be explained later in this chapter. This method of film formation, vacuum filtration, has its advantages. First,

FIGURE 4.5   Vacuum filtration system for CNT films.

the homogeneity of the CNT film is guaranteed by the process itself. As the CNTs are deposited on the filter, if thicker areas are developed, the filtration rate will also decrease in these localized areas, thus favoring the deposition of CNT in thinner areas where the permeation rate is higher, and so on. Therefore the resulting CNT thin film will be homogenous across the entire filtration area. The second advantage of this method comes from the fact that CNTs are extremely rigid for their size. Therefore they will tend to lie onto the filter straight, thus maximizing overlap and interpenetration with other CNTs within the film. This is the morphology required to maximize the electrical conductivity and mechanical integrity within the CNT thin film. Lastly, film thickness can be precisely controlled by the amount and concentration of CNT solution filtered through the membrane filter. Figure 4.6 shows resulting CNT thin films on cellulose ester membrane filters of different thicknesses. The amount of CNT filtered is proportional to the resulting conductivity of the film. However, the transparency of the film will suffer as the conductivity is increased. Therefore one needs to understand the trade-offs involved between transparency and conductivity.

4.4.2.2   Transfer of CNT Thin Film onto a Substrate

The final step in the processes of CNT film fabrication involves the transfer of the CNT film from the membrane filter onto the desired substrate. The original method developed by Wu et al. [37] uses a cellulose ester membrane filter that can be dissolved in organic solvents such as acetone. An illustration of the transfer and patterning process can be seen in Figure 4.7.

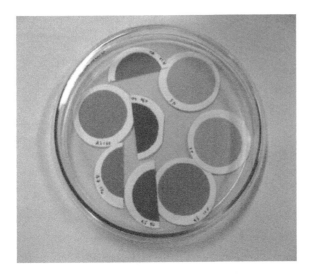

FIGURE 4.6 Different thicknesses of CNT thin films on mixed-cellulose ester membrane filters.

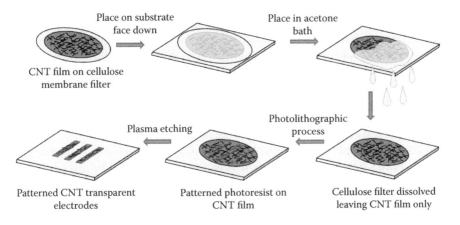

FIGURE 4.7 Transfer of CNTs from cellulose membrane filter to substrate by dissolving filter.

First, the membrane filter with the CNT film is placed face down onto the desired substrate. The cellulose ester membrane can then be dissolved by immersing the substrate with the filter in an organic solvent or just placed in a vapor bath of the organic solvent for a couple of hours. A drawback of this method is that the substrate needs to be able to resist organic solvents, and the device structure is limited by having CNT electrodes as the first structure on the device so that subsequent process steps are not affected; furthermore, in order to obtain the desired pattern of CNTs, one needs to use a traditional photolithographic process or a precisely cut shadow mask. A mask is required to protect wanted portions of the CNT film when the substrate is subjected to an oxygen plasma reactive ion etching to etch away CNTs from unprotected

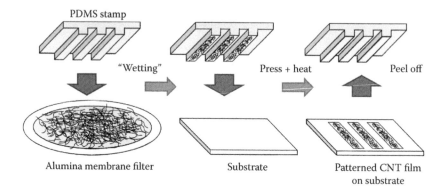

FIGURE 4.8 Transfer of CNTs from alumina membrane filter to substrate using a PDMS stamp.

parts. If a photolithographic technique is used to create the desired electrode pattern, given the nature of this process, there is a very high probability of contamination of the substrate throughout the process. Furthermore, the porosity of CNTs makes the removal of residual photoresist a laborious task.

A modified method for transferring CNT films from vacuum filtration was demonstrated by Zhou et al. [38] with the use of alumina membrane filters instead of cellulose ester membrane filters. The CNT film on alumina membrane filters can be transferred by picking up CNTs from the filter with a poly(dimethyl siloxane) (PDMS) stamp and then stamping the CNTs directly onto the desired surface. The latter method has the advantages of being a direct patterning and a "cleaner" method, and also of being able to deposit CNTs as intermediate or final layers of the device. This method is versatile and fast, as the deposition surfaces or substrates do not need to be solvent resistant.

A commonly used PDMS is provided by Dow Corning, Sylgard 184. The fabrication of a PDMS stamp is straightforward. A master mould needs to be first fabricated using traditional photolithography on a substrate, e.g., a silicon wafer. The PDMS base and curing agent are then mixed and poured onto the mold and cured in an oven. Once cured, the resulting PDMS stamp is peeled off the master mold and is ready to be used. Figure 4.8 illustrates the transfer process using a PDMS stamp. The transfer of the CNTs on the alumina membrane filter begins by pressing the PDMS stamp against the CNT film on top of the alumina filter. CNTs will adhere only to the protrusions of the PDMS stamp. Finally, the stamp is pressed against the substrate where CNT electrodes should be deposited, and CNTs are transferred from the stamp to the substrate, completing the electrode fabrication process. The curing time and temperature of the PDMS stamp are crucial for achieving a successful transfer of CNTs onto the substrate. Changing the curing time and curing temperature will change the adhesion of the PDMS surface. The right conditions must be found in order for the stamp to pick up the CNTs from the membrane filter and be able to release them onto the desired substrate. For example, curing for too long or raising the temperature too much will render the stamp too hard and it will lose its flexibility, and therefore will be unable to transfer the CNT films properly [39]. Figure 4.9

FIGURE 4.9 Photograph of SWCNT electrodes deposited on a PET substrate by PDMS stamping.

shows a sample of SWCNT electrodes transferred onto a poly(ethylene terephthalate) (PET) substrate using the above-mentioned transfer method.

4.4.2.3 SWCNT Electrode OLED Devices

OLEDs with SWCNT transparent electrodes were first reported by Aguirre et al. [40] in 2006. This group fabricated SWCNT electrodes on glass substrates using the vacuum filtration method and transfer from mixed-cellulose ester membrane filters.

The SWCNTs used in this experiment were produced in-lab by a pulsed laser vaporization technique and purified following a standard procedure. The purification process aims at eliminating amorphous carbon and metal catalyst impurities, and results in p-type charge transfer doping of the CNTs. The purified CNT powder was then dispersed in a 2% sodium cholate solution and centrifuged at 5000 g for 2 h.

The centrifuged solution was then used to filter several CNT sheets of various thicknesses and transferred onto clean glass slides by dissolving the cellulose filter in acetone. Figure 4.10 shows the resulting sheet resistance of the CNT thin film measured by a four-point probe versus the thickness and transmittance of the film for a wavelength of 520 nm. Electrical contacts to the CNT electrodes were made by evaporating 50 nm of Ti to one end of the CNTs.

The device structure of the SWCNT OLED is shown in Figure 4.11a. An ultra-thin (~1 nm) buffer layer of parylene was also deposited by CVD between the CNTs and the stacked OLED layers, which was absent in the control ITO OLEDs present. The addition of this layer claims to improve the wetting and adhesion of the evaporated organic semiconductor layer to the CNTs. The organic semiconductor layers that formed the OLED consisted of a 10 nm copper phthalocyanine (CuPc) hole injection layer (HIL), a 50 nm N,N-bis-1-naphthyl-N,N-diphenyl-1,1-biphenyl-4,4-diamine (NPB) hole transport layer (HTL), and a 50 nm tris-8-hydroxyquinoline aluminum (Alq3) electron transport layer (ETL) and emissive layer (EL) deposited in a thermal evaporator. The surface roughness of the CNT film was measured using

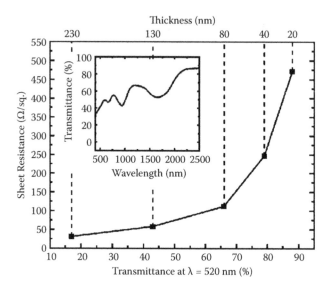

FIGURE 4.10   Sheet resistance versus thickness and transmittance of SWCNT thin film. (From Aguirre, C.M., et al., *Applied Physics Letters*, 2006; 88: 183104-1. With permission.)

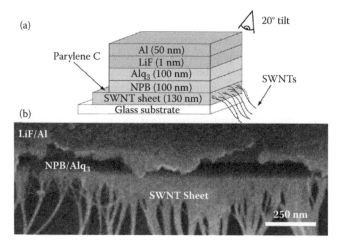

FIGURE 4.11   (a) Device structure of SWCNT OLED. (b) Corresponding cross-sectional scanning electron microscopy image at a broken edge taken at a 20° angle from the surface normal. (From Aguirre, C.M., et al., *Applied Physics Letters*, 2006; 88: 183104-1. With permission.)

atomic force microscopy (AFM) at 12 nm rms. The considerable roughness of the SWCNT films imposes a lower limit to the thickness of the organic layers. Therefore 100 nm layers of NPB and Alq$_3$ were used instead for SWCNT OLEDs. The cathode in both ITO OLED and SWCNT OLED was made by evaporating 1 nm of lithium fluoride and 50 nm of aluminum. Worth noting, Figure 4.11b shows a scanning electron microscope (SEM) image of the cross section of the SWCNT OLED. The sample for the SEM image was prepared by cleaving the glass substrate at the center

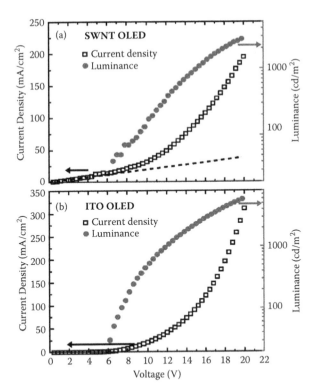

FIGURE 4.12 Current density (squares) and luminance (circles) as a function of applied voltage for OLEDs fabricated (a) on carbon nanotube anodes (SWNT OLED) and (b) on oxygen-plasma-treated ITO anodes (ITO OLED). (From Aguirre, C.M., et al., *Applied Physics Letters*, 2006; 88: 183104-1. With permission.)

of the emissive area. It can be seen that the SWCNT film network is overhanging across the glass substrate where the break was made, thus illustrating the flexibility and fabric quality of the SWCNT thin film.

The maximum luminescence reported for these devices was 2800 cd/m² with 1.2 cd/A current efficiency at 20 V. These results are said to be comparable to ITO devices with the same structure with maximum luminescence of 6000 cd/m² and 1.9 cd/A current efficiency. Detailed I-V curve characteristics are shown in Figure 4.12. The measured turn-on voltage of the SWCNT OLED was 6.6 V, only slightly higher than the ITO OLED, 6.2 V, despite the much thicker organic semiconductor layers. Furthermore, taking into consideration that the SWCNT film used in the maximum luminescence reported has a transmittance of only 44%, compared to ITO's 90%, it is possible to conclude that SWCNT injection properties are very similar to those of ITO.

Two other publications involving SWCNT OLEDs were presented during the same year. These two research groups used commercially available SWCNTs instead to carry out their experiments. Li et al. [41] fabricated their CNT electrodes using the vacuum filtration method and PDMS stamping on PET substrates. Multiple combinations of spin-coated HILs, HTLs, and ELs were explored. Devices based on ITO electrodes on PET substrate were also fabricated as reference. The maximum

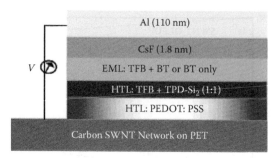

FIGURE 4.13 SWCNT electrode OLED structure on PET. (From Li, J., et al., *Nano Letters*, 2006; 6(11): 2472–2477. With permission.)

luminescence and current efficiency for CNT devices were 3500 cd/m^2 and 1.6 cd/A, respectively, compared to 20,000 cd/m^2 and 6 cd/A for ITO on PET samples. However, the lifetimes of those devices were measured to be similar at ~50 h with an initial brightness of 1400 cd/m^2.

SWCNT devices that were fabricated included PEDOT:PSS as HIL, PEDOT:SS with MeOH as HIL, and no HIL (see Figure 4.13). Also, devices with poly(9,9-dioc-tylfluorene-co-N-(4-butylphenyl)diphenylamine) (TFB) + poly(9,9-dioctylfluorene-co-benzothiadiazole) (BT) and BT as EL were compared. TFB is a hole transport polymer, and BT is an electron-dominated emission material. Additionally, all the devices included a layer of TFB + 4,4′-bis[(p-trichlorosilylpropylphenyl)phenylamino] (TPD-Si2) as HTL. The following observations were found. First, devices with HIL and HTL performed better, which could indicate two things: PEDOT:PSS is effective in planarizing the CNTs and lowering the barrier for hole injection. Furthermore, devices with PEDOT:PSS with MeOH performed better than with PEDOT:PSS alone, which suggests that MeOH help reduce the surface roughness, and thus minimize leakage current. Lastly, devices with BT only as EL showed the best performances, indicating that SWNTs are more efficient hole-injecting electrodes than ITO since ITO devices show the opposite with better devices using TFB + BT as EL.

In the experiment from Zhang et al. [42], different SWCNT sources obtained from different growth methods are compared. It was found that P3 nanotubes (arc discharge from Carbon Solutions, Inc.) exhibit much greater surface smoothness and lower resistivity than HiPCO nanotubes (Carbon Nanotechnology, Inc.) when the CNT film is formed using the PDMS stamping method. The OLED devices were fabricated on glass substrates with thermally evaporated HTL (NPD) and EL (Alq$_3$) layers and a spin-coated HIL layer (PEDOT:PSS). However, the performances obtained are below those of other reports using a similar structure. The maximum luminescence was only 17 cd/m^2 at 20 V with a turn-on voltage of 5 V for P3 electrode devices. The authors attribute this poor performance to the surface roughness, high resistivity, and low work function of SWCNTs compared to ITO electrodes.

4.4.2.4 Top Emission SWCNT Electrode OLED

So far, only bottom emission devices were discussed. However, in 2010, Chien et al. [39] successfully demonstrated the first top emission SWCNT electrode OLED based on solution-processed HIL and EL.

In this report, raw SWCNTs were purchased commercially from Carbon Solutions, Inc. (P2-SWNT). P2-SWNTs are purified CNTs (>90%) with low functionality and used as received without further treatment. The SWCNTs in powder form were prepared into a colloidal aqueous solution of 0.1 wt% of CNT in a 1 wt% surfactant solution in deionized water. The surfactant used was sodium dodecyl sulfate (SDS). The CNT/SDS solution was left in an ultrasonic bath for 24 h and was followed by a centrifugation at 30,000 rpm (154,000 g) for 1 h. Only the top portion of the CNT solution was extracted for use in the following steps, as the centrifugation process removes the larger bundles of CNTs to the bottom of the tube that the ultrasonic bath was not able to disperse. The result is a highly homogenous solution of CNT dispersed in deionized water.

In order to build a top-emitting device over organic semiconductor layers, the process needs to be carried out at low temperature to avoid damaging the organic layers underneath. The top transparent electrode was therefore deposited using PDMS stamping of the SWCNT method on top of the PEDOT:PSS HIL, which was previously spin-coated over the EL. Such a process offers the possibility of implementing CNT OLED devices onto any type of the substrates, including nontransparent surfaces. Vacuum filtration through 0.1 μm porosity alumina filters (Whatman, Inc.) was carried out to obtain a uniform CNT film, and the transfer process was carried out as described earlier for PDMS stamping.

The top emission SWCNT OLEDs were fabricated on glass substrates in the initial investigation to determine the feasibility of such a procedure. Figure 4.14 shows the cross-sectional structure of the top emission OLED. The bottom cathode consists of a layer of Al (100 nm) and LiF (1 nm) evaporated through a shadow mask in a thermal evaporator at around 1×10^{-6} Torr. The organic emissive material was then spin-coated from a solution, and heat-treated in ambient air at 100°C for 5 min. The chosen organic material was prepared according to Park et al. [43] and consisted of a blend of poly(vinylcarbazole) (PVK), 2-(4-biphenylyl)-5-(4-*tert*-butylphenyl)-1,3,4 oxadiazole (PBD), tris(2-phenyl-pyridinato) iridium (Ir(ppy)$_3$), and N,N'-diphenyl-N,N'-bis(3-methylphenyl)-1,1'-biphenyl-4,4'-diamine (TPD) into mixed solvent of 1,2-dichloroethane and chloroform. PEDOT:PSS (CLEVIOS P VP AI 4083 from H.C. Starck) was then spin-coated on top of the emission layer. The PEDOT:PSS was diluted with 10 wt% of 2-propanol to promote the wetting on the emission layer in order to obtain a more uniform buffer layer. The PDMS stamp was then used to transfer the CNTs from the alumina membrane filters onto the PEDOT:PSS layer.

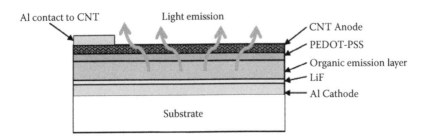

FIGURE 4.14 Top emission SWCNT OLED device structure.

This process takes place on a hot plate. The appropriate temperature to transfer CNTs on top of a PEDOT substrate was determined to be 100°C. The PDMS "wetted" with CNTs is put in conformal contact with the PEDOT:PSS layer with light pressure from a fingertip for 30 s. The PDMS stamp is then carefully peeled off, leaving the CNT film on the PEDOT:PSS layer. Heating the substrate while stamping the CNTs serves two purposes. It helps the release of the CNTs from the PDMS and also helps the adhesion of the CNTs to the PEDOT:PSS layer. The top emission OLED is then completed by thermally evaporating through another shadow mask a layer of Al on the CNT film outside the active region in order to make the electrical contact of the CNTs. Some research groups [37,42,44] have deposited additional layers of Ti and Pd on the CNTs to obtain true ohmic contacts with the CNTs; however, Chien et al. did not find this to be necessary as the devices performed adequately to be able to take reliable measurements.

CNT films of various thicknesses were also fabricated and characterized. Resistivity (R_s) and transmittance (%T) of the CNT films were measured with a four-point probe and a UV-visible spectrometer, respectively. For these measurements, the CNTs were transferred on a plain glass substrate spin-coated with PEDOT:PSS to simulate the surface conditions of a real device. As mentioned previously, resistivity and transmittance are affected by the thickness of the CNT film. As the thickness of the CNTs is decreased, the film becomes less conductive and more transparent. Figure 4.15 shows the resistivity versus the transmittance of the film at 510 nm corresponding to the peak emission of a green emission OLED. For an R_s: 100 Ω/sq CNT film, the transmittance is about 60%. At 90% transmittance, the film resistivity reaches 500 Ω/sq. These values are on par with numbers reported by other groups with films transferred from cellulose or alumina filters and on glass or plastic substrates. The devices in this report were fabricated using the 90% transmittance films. Figure 4.16 shows an AFM image of the CNT on a complete top emission device. The surface RMS roughness is about 7 nm, which is equivalent to the roughness of

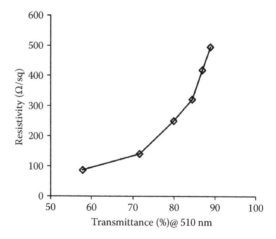

FIGURE 4.15   Sheet resistance as a function of transmittance of CNT films transferred onto the device measured for a 510 nm emission.

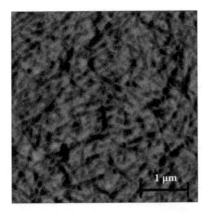

FIGURE 4.16 AFM image of CNT film already transferred onto an actual device (R_s: 500 Ω/sq, %T: 90%). Film RMS roughness is about 7 nm.

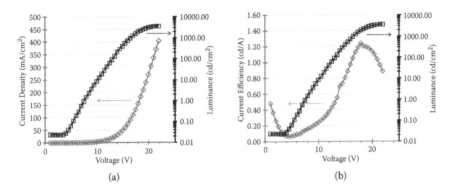

FIGURE 4.17 I-V curve characteristics of top emission SWCNT OLED.

CNT films alone, showing that the transfer on top of the OLED devices does not affect the characteristics of the CNT films.

OLED devices were encapsulated in a nitrogen glove box immediately after fabrication and were characterized with Keithley 2400 and Keithley 2610 DC source monitors. Luminance measures were taken with a photodiode with an infrared (IR) cutoff filter, which was calibrated with a Delta OHM HD 2102.1 photometer for their green OLEDs. About 30 top emissions with CNT anode OLEDs were fabricated in total. However, only half the devices had complete CNT film transfers. For these devices, the performance characteristics were very similar. Figure 4.17 shows the current density, luminance, and current efficiency as a function of voltage applied of a typical device. Maximum luminance and current efficiency obtained were 3588 cd/m^2 and 1.24 cd/A, respectively. These numbers are comparable to devices reported by other groups for bottom emission and evaporated organic material devices and CNT electrodes. Figure 4.18 shows a photograph of an actual solution-processed top emission OLED with a CNT transparent anode. As can be seen, the light emission of the device is very uniform and comparable to typical OLEDs with ITO electrodes.

FIGURE 4.18   Photograph of a working top emission SWCNT-OLED.

However, even though these top emission OLEDs with CNT electrodes have excellent results compared to other CNT electrode devices reported in the literature, bottom emission devices with ITO electrodes fabricated in our lab, with the same spin-coated organic material, still have better performance than the top emission CNT devices. Optimized ITO electrode devices with luminance around 30,000 cd/m² were achieved. It is believed that if further optimization is conducted, better performance should be expected from CNT top emission devices. For example, by reducing the resistivity of CNT films, better performances should be expected. This can be achieved by optimizing the CNT solution dispersion process and through chemical treatment of the prepared films. Additionally, charge injection properties from CNTs into the organic materials is of great importance in order to obtain better hole injection and PEDOT, which is appropriate for ITO electrodes, should be replaced by a more appropriate material for use with CNT electrodes. Finally, better control of the transferring process of CNTs with the help of a mechanical instrument is desired to obtain more reliable and consistent results.

4.4.2.5 MWCNT Electrode OLED Devices

The first reported OLED devices using CNTs only as transparent hole-injecting electrodes were fabricated using MWCNTs on glass substrate [45]. This publication reported devices with spin-coated emission layers and hole injection layers, MEH-PPV and PEDOT:PSS, respectively. The achieved performances were 500 cd/m² for maximum luminescence, and a low turn-on voltage of 2.4 V. There was no mention of the efficiencies obtained. Additionally, this report uses a unique method in fabricating thin MWCNT films. The authors call it a dry solid-state process where transparent CNT sheets are drawn from a sidewall of MWCNT forests. Figure 4.19 illustrates

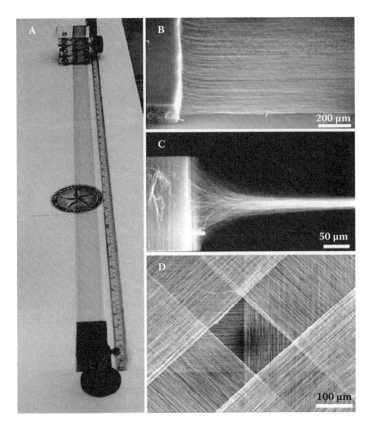

FIGURE 4.19 MWCNT forest conversion into sheets (dry solid-state process). (From Zhang, M., et al., *Science*, 2005; 309(5738): 1215–1219. With permission.)

such a process. Figure 4.19A shows a photograph of a self-supporting 3.4 cm wide and 1 m long MWCNT sheet that was hand drawn from a CNT forest. Figure 4.19B and C shows SEM images of the process at 35° and 90° angles with respect to the forest plane, respectively. Figure 4.19D shows a two-dimensionally reinforced structure of four overlaying MWCNT sheets with 45° shift in orientation between them. In order to make useful transparent conducting electrodes for organic devices, such sheets can also be directly placed onto a substrate and immersed momentarily in ethanol and pulled out vertically, which increases the density and transparency of the MWCNT sheets. One of the major advantages of such a process is the absence of sonication, which is known to shorten CNTs, thus decreasing electrical and thermal conductivities and mechanical properties.

Another paper on MWCNT OLEDs was presented by Williams et al. [46] as a continuation to the previous publication focuses on improving the surface roughness of the MWCNT sheet by spin coating up to nine layers of PEDOT:PSS. Maximum luminance with 4500 cd/m² with maximum current efficiency of 2.3 cd/A was achieved with this approach. This time, however, vacuum-evaporated active layers of α-NPB and Alq$_3$ were deposited as active OLED layers.

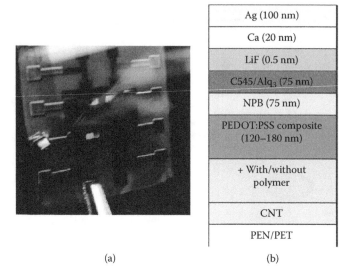

Ag (100 nm)
Ca (20 nm)
LiF (0.5 nm)
C545/Alq$_3$ (75 nm)
NPB (75 nm)
PEDOT:PSS composite (120–180 nm)
+ With/without polymer
CNT
PEN/PET

(a)   (b)

FIGURE 4.20 (a) Photograph of working OLED with CNT electrode. (b) Structure of CNT OLED. (From Ou, E.C.W., et al., *ACS Nano*, 2009; 3(Compendex): 2258–2264. With permission.)

4.4.2.6 Summary of CNT Electrode OLED Performance Criteria

A recent report by Ou et al. [47] provided a good insight into the key elements that are important to the fabrication of a CNT-only transparent anode for OLEDs that performs well.

In this experiment, CNT thin films were deposited on PEN and PET substrates using a high-speed roll-to-roll slot die method. This report shows the highest-performing OLEDs with CNT electrodes to date with a luminescence of 9000 cd/m^2 and a current efficiency of 10 cd/A. The active emission layer consisted of *N,N*-diphenyl-1,1-bihyl-4,4-diamine (NPB) and tris-(8-ydroxquinoline) aluminum (Alq$_3$) coevaporated with coumarin 545. Figure 4.20 shows a photograph of a working CNT OLED and its corresponding structure. The experiments were centered around the use of a proprietary HIL that contains high-conductivity PEDOT:PSS and poly(ethylene glycol) (PEG), referred to as PS^C in the report. Key elements to obtaining high-performance OLEDs with CNT electrodes discussed were surface roughness, sheet resistance, and work function of the injecting electrodes.

First, the surface roughness of the anode is important because protruding CNTs (Figure 4.21) generate very high local electrical fields, which can in turn cause local device failure and lead to a shorted device. Ou et al. reduced the surface roughness of their electrodes by more than half in a two-step process consisting of the deposition of a 5 nm polyvinylpyrrolidone (PVP) polymer with the Meyer rod method and ozone plasma treating the samples, followed by the spin coating of PS^C (HIL). The resulting surface roughness can be seen in Table 4.1, comparing with the initial CNT film roughness.

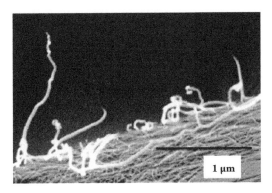

FIGURE 4.21 SEM image of a SWCNT surface with occasional protruding CNT. (From Hu, L., et al., *Nanotechnology*, 2010; 21(Compendex). With permission.)

TABLE 4.1
Surface Roughness Change after PS^C Coating on CNTs

Sample	Description	Initial Roughness (nm)	Final Roughness after Spin-Coated with PS^C (nm)
2	PEN/CNT/PS^C	14.55	6.0
3	PEN/CNT/5 nm polymer/PS^C	14.65	4.6
4	PET/CNT/PS^C	9.3	5.9
5	PET/CNT/5 nm polymer/PS^C	9.72	5.2
6	PET/Doped CNT/PS^C	10.85	6.0

Source: Ou, E.C.W., et al., *ACS Nano*, 2009; 3(Compendex): 2258–2264. With permission.

Second, the sheet resistance and work function of the hole-injecting electrode were also improved by doping the CNT with HNO_3, which resulted in a 50% sheet resistance drop and an increase of 0.13 eV for the work function. CNTs doped with HNO_3 showed higher luminescence and lower turn-on voltage than undoped ones (see Figure 4.22).

Finally, to further illustrate the importance of the sheet resistance of the electrode in an OLED, PEN/PS^C anode (310.5 ohms/sq) and anode (102.9 ohms/sq) devices were compared. As shown in Figure 4.23, the PEN/CNT/PS^C anode OLED was a huge improvement over the PEN/PS^C anode OLED, with as much as three times the luminescence for the same driving voltage.

4.5 CONCLUSION

The development of a true high-performance transparent flexible electrode to be used for OLED fabrication is still a challenge to be resolved. Among various alternatives the use of CNT electrodes is one of the more promising approaches. CNT electrodes

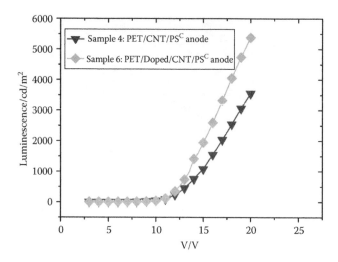

FIGURE 4.22 Luminescence versus voltage of OLEDs for PET/CNT/PEDOT:PSS versus PET/doped CNT/PEDOT:PSS. (From Ou, E.C.W., et al., *ACS Nano*, 2009; 3(Compendex): 2258–2264. With permission.)

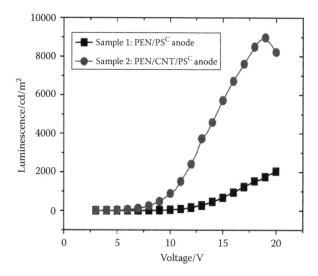

FIGURE 4.23 Luminescence characteristics of PEN/PS^C and PEN/CNT/PS^C devices. (From Ou, E.C.W., et al., *ACS Nano*, 2009; 3(Compendex): 2258–2264. With permission.)

with sheet resistivities around 100 ohms/sq and transparencies around 60% have been produced. OLEDs made with those electrodes have shown high luminescence and current efficiency. Even though those performances are still inferior to the ones obtained with standard ITO electrodes, various approaches for reducing resistivity of the CNT films without losing transparency are presently under study and should lead to the development of real flexible displays in the near future.

REFERENCES

1. Yilun, L., et al. Engineering the band gap of carbon nanotubes. In *8th IEEE Conference on Nanotechnology 2008 (NANO '08)*. 2008.

2. Tang, C.W., and S.A. VanSlyke. Organic electroluminescent diodes. *Applied Physics Letters*, 1987; 51(12): 913–915.

3. Kim, H., et al. Transparent conducting electrode materials grown by pulsed laser deposition for organic light-emitting devices. In *Organic Light-Emitting Materials and Devices III*, July 19–21, 1999.

4. Vaufrey, D., et al. Sol gel deposited Sb doped SnO_2 as transparent anode for OLED: Process, patterning and hole injection characteristics. In *Organic Light-Emitting Materials and Devices V*, July 30–August 1, 2001. 2002.

5. Besbes, S., et al. Effect of surface treatment and functionalization on the ITO properties for OLEDs. *Materials Science and Engineering C*, 2006; 26(2–3): 505–510.

6. Gardonio, S., et al. Degradation of organic light-emitting diodes under different environment at high drive conditions. *Organic Electronics*, 2007; 8(1): 37–43.

7. Melpignano, P., et al. Mechanism of dark-spot degradation of organic light-emitting devices. *Applied Physics Letters*, 2005; 86(4): 041105-3.

8. de Jong, M.P., L.J. van Ijzendoorn, and M.J.A. de Voigt. Stability of the interface between indium-tin-oxide and poly(3,4-ethylenedioxythiophene)/poly(styrenesulfonate) in polymer light-emitting diodes. *Applied Physics Letters*, 2000; 77(14): 2255–2257.

9. Tolcin, A.C. *Mineral commodity summaries*. U.S. Geological Survey, January 2010.

10. Lin, Y.C., W.Q. Shi, and Z.Z. Chen. Effect of deflection on the mechanical and optoelectronic properties of indium tin oxide films deposited on polyethylene terephthalate substrates by pulse magnetron sputtering. *Thin Solid Films*, 2009; 517(5): 1701–1705.

11. Shirakawa, H., et al. Synthesis of electrically conducting organic polymers: Halogen derivatives of polyacetylene, (CH)x. *Journal of the Chemical Society, Chemical Communications*, 1977; 1977(16): 578–580.

12. Gustafsson, G., et al. Flexible light-emitting diodes made from soluble conducting polymers. *Nature*, 1992; 357(6378): 477–479.

13. Yang, Y., and A.J. Heeger. Polyaniline as a transparent electrode for polymer light-emitting diodes: Lower operating voltage and higher efficiency. *Applied Physics Letters*, 1994; 64(10): 1245–1247.

14. Karg, S., et al. Increased brightness and lifetime of polymer light-emitting diodes with polyaniline anodes. *Synthetic Metals*, 1996; 80(2): 111–117.

15. Fung, M.K., et al. Anode modification of polyfluorene-based polymer light-emitting devices. *Applied Physics Letters*, 2002; 81(8): 1497–1499.

16. Ghosh, S., and O. Inganäs. Nano-structured conducting polymer network based on PEDOT-PSS. *Synthetic Metals*, 2001; 121(1–3): 1321–1322.

17. Makinen, A.J., et al. Hole injection barriers at polymer anode/small molecule interfaces. *Applied Physics Letters*, 2001; 79(5): 557–559.

18. Granlund, T., L.A.A. Pettersson, and O. Inganas. Determination of the emission zone in a single-layer polymer light-emitting diode through optical measurements. *Journal of Applied Physics*, 2001; 89(11): 5897–5902.

19. Kim, J.Y., et al. Enhancement of electrical conductivity of poly(3,4-ethylenedioxythiophene)/poly(4-styrene sulfonate) by a change of solvents. *Synthetic Metals*, 2002; 126(2–3): 311–316.

20. Kim, W.H., et al. Molecular organic light-emitting diodes using highly conducting polymers as anodes. *Applied Physics Letters*, 2002; 80(20): 3844–3846.

21. Kim, W., et al. Efficient silole-based organic light-emitting diodes using high conductivity polymer anodes. *Chemistry of Materials*, 2004; 16(Compendex): 4681–4686.

22. Fehse, K., et al. Highly conductive polymer anodes as replacements for inorganic materials in high-efficiency organic light-emitting diodes. *Advanced Materials*, 2007; 19(3): 441–444.

23. Kajii, H., et al. Organic light-emitting diodes with highly conductive polymer electrodes as anode and their stress tolerance. *Japanese Journal of Applied Physics*, 2008; 47(1 Part 2): 460–463.

24. Curran, S.A., et al. A composite from poly(m-phenylenevinylene-co-2,5-dioctoxy-p-phenylenevinylene) and carbon nanotubes: A novel material for molecular optoelectronics. *Advanced Materials*, 1998; 10(14): 1091–1093.

25. Woo, H.S., et al. Hole blocking in carbon nanotube-polymer composite organic light-emitting diodes based on poly (m-phenylene vinylene-co-2, 5-dioctoxy-p-phenylene vinylene). *Applied Physics Letters*, 2000; 77(9): 1393–1395.

26. Woo, H.-S., et al. Tailoring hole transport in organic light-emitting devices using carbon nanotube-polymer nanocomposites. *Journal of the Korean Physical Society*, 2004; 45: 507–511.

27. Woo, H.S., et al. Organic light emitting diodes fabricated with single wall carbon nanotubes dispersed in a hole conducting buffer: The role of carbon nanotubes in a hole conducting polymer. In *Fourth International Topical Conference on Optical Probes of pi-Conjugated Polymers and Photonic Crystals*, February 2000. Switzerland: Elsevier, 2001.

28. Wang, G.-F., et al. Improvement in performance of organic light-emitting devices by inclusion of multi-wall carbon nanotubes. *Journal of Luminescence*, 2007; 126(Compendex): 602–606.

29. Wang, G.-F., X.-M. Tao, and R.-X. Wang. Fabrication and characterization of OLEDs using PEDOT:PSS and MWCNT nanocomposites. *Composites Science and Technology*, 2008; 68(Compendex): 2837–2841.

30. Su, S.-H., et al. Enhancing efficiency of organic light-emitting diodes using a carbon-nanotube-doped hole injection layer. *Japanese Journal of Applied Physics*, 2010; 49(Compendex).

31. Wang, G.-F., X.-M. Tao, and R.-X. Wang. Flexible organic light-emitting diodes with a polymeric nanocomposite anode. *Nanotechnology*, 2008; 19(14): 145201-1.

32. Duggal, R., F. Hussain, and M. Pasquali. Self-assembly of single-walled carbon nanotubes into a sheet by drop drying. *Advanced Materials*, 2006; 18(Compendex): 29–34.

33. Southard, A., et al. Solution-processed single walled carbon nanotube electrodes for organic thin-film transistors. *Organic Electronics: Physics, Materials, Applications*, 2009; 10(Compendex): 1556–1561.

34. Li, J., and Y. Zhang. Langmuir-Blodgett films of single-walled carbon nanotubes. *Carbon*, 2007; 45(3): 493–498.

35. Jong Hyuk, Y., et al. Fabrication of transparent single wall carbon nanotube films with low sheet resistance. *Journal of Vacuum Science and Technology B (Microelectronics and Nanometer Structures)*, 2008; 26: 851–855.

36. Vichchulada, P., et al. Sonication power for length control of single-walled carbon nanotubes in aqueous suspensions used for 2-dimensional network formation. *Journal of Physical Chemistry C*, 2010; 114(29): 12490–12495.

37. Wu, Z., et al. Transparent, conductive carbon nanotube films. *Science*, 2004; 305(5688): 1273–1276.

38. Zhou, Y., L. Hu, and G. Gruner. A method of printing carbon nanotube thin films. *Applied Physics Letters*, 2006; 88(12): 123109-3.

39. Chien, Y.-M., et al. A solution processed top emission OLED with transparent carbon nanotube electrodes. *Nanotechnology*, 2010; 21(13): 134020.

40. Aguirre, C.M., et al. Carbon nanotube sheets as electrodes in organic light-emitting diodes. *Applied Physics Letters*, 2006; 88: 183104-1.

41. Li, J., et al. Organic light-emitting diodes having carbon nanotube anodes. *Nano Letters*, 2006; 6(11): 2472–2477.
42. Zhang, D., et al. Transparent, conductive, and flexible carbon nanotube films and their application in organic light-emitting diodes. *Nano Letters*, 2006; 6(9): 1880–1886.
43. Park, J.H., et al. Double interfacial layers for highly efficient organic light-emitting devices. *Applied Physics Letters*, 2007; 90(15): 153508-3.
44. Nosho, Y., et al. Relation between conduction property and work function of contact metal in carbon nanotube field-effect transistors. *Nanotechnology*, 2006; 17(14): 3412–3415.
45. Zhang, M., et al. Strong, transparent, multifunctional, carbon nanotube sheets. *Science*, 2005; 309(5738): 1215–1219.
46. Williams, C.D., et al. Multiwalled carbon nanotube sheets as transparent electrodes in high brightness organic light-emitting diodes. *Applied Physics Letters*, 2008; 93(Compendex).
47. Ou, E.C.W., et al. Surface-modified nanotube anodes for high performance organic light-emitting diode. *ACS Nano*, 2009; 3(Compendex): 2258–2264.
48. Hu, L., et al. Flexible organic light-emitting diodes with transparent carbon nanotube electrodes: Problems and solutions. *Nanotechnology*, 2010; 21(Compendex).

5 Direct Graphene Growth on Dielectric Substrates

Jeffry Kelber

CONTENTS

5.1 INTRODUCTION

Graphene consists of a single atomic layer of graphite—C atoms arranged in a hexagonal lattice, with sp^2 hybridization. Measurements on isolated single- and few-layer graphene sheets have demonstrated a number of properties with great potential for novel device applications, including electron or hole mobilities orders of magnitude greater than in Si [1,2], electron spin diffusion lengths on the order of microns [3], and predicted [4] ferromagnetism in the vicinity of a ferromagnet. The initial demonstration, by Novoselov et al., of some of these properties on isolated single- and few-layer graphene sheets [5,6] has stimulated intense interest in the electronic and spintronic properties of graphene.

A key step toward the industrial development of graphene-based devices is the ability to directly grow graphene on dielectric substrates. To date, however, the great majority of published accounts employ one of two approaches: (1) the physical transfer of graphene sheets grown on metal substrates [7–9] or from highly oriented pyrolytic graphite (HOPG) [5,10] onto patterned SiO_2/Si substrates, or (2) the graphitization of SiC(0001) by high-temperature evaporation of Si [11–13]. The former approach has been successful in terms of elucidating fundamental physical properties of the graphene system, but obviously presents severe problems to device fabrication on an industrial scale, not the least of which is the formation of nanoscale interfacial inhomogeneities [14]. The latter approach has been used successfully to fabricate high-frequency field effect transistors (FETs) [15,16], but the absence of a band gap (at least in zero applied field) leads to low device on/off current ratios (see below). Additionally, this approach is limited to SiC.

Recently, macroscopically large and uniform graphene films (as opposed to nanoflakes) have been grown, without metal catalysts, on h-BN(0001) monolayers on Ru(0001) [17], on MgO(111) [18,19], on mica [20,21], and on $Co_3O_4(111)$/ Co(111) [22]. Preliminary accounts [23] also suggest that continuous graphene films can be grown on α-$Al_2O_3(0001)$. These studies have employed scalable methods, including chemical vapor deposition (CVD), physical vapor deposition (PVD), and molecular beam epitaxy (MBE). The results, reviewed below, present exciting possibilities for the development of graphene-based electronic and spintronic devices. These results, however, also demonstrate that the properties of graphene sheets are strongly impacted by interactions with the substrate. This is generally true even for isolated graphene sheets physically transferred to high dielectric constant substrates [15,24], or to boron nitride (BN) nanocrystals [25], where dramatic increases in carrier mobility have been reported, relative to mobilities for graphene on SiO_2. In the case of high dielectric substrates, screening of electron scattering from charged impurities may be responsible for the observed mobility enhancement [24], while for graphene on BN, substrate flatness appears to play a role [25]. For graphene grown directly on substrates, such issues as graphene/substrate charge transfer [17,18], interfacial chemical reaction or reconstruction [19,26], and the relative orientations and interactions of graphene sheets [12,27] also present themselves. Such issues not only are important in a fundamental, scientific sense, but directly impact practical device applications.

The primary focus of this review is on direct graphene growth on dielectric substrates, specifically h-BN(0001), and certain oxides—MgO(111), mica, and $Co_3O_4(111)$. There is a growing literature on the use of metal catalysts to form graphene sheets on insulating substrates [28–30], but the chemistry involved appears to be essentially that of graphene growth on transition metal substrates [8,31]. Additionally, the presence of transition metal impurities presents potentially serious difficulties for device applications because even low concentrations of transition metal impurities in Si or other semiconductors can significantly degrade electronic device properties.

Finally, a section of this review is devoted to exploring potential device applications—including electronic applications made possible by the controlled growth of graphene on MgO, which results in a band gap, or spintronics applications for graphene grown on magnetically polarizable oxides, such as Co_3O_4. Graphene conduction electrons are predicted to become spin-polarized in proximity to a ferromagnet [4]. Preliminary calculations [32] suggest the possibility that such a system may exhibit nonlocal magnetoresistance values orders of magnitude higher than spin valves or spin

FETs that are based on the injection and diffusion of individual spins through a graphene sheet. Such effects, if demonstrated, could yield manufacturable, low-power, and nonvolatile spintronics devices operating at or above room temperature [32].

5.2 GRAPHENE SUBSTRATE INTERACTIONS AND BAND GAP FORMATION

The lattice of a single graphene sheet (Figure 5.1a) consists of a two-atom unit cell with crystallographically distinct sites A and B that are chemically identical in the isolated graphene lattice. This chemical identity results in the highest occupied molecular orbital (HOMO) and lowest unoccupied molecular orbital (LUMO) being degenerate at the Dirac point (Figure 5.1b) [33]. Thus an isolated graphene sheet is a zero-band gap semiconductor. This band structure—including the singularity at the Dirac point and linear dispersion in its vicinity—is responsible for many of graphene's unusual properties, including measured DC mobilities for an isolated graphene sheet in excess of 100,000 V $cm^{-2} s^{-1}$ (at low carrier densities), several orders of magnitude greater than that of an electron in Si [1]. The band structure shown in Figure 5.1b, however, presents a major problem for applications in such devices as FETs. Specifically, the "off state" in such a device would, at room temperature or

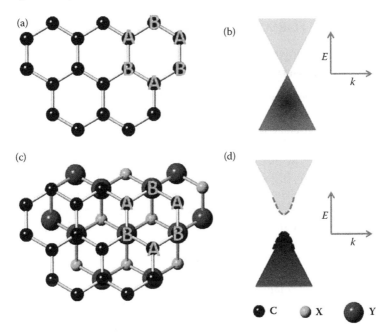

FIGURE 5.1 Substrate-induced band gap formation in graphene. (a) Structure of the isolated graphene lattice. A sites and B sites are crystallographically distinct but chemically identical, leading to (b) HOMO/LUMO degeneracy at the Dirac point. (c) Interaction of graphene with a commensurate substrate (atoms arbitrarily labeled X and Y, as shown) can, depending on the relative orientation of the two layers, destroy the A site/B site chemical equivalence, lifting HOMO/LUMO degeneracy at the Dirac point and (d) resulting in a band gap.

above, still exhibit a considerable current, leading to very low on/off ratios. FETs formed with graphene grown on the Si face of SiC(0001), for example, exhibit on/off ratios of ~30 [34]. A band gap of 0.26 eV for single-layer graphene on Si-terminated SiC(0001) has been reported on the basis of angle-resolved photoemission measurements [35]. The magnitude of the gap, however, decreases with additional graphene layer formation [35], and a band gap of this magnitude may in any case be too small for conventional (nontunneling) FET applications [36,37]. Importantly, there is a lack of transport measurements pointing to the formation of such a band gap.

The formation of a ~0.25 eV band gap in bilayer graphene in the presence of an external electric field has been demonstrated [38]. The required presence of an external field, however, as well as the apparent limiting value of only 0.25 eV, presents obvious difficulties for attaining a high on/off ratio in a graphene-based nontunneling [36,37] FET, even assuming the ability to precisely form a bilayer graphene channel. Band gap formation has also been reported [39] for graphene nano-ribbons, but this effect is sensitive to the width of the ribbon on the nanoscale, and perhaps as well to the termination of the nano-ribbon edge sites [40]. Thus the most robust way forward in introducing a band gap to graphene for device applications would appear to be the growth of graphene on a suitable substrate to induce such interactions as shown in Figure 5.1c and d.

5.3 GRAPHENE FORMATION ON h-BN(0001)

5.3.1 GRAPHENE GROWTH ON H-BN MONOLAYERS FORMED ON METAL SUBSTRATES

The basal plane of hexagonal boron nitride (h-BN(0001)) is in many respects an ideal dielectric surface for graphene heteroepitaxy. Isoelectronic with graphene, h-BN(0001) has an in-plane lattice constant of 2.51 Å—less than a 2% lattice mismatch to graphene. h-BN(0001) monolayers have been formed on Ni(111), Pt(111), and Pd(111) substrates by thermal decomposition of borazine ($B_3N_3H_6$) [41,42]. The inertness of the initial BN surface layer to further thermal reaction and precursor decomposition results in a self-limiting monolayer growth process [42]. Such monolayers have also been formed by borazine decomposition on Ru(0001), with observation of a puckered "nanomesh" structure [43]. An R30 ($\sqrt{3} \times \sqrt{3}$) structure was observed (Figure 5.2a), however, for a monolayer (1 ML) BN film on Ru(0001) formed by atomic layer deposition (ALD) using two cycles of BCl_3 and NH_3 precursors [17]. This structure is consistent with a flat BN layer rotated 30° with respect to the Ru(0001) lattice (Figure 5.2a, inset). Bulk h-BN(0001) has a band gap of ~6 eV [44]. Scanning tunneling microscopy (STM) dI/dV data, however (Figure 5.2b), proportional to the local density of states, indicate a band gap of ~2 eV, demonstrating substantial Ru/BN orbital hybridization. Such hybridization has also been reported for epitaxial h-BN layers on certain other transition metal surfaces [41,45]. Subsequent exposure of h-BN(0001)/Ru(0001) to C_2H_4 at 1000 K (60 s, 0.1 Torr) yields an sp^2-characteristic C(KVV) Auger feature (not shown), the low-energy electron diffraction (LEED) image shown in Figure 5.2c, and the STM dI/dV data in Figure 5.2d, demonstrating formation of a graphene overlayer.

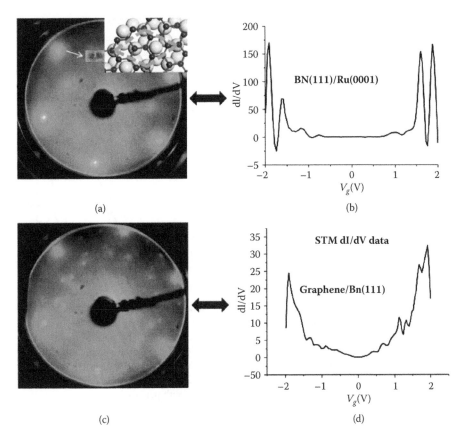

(a)   (b)

(c)   (d)

FIGURE 5.2 Formation of graphene/BN heterojunction on Ru(0001): (a) LEED image for BN(0001) monolayer deposited by ALD. Bifurcation of outer spots due to Ru substrate is shown (arrow). Inset is the model for unit cell orientation relative to the Ru lattice. (b) STM dI/dV data indicating BN/Ru orbital hybridization. (c) LEED after CVD of graphene using C_2H_4. (Note satellite features attributed to multiple scattering.) (d) Corresponding STM dI/dV data indicating graphene-like density of states near the Fermi level. (From C. Bjelkevig et al., *J. Phys. Cond. Matt.*, 22 (2010) 302002. With permission.)

The well-defined graphene/monolayer h-BN(0001)/Ru(0001) LEED image shown in Figure 5.2b contrasts with the diffuse nature of the LEED image reported [46] for graphene/monolayer h-BN(0001) heterojunction formation on Ni(111). One obvious cause for the LEED spot broadening in the latter case could be very small domain sizes. The authors suggest, however, on the basis of high-resolution electron energy loss data, that graphene interaction with BN results in the BN layer undergoing a slight relaxation, due to a weakening of BN/substrate interactions upon formation of the graphene layer [46]. In either case, the differing LEED data for graphene/monolayer h-BN(0001)/Ru(0001) (Figure 5.2a) and for graphene/monolayer h-BN(0001)/Ni(111) demonstrate that interlayer interactions between graphene and the BN/metal substrate depend strongly on the nature of the metal, and perhaps the degree of BN/metal lattice matching.

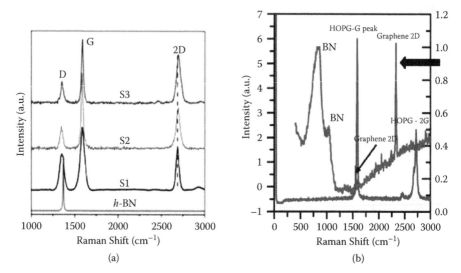

FIGURE 5.3 Raman spectra (a) for few-layer graphene grown on BN nanoflakes by flame pyrolisis (S1–S3 refer to different samples grown by the same method) and (b) for monolayer graphene grown by CVD on monolayer BN/Ru(0001). Graphene/BN data in upper trace. Lower trace is a reference spectrum acquired from HOPG. Note the significant red-shift of the 2D feature for graphene/BN/Ru (arrow) relative to graphene/BN nanoflake. (Data in (a) adopted from T. Lin et al., *J. Mater. Chem.*, 22 (2012) 2859. Data in (b) adopted from C. Bjelkevig et al., *J. Phys. Cond. Matt.*, 22 (2010) 302002. With permission.)

5.3.2 GRAPHENE/BORON NITRIDE INTERFACIAL INTERACTIONS

Significant interactions occur between graphene and BN monolayers on metal substrates. The nature of such interactions can be discerned in part by a comparison of Raman spectra for graphene/monolayer-BN(0001)/Ru(0001) [17] with similar data for multilayer graphene grown on BN nanoflakes (50–250 nm) by flame pyrolysis of a glassy carbon source [47] (Figure 5.3).

The Raman spectra for graphene grown directly on a BN nanoflake (Figure 5.3a) exhibit a significant D feature, indicative of a high concentration of edge or defect sites, as expected for a small domain film grown on a nanoflake, and a 2D feature at a typical energy of 2687 cm^{-1} [47]. The Raman spectrum for graphene/BN/Ru(0001) (Figure 5.3b, red trace) contains normally IR-active BN modes, due to the broken symmetry of the sample, no discernable D feature, and a G/2D intensity ratio consistent with single-layer graphene [17]. The 2D mode, however (Figure 5.3b, arrow), is red-shifted by over 300 cm^{-1} from the usual value, as demonstrated by comparison with the 2D mode for HOPG (Figure 5.3b, lower trace). Since the 2D mode in graphene is a mixed vibrational-electronic two-phonon mode [48], such a red-shift is strongly suggestive of charge transfer into the graphene π^* band, and mode softening. Thus Raman data indicate negligible substrate $\rightarrow$ graphene charge transfer for graphene grown on BN nanoflakes, and substantial substrate $\rightarrow$ graphene charge transfer for graphene/monolayer h-BN(0001)/Ru(0001).

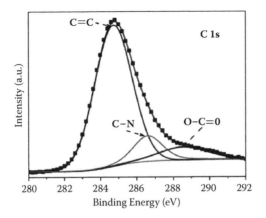

FIGURE 5.4 C(1s) XPS data for graphene grown directly on BN nanoflakes. The peak maximum is at a bindng energy of 284.7 eV, indicative of some graphene charge transfer to the substrates. (Data adapted from T. Lin et al., *J. Mater. Chem.*, 22 (2012) 2859. With permission.)

The absence of substantial substrate → graphene charge transfer for graphene/BN nanoflake heterojunctions is corroborated by x-ray photoelectron spectroscopy (XPS) C(1s) data (Figure 5.4). The data indicate a C(1s) peak maximum at 284.7 eV binding energy. While the figure also indicates some C-N and C=O bonding, which may occur at edge sites, a main peak binding energy of 284.7 eV is somewhat higher than the 284.5 eV commonly reported for graphite. Since shifts in core level binding energy usually correspond to changes in the valence charge density around an atom—the ground state atomic charge population [49]—a slight shift toward higher binding energy indicates some charge transfer from graphene to the BN nanoflake substrate. By comparison, multilayer sheets comprised of mixed BN and graphene domains formed by CVD exhibit a C(1s) binding energy of 284.4 eV [50], close to that of bulk graphite. These data therefore indicate that graphene grown directly on multilayer BN nanoflakes exhibits some charge transfer from the graphene to the BN substrate.

In contrast, there is pronounced charge transfer in the other direction (i.e., to the graphene film from the substrate) in the case of graphene/monolayer h-BN(0001)/Ru(0001), nearly filling the π* band [17]. This is demonstrated by the data displayed in Figure 5.5, which compare angle-integrated valence band photoemission and *k*-vector-resolved inverse photoemission (conduction band) spectra (Figure 5.5a) with STM dI/dV data for graphene/monolayer h-BN(0001)/Ru(0001) (Figure 5.5b and Figure 5.2d), with monolayer h-BN(0001)/Ru(0001) (Figure 5.5c and Figure 5.2b), and with corresponding inverse photoemission data for multilayer graphene on SiC(0001) (Figure 5.5e) [51]. Specifically, the feature closest to the Fermi level in the inverse photoemission data (Figure 5.5a)—that is, the lowest unoccupied level—is the σ* feature, as shown by the dispersion of this feature in *k*-space (Figure 5.5d). In contrast, an empty π* feature is the lowest unoccupied level for corresponding measurements on multilayer graphene/SiC(0001) [51], as well as for graphene on Ni or Cu [18]. These comparisons demonstrate that for graphene/monolayer h-BN(0001)/Ru(0001), the graphene π* band is effectively filled, consistent with Raman data

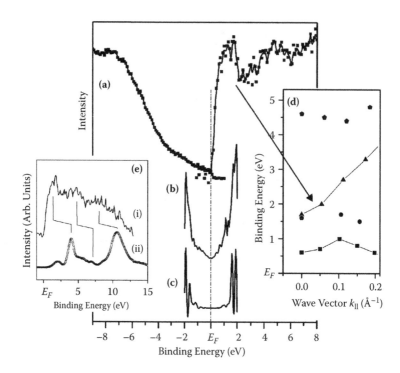

FIGURE 5.5 Density of states data for graphene/monolayer h-BN(0001)/Ru(0001). All data plotted relative to the Fermi level (E_F): (a) Valence band-integrated photoemission (left) and k-vector-resolved inverse photoemission data (right), (b) STM dI/dV data for graphene/BN/Ru (as in Figure 5.2d), (c) corresponding data for BN/Ru (as in Figure 5.2b), (d) dispersion data for inverse photoemission feature closest to Fermi level, and (e) inverse photoemission data (i) compared to those of (ii) multilayer graphene/SiC(0001). (Data in (e)(ii) from I. Forbeaux et al., *Phys. Rev. B*, 58 (1998) 16396. Figure adopted from C. Bjelkevig et al., *J. Phys. Cond. Matt.*, 22 (2010) 302002. With permission.)

(Figure 5.4b) indicating vibrational mode softening for the 2D mode. More detailed analysis [18] indicates that this charge transfer amounts to 0.07 e$^-$ per carbon atom. The dispersion of the σ* band, however, yields an estimate for the effective mass of 0.05 m$_e$ [17], in good agreement with transport measurement-derived values of 0.4–0.6 m$_e$ [14,52,53] on transferred graphene sheets. These data demonstrate that the charge transfer results from graphene band filling, rather than from interfacial carbon rehybridization. Therefore this charge transfer does not significantly alter the basic graphene band structure.

The data presented in Figures 5.3–5.5 thus demonstrate a significant difference in interfacial interactions for graphene/BN(nanoflakes) vs. graphene/monolayer h-BN(0001)/Ru(0001). For graphene grown directly on a BN nanoflake, there is some graphene → substrate charge transfer. In contrast, for graphene/monolayer h-BN(0001)/Ru(0001), there is strong substrate → graphene charge transfer, almost completely filling the graphene π* band. The ability to grow multilayer epitaxial h-BN(0001) with layer-by-layer precision on a transition metal substrate should

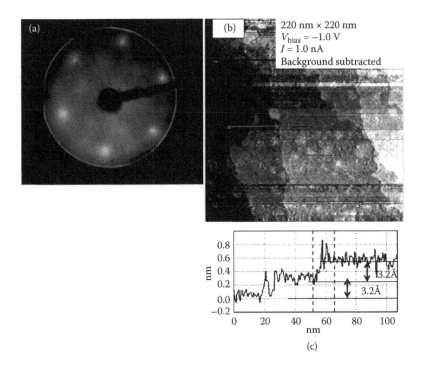

FIGURE 5.6 Trilayer h-BN(0001) on Ru(0001) by ALD: (a) LEED of a 3 ML film produced by 6 BCl₃/NH₃ cycles at 550 K, followed by annealing to 1000 K in UHV (beam energy = 75 eV), and (b) STM constant current image for a BN(0001) film grown epitaxially on a Ru(0001) substrate. STM line trace (c) shows two steps with an individual step height of 3.2 Å, within experimental error of the nominal BN interlayer distance of 3.3 Å.

therefore present a route toward "tuning" graphene/transition metal interactions. Such precision growth is possible using an atomic layer deposition (ALD) process incorporating cycles of BCl_3 and NH_3, at ~550 K, which has been used to grow conformal polycrystalline BN films on a variety of substrates [54]. This process has been adapted to the layer-by-layer growth of epitaxial h-BN layers on Ru(0001), as shown by the LEED and STM data in Figure 5.6. The ability to form a h-BN(0001) trilayer on Ru(0001) (Figure 5.6; six BCl_3/NH_3 cycles at 550 K, followed by annealing to ~1000 K in ultrahigh vacuum (UHV)) suggests that this ALD process is a viable route toward forming varying thicknesses of BN on appropriate metal substrates for systematic variation of graphene/metal interactions. The data shown in Figures 5.3–5.5 also suggest that the chemical reactivity of a monolayer of BN on a transition metal substrate with strong BN/metal hybridization (e.g., Figure 5.2b) could differ substantially from that of a multilayer film that approaches the electronic structure of bulk h-BN. Thus a graphene film grown directly on the surface of monolayer h-BN(0001)/Ru(0001) may well display different electronic properties than a graphene layer grown on a multilayer h-BN(0001) film. An obvious issue is whether a multilayer film will be as reactive toward C-H bond scission during CVD as the monolayer BN film on Ru(00001). Recent studies, however, suggest that processes such as flame pyrolysis [47], or perhaps MBE, will permit direct graphene growth on multilayer h-BN(0001) films.

Finally, the data in Figure 5.2d and Figure 5.5 demonstrate that graphene/monolayer h-BN(0001)/Ru(0001) does not exhibit an observable band gap at 300 K. This is somewhat in contrast to recent density functional theory (DFT) calculations predicting the formation of a small band gap of ~0.05 eV in graphene/BN bilayers, due to adoption of a relative orientation similar to that in Figure 5.1c [55]. Aside from well-known difficulties in the use of DFT methods to predict band gap size [56], there are several factors that could contribute to this apparent absence of an experimental graphene band gap, the most obvious being that a band gap so small would be difficult to observe by room temperature STM dI/dV or photoemission/inverse photoemission (PES/IPES) measurements. Other factors include the possible orientation of the graphene and BN layers being such as to lead to equivalent average bonding environments for graphene lattice A sites and B sites. Indeed, an analysis of graphene-related LEED intensity spots on h-BN(0001)/Ru(0001) (Figure 5.2c) revealed true C_6v symmetry in the LEED pattern [19], thus indicating graphene A site/B site chemical equivalence, and implying either that the relative orientation of the graphene and BN layers differed from that predicted by the DFT calculations [55], or that the charge transfer from the Ru substrate was sufficiently great as to overwhelm and obscure effects resulting from graphene/BN interfacial interactions. In any event, the LEED, STM dI/dV, and PES/IPES data all indicate the absence of a band gap at room temperature for graphene grown directly on monolayer h-BN(0001)/Ru(0001).

5.4　GRAPHENE GROWTH ON MgO(111)

5.4.1　Growth Studies

Graphene has been grown on the surface of bulk MgO(111) single crystals, using either chemical vapor deposition (thermally dissociated C_2H_4) [18], annealing of adventitious carbon layers in the presence of C_2H_4 (Figure 5.7) [19], or magnetron sputter deposition from a graphite source at room temperature, followed by annealing in UHV [26]. In all these cases, annealing in UHV to 1000 K appears to be sufficient to induce (111)-ordered sp^2-hybridized C overlayers—graphene. That these overlayers are macroscopically continuous is evident from the fact that the samples can be repeatedly exposed to ambient without significant change in, e.g., LEED or other surface spectra [18]. XPS and LEED data (Figure 5.7) [19] demonstrate that formation of an ordered overlayer is accompanied by an interfacial chemical reaction that results in at least some of the carbon atoms in a higher oxidation state, as evidenced by a C(1s) feature at >288 eV binding energy (Figure 5.7d), and as observed for graphene deposited by other deposition methods [18,26]. Thus at least part of the interfacial carbon layer is strongly oxidized.

The existence of a highly oxidized component in the C(1s) XPS spectrum (Figure 5.7d) is strong evidence that the graphene layer (perhaps partially oxidized) is in contact with and reacting with an oxygen-terminated or hydroxyl-terminated MgO substrate surface. The bulk-terminated (111) layers of MgO and other oxides with the rock salt structure are composed entirely of either oxygen anions or metal cations, as

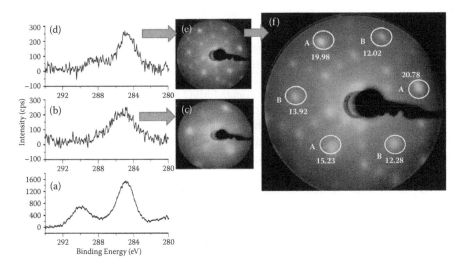

FIGURE 5.7 Graphene formation on a MgO(111) single crystal. (a) C(1s) spectrum of an adventitious C overlayer upon sample insertion into UHV. (b) After annealing at 700 K in the presence of 10^{-6} Torr O_2. The XPS-derived average C thickness is 1 ML. (c) Corresponding LEED pattern exhibits C_3v symmetry. (d) C(1s) XPS after additional annealing to 1000 K in the presence of 5×10^{-7} Torr. (e) Corresponding complex LEED pattern. (f) Close-up of the pattern in (e) with integrated, background-subtracted intensities (arbitrary units) for A and B spots (circled). The A spots have an average intensity of 18.7 ± 3, while the B spots have an average intensity of 12.9 ± 1. The uncertainties are the standard deviations. Other spots in the image are weaker and are attributed to multiple diffraction. The LEED patterns were acquired at 80 eV beam energy. The XPS spectra binding energies are referenced to a MgO lattice oxygen O(1s) binding energy of 530.0 eV. (Adopted from S. Gaddam et al., *J. Phys. Cond. Matt.*, 23 (2011) 072204.)

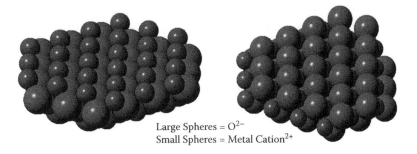

Large Spheres = O^{2-}
Small Spheres = Metal Cation^{2+}

FIGURE 5.8 Schematic of the bulk-terminated (111) layers of an oxide with the rock salt structure: (left) metal cation terminated; (right) oxygen anion terminated.

shown schematically in Figure 5.8. This composition—either 100% cation (Figure 5.8 left) or 100% anion (Figure 5.8 right)—results in a net surface charge and instability in the Madelung potential, which is often stabilized by hydroxyl termination of the surface layer [57–60]. For this reason, the (111) surfaces of rock salt oxides such as MgO are prone to reconstruction upon, e.g., the adsorption of transition metals [61].

The data in Figure 5.7 also strongly suggest that, in addition to reaction with carbon, the oxide surface undergoes a reconstruction to yield an interface commensurate with the (partially oxidized) graphene overlayer. Specifically, the LEED data for the ordered graphene overlayer (Figure 5.7e) exhibit C_3v symmetry, with the LEED spots thus exhibiting a strong (A site) and weak (B site) configuration. The O-O nearest-neighbor distance for bulk-terminated MgO is ~2.8 Å, and should therefore form an incommensurate interface with graphene (lattice constant 2.5 Å). In such an incommensurate structure, both graphene A sites and B sites will be in an ensemble of chemical environments, and one would expect the average A site and B site electron density to be the same. This would then give rise to a C_6v LEED pattern and (Figure 5.1b) the absence of a band gap at the Dirac point. To the contrary, a C_3v LEED pattern, as in Figure 5.7f, indicates that graphene lattice A sites have a higher electron density than the B sites—implying the formation of a commensurate interface [26] and rehybridization of the HOMO and LUMO molecular orbitals, lifting their degeneracy at the Dirac point [33], as in Figure 5.1c and d. Thus the data in Figure 5.7 indicate that graphene on MgO(111) should exhibit a band gap.

5.4.2 BAND GAP FORMATION

The formation of a 0.5–1 eV band gap for a 2.5 ML (average thickness) graphene film formed on MgO(111) by thermally dissociated C_2H_4 has been reported [18]. Valence band photoemission (angle-integrated PES) and conduction band inverse photoemission (k-vector-resolved IPES) data for this film are shown in Figure 5.9. The PES data (He I source) are in excellent agreement with those reported for high-quality graphene films grown on transition metal surfaces [62], while the IPES data exhibit the expected π^* and σ^* features in good agreement with IPES data for multilayer graphene/SiC(0001) [51]. Between the PES/valence band and IPES/conduction band,

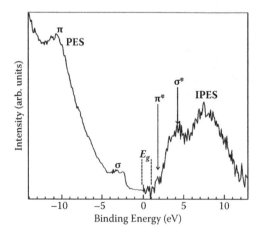

FIGURE 5.9 Band gap formation for 2.5 ML graphene on MgO(111). (Left) Angle-integrated valence band photoemission (He I source): PES. (Right) K-vector-resolved inverse photoelectron data: IPES. Energies are plotted relative to the Fermi level (0 eV). (Adopted from L. Kong et al., *J. Phys. Chem. C*, 114 (2010) 21618. With permission.)

there is a feature-free region—a band gap. The magnitude of this band gap is ~0.5–1 eV, with the uncertainty being due to possible charging effects and the inherent limiting resolution (400 meV) in the IPES measurements [18].

Recent charge transport measurements [63] for a single-layer graphene film on MgO(111) confirm semiconducting behavior, with an estimated activation energy for carrier transport of 0.64(± 0.05) eV derived from the temperature-dependent conductivity data. Although the activation energy can differ from the band gap energy by the final state exciton binding energy and other factors, an activation energy of 0.6 eV for the monolayer film is in very good agreement with the magnitude of the band gap indicated by the photoemission/inverse photoemission data (Figure 5.9) for a few-layer film. Importantly, LEED data for both films exhibit C_3v symmetry [19].

5.4.3 INTERFACIAL CHARGE TRANSFER

Although substrate → graphene charge transfer occurs readily for graphene on transition metals and on monolayer h-BN(0001)/Ru(0001) [18] graphene/MgO exhibits charge transfer in the opposite direction: the graphene becomes p-type due to charge transfer to the substrate [18]. This is demonstrated by the data shown in Figure 5.10, comprising inverse photoemission spectra for graphene grown on a variety of substrates, compared to multilayer graphene/SiC(0001) [51]. Since inverse photoemission spectra reveal only the unoccupied states, the energy of a specific feature, such as the σ^*, relative to the Fermi level, gives a quantitative measure of the charge transfer into the graphene conduction band from the substrate, or out of the graphene valence band to the substrate [18]. Taking the IPES data for multilayer graphene/SiC(0001) [51] as indicative of negligible graphene/substrate charge transfer [64], the relative energies of the π^* and σ^* features in the IPES spectrum for graphene/MgO(111) indicate charge transfer from graphene to the oxide. This finding is also consistent with XPS spectra [26,65].

In summary, the formation of single-layer graphene on MgO(111) yields an ordered graphene layer with C_3v symmetry, and a band gap of ~0.5–1 eV, as determined by DC transport measurements [63]. A 2.5 ML film on MgO(111) also exhibits

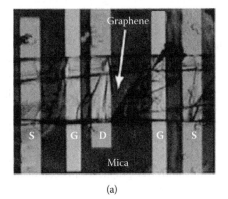

(a)

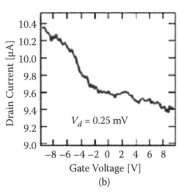

(b)

FIGURE 5.10 Multilayer graphene structure on mica. (a) Optical micrograph. (b) I-V behavior. (From G. Lippert et al., *Phys. Stat. Sol. B*, 248 (2011) 2619. With permission.)

a band gap of about the same magnitude, as determined by photoemission/inverse photoemission [18]. Both films exhibit C_3v symmetry in the LEED pattern, indicating that (1) interfacial reconstruction is probably occurring to yield a commensurate interface as well as (Figure 5.7d) a partially oxidized initial carbon layer, and (2) that such interactions are not completely screened in the second layer. A band gap of 0.5–1 eV is highly suitable for conventional FET-type logic devices [36,37]. Further, (111)-textured MgO films have been deposited on Si(100) by pulsed laser deposition [66], indicating a straightforward route toward the production of simple graphene-based FET structures integratable with Si CMOS.

5.5 GRAPHENE GROWTH ON MICA BY MBE

The growth of graphene by MBE on mica has recently been reported [20,21]. The particular sample, "black mica" or biotite, has a stoichiometry of $K(Mg,Fe)_3AlSi_3O_{10}(OH)_2$ (Fe content < 5% of Mg) [21] and a structure consisting of sheets of silica/alumina layers separated by potassium cations—permitting facile cleavage exposing ordered silica layers.

Raman mapping [21] of the G/2D intensities for graphene MBE growth at 1073 K indicates that graphene thickness is nonuniform, and that the quality of the graphene layer—as judged by the relative intensity of the D peak—decreases with increasing number of graphene layers [21].

The temperature sensitivity of the growth process, and the macroscopic variability in graphene thickness, strongly suggest that the growth process involves an initial carbon interaction with the substrate to form a buffer layer, and that succeeding sp^2 carbon-graphene layers grow in an island-like mode and coalesce. Ab initio growth simulations [20] indicate that growth may start at defect, or seed sites on the flat mica surface, with graphene flakes becoming immobilized when they coalesce. The studies further suggest that carbon oxidation occurs at the edge sites of the graphene flakes, which could also immobilize the separate domains. This last suggestion is certainly consistent with the finding of enhanced graphene quality with increased number of graphene layers, i.e., that graphene MBE growth on graphene occurs more readily than the growth of the initial layer on the mica.

Graphene/mica sheets grown in this way were fabricated into FETs, as shown in Figure 5.10a [20], and yielded I-V behavior characteristic of multilayer graphene [20] (Figure 5.13b). The ability to grow graphene directly on a SiO_2-like substrate suggests a path toward direct graphene integration with Si CMOS. However, the apparent increase in graphene quality with increasing graphene layers indicates that growth on such substrates involves formation of an interfacial layer, with perhaps some oxidation at edge sites, followed by a more regular growth process on the initial graphene layer. In other words, the growth of graphene on graphene apparently yields a higher-quality graphene sheet than graphene on silica. This simple fact indicates that the ability to grow few-layer or multilayer graphene, instead of just a single layer, may be of significant practical importance for device fabrication. In this respect, MBE would appear to have significant advantages over CVD, as the formation of the initial graphene (or graphene-like) layer results in a surface relatively inert toward further C-H bond scission [17].

5.6 GRAPHENE GROWTH ON Co₃O₄(111)/Co(111) BY MBE

5.6.1 GROWTH STUDIES

Direct graphene growth by MBE on Co_3O_4/Co(111) at 1000 K has been reported [22], and the ability to grow graphene directly on a magnetically polarizable oxide suggests new possibilities for spintronics applications. Further, graphene growth on Co_3O_4(111) exhibits important differences with graphene growth on MgO(111) or on mica with respect to both the nature of interfacial chemical interactions and the mode of growth. Co_3O_4(111) film formation occurred by the deposition, at 750 K, of Co(111) films ~40 Å thick, on Al_2O_3(0001), followed by annealing in UHV at 1000 K. The resulting oxygen segregation resulted in formation of a Co_3O_4(111) film ~3 ML thick [22]. Subsequent MBE from a solid carbon source yielded sp^2-hybridized carbon at average carbon coverages from 0.4 ML up to the thickest coverage examined, 3 ML (Figure 5.11a).

Importantly, the evolution of the Auger-derived average carbon thickness (Figure 5.11b) is well fit by a series of linear segments, with changes in slope corresponding to the completion of one carbon layer and the beginning of another [22] indicative of layer-by-layer growth [67]. LEED data acquired at 0.4 ML and 3.0 ML coverages (Figure 5.12) indicate the formation of a hexagonal carbon lattice with a 2.5 Å lattice constant, thus forming an incommensurate interface with the unreconstructed oxide surface, which has an O-O distance of 2.8 Å [22].

Additional analysis of the LEED data indicates that the graphene LEED pattern at both 0.4 and 3 ML graphene thickness displays C_6v symmetry [22], consistent with the formation of an incommensurate graphene/oxide interface [26], and indicating that the graphene lattice A sites and B sites are chemically equivalent at all coverages examined. Additional analysis of the full width at half maxima of LEED diffraction spots indicates an average graphene domain size of ~1800 Å, comparable

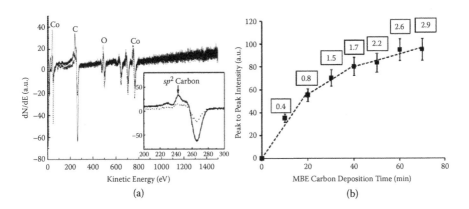

FIGURE 5.11 Auger spectra for MBE graphene growth at 1000 K on Co_3O_4(111)/Co(111). (a) Auger spectra at 0.4 ML (dashed trace) and 3 ML (solid trace) carbon coverage—insert shows expanded view of C(KVV) spectra; (b) evolution of average carbon overlayer thickness as a function of MBE deposition time. (From M. Zhou et al., *J. Phys. Cond. Matt.*, 24 (2012) 072201. With permission.)

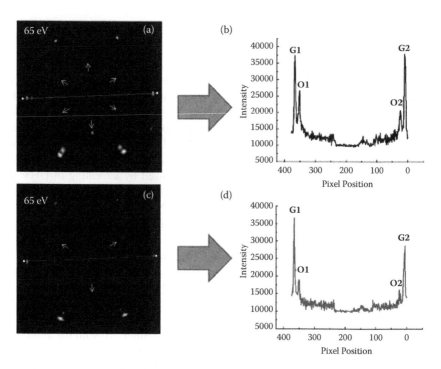

FIGURE 5.12 LEED and corresponding line scan data for (a, b) 0.4 ML graphene on $Co_3O_4(111)$, and (c, d) 3 ML graphene on $Co_3O_4(111)$. Arrows (a, c) mark diffraction spots associated with $Co_3O_4(111)$, as do inner spots in the outer ring of bifurcated features (e.g., O1, O2—b, d). Outer spots in the outer ring of bifurcated features (e.g., G1, G2—b, d) are graphene related. LEED beam energy is 65 eV. (From M. Zhou et al., *J. Phys. Cond. Matt.*, 24 (2012) 072201. With permission.)

to HOPG [22]. C(1s), Co(2p), and O(1s) XPS spectra acquired for the 3 ML film are displayed in Figure 5.13. The data indicate a characteristically asymmetric C(1s) feature (Figure 5.13a) with the expected $\pi \rightarrow \pi^*$ satellite feature (Figure 5.13a, inset). The Co(2p) and O(1s) spectra are as expected for a 3 ML $Co_3O_4(111)$ film on Co(111) [22,68]. In contrast to XPS spectra for graphene/mica (Figure 5.12a), or for graphene/MgO(111) (Figure 7d in [26]), there is no evidence of C chemical reaction with the oxide substrate. The observed C(1s) binding energy, 284.9(± 0.1) eV, however, is significantly higher than that of bulk graphite—284.5 eV. This observed binding energy is also quite close to that observed, 284.75 eV, for the first C(1s) layer on Si-terminated SiC(0001) [64]. This increase in binding energy is indicative of charge donation from the graphene layer to the oxide substrate [64,69]. Such graphene $\rightarrow Co_3O_4(111)$ charge transfer was also inferred from the blue-shift in the $\pi \rightarrow \pi^*$ absorption feature for spectroscopic ellipsometry data, relative to that of multilayer graphene/SiC or graphene physically transferred to SiO_2 [22]. Thus the XPS data in Figure 5.13a indicate significant graphene $\rightarrow Co_3O_4(111)$ charge transfer, as to also inferred from IPES spectra for graphene on MgO(111).

Macroscopic continuity and uniformity of the graphene sheets deposited by MBE on $Co_3O_4(111)$ is indicated by two pieces of evidence [22]:

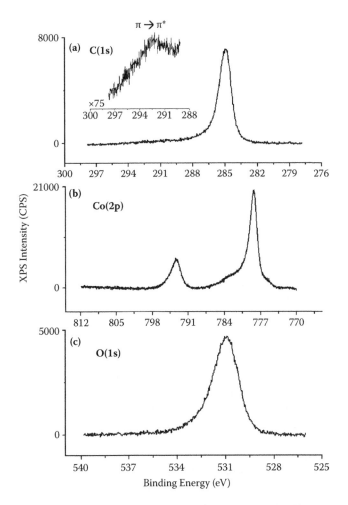

FIGURE 5.13 Core level XPS spectra acquired for a 3 ML graphene film grown on $Co_3O_4(111)/Co(111)$. (a) C(1s) (inset: π à π* satellite feature), (b) Co(2p), and (c) O(1s) spectra. (From M. Zhou et al., *J. Phys. Cond. Matt.*, 24 (2012) 072201. With permission.)

1. The sample with 3 ML graphene coverage, upon exposure to ambient, showed no change in Auger or LEED spectra. A 3 ML $Co_3O_4(111)/Co(111)$ film, unmodified, would be extremely sensitive to air exposure. This passivating effect of continuous graphene films has been observed for graphene on other air-sensitive substrates, including Ru(0001) [18] and Co(0001) [70].

2. Micro-Raman spectra (Figure 5.14). Micro-Raman spectra were acquired at spots separated by ~3 mm on a 1 × 1 cm sample with a 3 ML graphene film. The spectra acquired at full laser power (Figure 5.14a—upper trace, Figure 5.14b) display identical spectra, indicating macroscopic film uniformity. The observed G/2D intensities are consistent with multilayer graphene [48,71]. Importantly, however, both spectra display an intense D peak, generally associated with grain boundaries and edge sites [71].

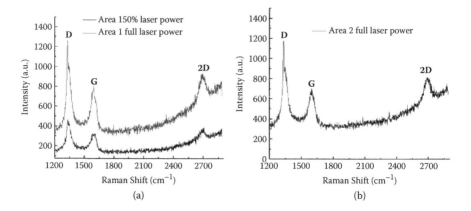

FIGURE 5.14 Micro-Raman spectra of a 3 ML graphene film on $Co_3O_4(111)/Co(111)$ acquired at two regions ~3 mm apart on a 1 cm × 1 cm sample. (a) Spectra acquired from area 1 at (top) full and (bottom) 50% laser power; (b) spectrum acquired at full laser power from area 2. (From M. Zhou et al., *Phys. Cond. Matt.*, 24 (2012) 072201 (supplementary data online). With permission.)

This is inconsistent with the determination, from LEED data, of an average domain size—1800 Å—comparable to HOPG [22], which typically yields a negligible D peak (e.g., Figure 5.3b, blue trace). The explanation for this inconsistency can be found in a comparison of two spectra acquired from the same spot at full and half laser power (Figure 5.14a, red and black traces, respectively). The data indicate that the intensities of the G and 2D features increase proportionally to the laser power, but that the increase in the D feature is much greater, roughly as (laser power)^2. This is indicative of an interaction between the D vibrational mode—the only mode in the spectrum with an out-of-plane vibrational component—and the dipole from the substrate oxide layer [72]. Thus the Raman data (Figure 5.14) indicate that the 3 ML graphene film is macroscopically uniform and continuous. However, the data also demonstrate that Raman spectroscopy is *not* a reliable metric for film quality (as determined by D peak relative intensity) on oriented polar oxide substrates.

In summary, the data in Figures 5.11–5.14 demonstrate that macroscopically continuous and uniform graphene sheets have been grown by MBE at 1000 K on $Co_3O_4(111)/Co(0001)$. Furthermore, the growth proceeds in a layer-by-layer manner, with no evidence of carbon reaction/oxidation at edge sites, or other evidence of interfacial chemical reaction. This is consistent with growth under MBE conditions on this surface being much more rapid at edge sites than at terrace sites. In this respect, growth of graphene by MBE on $Co_3O_4(111)/Co(0001)$ differs significantly from the reported results of graphene growth by the same method on silica-terminated mica [20,21], or of graphene growth on MgO(111) [18,19,26]. It would appear that the choice of oxide substrate rather than growth method determines the extent of interfacial interactions. However, the ability to grow graphene multilayers in a uniform and controlled manner, which is apparently feasible by MBE on $Co_3O_4(111)$ and (albeit at >1273 K) on SiC(0001) [73,74], may have important implications for controlling and limiting such effects as graphene/substrate charge transfer [64].

5.6.2 MAGNETIC PROPERTIES AND DEVICE APPLICATIONS

A principal motivation for direct graphene growth on dielectric substrates is, of course, the development of graphene-based electronic or spintronic devices. With respect to charge-based devices, the ability to induce a significant band gap for graphene/MgO(111) [18,19] suggests direct applications to graphene FETs, particularly as (111)-textured MgO films have been deposited on Si(100) by pulsed laser deposition [66]. While there is thus a direct potential route toward formation of graphene/MgO(111) FET structures, the formation of a band gap at the Dirac point leads to a change in the conduction and valence band dispersions in the vicinity, and therefore to an increase in the effective mass and decrease in the electron mobility. For graphene/MgO(111)-based FETs to be of real value, there must be a "sweet spot" in the trade-off between increased band gap (higher on/off ratio) and decreased mobility. Semiconducting behavior has been determined by transport measurements on single-sheet graphene on MgO(111) with a band gap of ~0.5–1 eV [63], consistent with spectroscopic measurements on few-layer graphene [18]. Band gaps of this magnitude are generally suitable for conventional (e.g., nontunneling) FETs [36,37].

The ability to deposit graphene on magnetically polarizable substrates raises the possibility of developing spintronic devices based on the proximity effect [4], in which the spin polarization of graphene electrons is predicted to be induced by proximity to a ferromagnetic layer. Such applications, because of the nature of the spin transport, would be distinctly different from conventional graphene spin valves or spin FETs based on injection and diffusion of discrete spins in graphene. Originally proposed by Semenov et al. [75], the spin FET involves injection of spin-polarized electrons at a source (Figure 5.15, S) into a graphene layer, with the diffusion of such spins through a graphene layer, and transport into a drain (Figure 5.15, D). A ferromagnetic dielectric (Figure 5.15, FMD) controls spin precession. This causes a difference in conductance (C) and magnetoresistance (MR) when the source and drain are in parallel (P) or antiparallel (AP) configurations:

$$MR(\%) = ([C_P - C_{AP}]/C_{AP}) \times 100 \qquad (5.1)$$

The long spin diffusion lengths and small spin-orbit coupling in graphene make such devices attractive [3], and variants have been constructed and tested by several groups [3,76,77].

The actual performance of such devices has been unexciting, however, at least with respect toward practical device applications, as typical MR values (Equation 5.1) are ~10%, and then only at cryogenic temperatures [3,76,77]. There are several reasons for this rather disappointing result, one of which is the necessity of using spin-polarized tunneling, typically through an oxide barrier, to maximize the low efficiency of spin injection/detection [78]. A second reason is that such devices involve the *diffusion* of individual spins through the graphene layer, with the result that the net spin

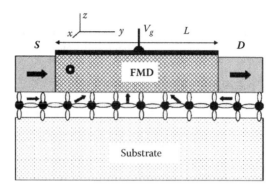

FIGURE 5.15 Proposed spin FET, consisting of a ferromagnetic source and drain (S, D) separated by a ferromagnetic dielectric (FMD) gate on top of a graphene channel, lying in turn on top of a graphene substrate. The external applied field from the FMD affects spin of the electron diffusing through the graphene channel, and therefore the conductance. (From Y.G. Semenov et al., *Appl. Phys. Lett.*, 91 (2007) 153105. With permission.)

polarization, and therefore MR, decreases significantly with source/drain distance [3] and possibly also with spin scattering at impurities and grain boundaries.

A second, somewhat different type of spin FET has been proposed [32], based on graphene in intimate contact with a FM substrate, as in an intercalated graphene/ FM/graphene structure (Figure 5.16, left). In such a structure, exchange interactions with the unpaired electrons on the ferromagnetic centers (Figure 5.16, left—middle spheres) induce spin polarization of the graphene conduction electrons, as previously predicted for graphene in proximity to EuO [4]. In the case of graphene/FM/graphene structures, where FM = Co, Ni, or Fe, DFT calculations predict AB stacking (graphene lattice A sites in one layer corresponding to lattice B sites in the next layer), with Fe atoms located over C-C bonds [22]. Substantial graphene conduction band polarizations (>60%) are also predicted for Co and Fe (but not for Ni) with quantum transport calculations suggesting MR values of ~200% at low source/drain bias and at room temperature (Figure 5.16, right) Notably, even at bias voltages of >0.1 V, the calculated MR at 300 K is several times better than reported values [3,76,77] for spin valves based on the tunneling injection and diffusion of discrete spins.

An important issue is whether the proximity effect, if it occurs, relies simply on the presence of a local magnetic mean field, as predicted [4], or whether induced magnetization is a sensitive function of C/FM orbital hybridization [32]. The graphene/Co_3O_4(111) interface is incommensurate [22] (Figure 5.15), indicating minimal graphene/Co orbital hybridization. The formation and testing of spin FETs on graphene/Co_3O_4(111)/Co(111) substrates would provide a rigorous test of these competing models, as well as perhaps open the way toward spin FETs with much enhanced magnetoresistance values at room temperatures.

5.7 SUMMARY AND CONCLUSIONS

The results presented in this chapter demonstrate that graphene can be deposited directly on h-BN(0001), MgO(111), mica, and Co_3O_4(111) by a variety of scalable,

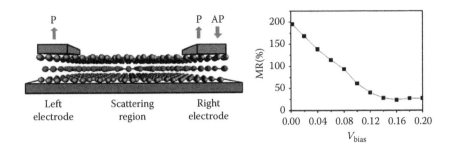

FIGURE 5.16 Proposed spin FET based on (left) graphene sheets (top and bottom layers of spheres) intercalated with an atomic layer of a ferromagnetic material (middle layers of spheres). (Right) Predicted MR values vs. source/drain voltage difference. (From J. Zhou et al., *J. Phys. Chem. C*, 115 (2011) 25280. With permission.)

industrially practical methods, including CVD, PVD, and MBE. These data also demonstrate that the type of substrate can significantly alter the electronic structure of graphene:

- *Graphene/BN.* For graphene/monolayer h-BN(0001)/Ru(0001), deposited by CVD using C_2H_4, there is substantial charge transfer (0.07 e⁻/carbon atom) from the substrate to the graphene, filling the graphene π* band, without impacting the dispersion of the σ* band. This coincides with red-shifting (by more than 300 cm⁻¹) of the graphene 2D Raman band, indicative of mode softening. In contrast, no such red-shift is observed for graphene/BN(nanoflakes). This and C(1s) spectra indicate that in fact there is some charge transfer from graphene to the substrate nanoflakes, and suggest that the formation of epitaxial BN multilayers by ALD can systematically decrease substrate → graphene charge transfer. Although a commensurate interface is observed for graphene/monolayer h-BN(0001)/Ru(0001), LEED images display C_6v symmetry, indicating the chemical equivalence of A and B sites, and consistent with the lack of a discernible band gap in STM dI/dV or PES/IPES measurements.

- *Graphene/MgO(111).* Single- or few-layer films of graphene have been deposited on MgO(111) by CVD (thermally predissociated C_2H_4), by PVD at room temperature followed by annealing to 1000 K, and by annealing a monolayer of adventitious carbon in the presence of C_2H_4. Both single- and few-layer films exhibit C_3v LEED symmetry, indicative of interfacial reconstruction resulting in a commensurate interface, and of graphene lattice A site/B site chemical inequivalence. XPS also indicates partial oxidation of the carbon overlayer, possibly at edge sites. Consistent with the LEED data, conductivity measurements indicate a band gap of ~0.6 eV for a single-layer film, and photoemission/inverse photoemission indicates a band gap of 0.5–1 eV for a few-layer film. Inverse photoemission data also indicate some graphene → oxide charge transfer.

- *Graphene/mica.* Graphene multilayers have been deposited on mica at 1000 K by MBE from a solid carbon source. XPS data and growth simulations suggest initial graphene nucleation at oxide defect sites, with oxidation of carbon edge sites. The growth process is highly temperature-sensitive, with graphene growth occurring at 1073 K, but not at 473 K or 1273 K. Raman data indicate macroscopic non-uniformity of the graphene films, with higher quality graphene growth (as determined by a decrease in the Raman D peak). XPS data indicate that, in addition to graphitic carbon, smaller amounts of carbidic carbon and carbon in a partially oxidized state are present.

- *Graphene/Co3O4(111).* Graphene layer-by-layer growth on $Co_3O_4(111)$ occurs at 1000 K, by MBE from a solid carbon source. The oxide/graphene interface is incommensurate, and consistent with this, the graphene LEED images exhibit C_6v symmetry at low and high graphene coverage, indicative of the chemical equivalency of the graphene A sites and B sites. The LEED data also indicate an average domain size of ~1800 Å, comparable to HOPG. Both inertness upon exposure to ambient and micro-Raman data demonstrate film continuity and uniformity over macroscopic distances. The intensity of the graphene D peak varies in a nonlinear manner with laser power, indicating strong interaction between the out-of-plane D vibrational mode and the oxide substrate dipole, thereby suggesting that Raman spectroscopy is not a reliable fingerprint for graphene domain/edge site density on polar substrates. XPS and spectroscopic ellipsometry data indicate significant graphene → oxide charge transfer.

These results point the way toward direct graphene growth on a variety of oxides and nitride substrates, for device applications ranging from graphene-based FETs to nonlocal spin valves and magnetic tunnel junctions. Further, some of these substrates (e.g., MgO(111), $Co_3O_4(111)$) can be formed directly on Si(100), indicating routes toward direct integration with Si CMOS. What is clear, however, is that practical development of graphene-based devices must consider not only the issues of direct graphene growth, but how interfacial chemistry modifies graphene electronic and magnetic properties, and how such interactions can be systematically varied by, e.g., the use of few-layer, rather than single-layer, graphene. Such basic considerations strongly suggest that studies of graphene growth will expand to focus not only on substrate interactions, but on control of the relative orientations and interactions of graphene sheets, in the transition from single- to few-layer graphene.

ACKNOWLEDGMENTS

This work was supported in part by the Semiconductor Research Corporation/Global Research Consortium under Task ID 2123.001 and Task ID 1770.001. Günther Lippert, Peter Dowben, Jincheng Du, and Paul Bagus are acknowledged for many stimulating conversations. Mi Zhou, Cao Yuan, Sneha Gaddam, and Frank Pasquale are gratefully acknowledged for their assistance in preparation of the manuscript.

REFERENCES

1. A.H.C. Neto, F. Guineau, N.M.R. Peres, K.S. Noveoselov, A.K. Geim. *Rev. Mod. Phys.* 81 (2009) 109.
2. E.V. Castro, K.S. Novoselov, S.V. Morozov, N.M.R. Peres, J.M.B. Lopes dos Santos, J. Nilsson, F. Gineau, A.K. Geim, A.H. Castro Neto. *Phys. Rev. Lett.* 99 (2007) 216802.
3. N. Tombros, C. Jozsa, M. Popinciuc, H.T. Jonkman, B.J. van Wees. *Nature* 448 (2007) 571.
4. H. Haugen, D. Huertas-Hernando, A. Brataas. *Phys. Rev. B* 77 (2008) 115406.
5. K.S. Novoselov, A.K. Geim, S.V. Morozov, D. Jiang, Y. Zhang, S.V. Dubonos, I.V. Grigorieva, A.A. Firsov. *Science* 306 (2004) 666.
6. K.S. Novoselov, A.K. Geim, S.V. Morozov, D. Jian, M.I. Katsnelson, I.V. Grigorieva, S.V. Dubonos, A.A. Firsov. *Nature* 438 (2005) 197.
7. X. Li, W. Cai, J. An, S. Kim, J. Nah, D. Yang, R. Piner, A. Velamakanni, I. Jung, E. Tutuc, S.K. Banerjee, L. Colobmo, R.S. Ruoff. *Science* 324 (2009) 1312.
8. A. Reina, S. Thiele, X. Jia, S. Bhaviripudi, M.S. Dresselhaus, J.A. Schaefer, J. Kong. *Nano Res.* 2 (2009) 509.
9. P.W. Sutter, J. Flege, E.A. Sutter. *Nature Mater.* 7 (2008) 406.
10. X. Liang, Z. Fu, S.Y. Chou. *Nano Lett.* 7 (2007) 3840.
11. C. Berger, Z. Song, T. Li, Z. Li, A.Y. Ogbazghi, R. Feng, Z. Dai, A.N. Marchenkov, E.H. Conrad, P.N. First, W.A. de Heer. *J. Phys. Chem. B* 108 (2004) 19912.
12. C. Berger, Z. Spong, X. Li, X. Wu, N. Brown, C. Naud, D. Mayou, T. Li, J. Hass, A. Marchenkov A.N., E.H. Conrad, P.N. First, W.A. de Heer. *Science* 312 (2006) 1191.
13. W.A. de Heer, C. Berger, X. Wu, P.N. First, E.H. Conrad, X. Li, T. Li, M. Sprinkle, J. Hass, M.L. Sadowski, M. Potemski, G. Martinez. *Sol. St. Commun.* 143 (2007) 92.
14. Y. Zhang, V.W. Brar, C. Girit, A. Zettl, M.F. Crommie. *Nature Physics* 5 (2009) 722.
15. F. Chen, J. Xia, D.K. Ferry, N. Tao. *Nano Lett.* 9 (2009) 2571.
16. Y.-M. Lin, C. Dimitrakopoulos, K.A. Jenkins, D.B. Farmer, H.-Y. Chiu, A. Grill, P. Avouris. *Science* 327 (2010) 662.
17. C. Bjelkevig, Z. Mi, J. Xiao, P.A. Dowben, L. Wang, W. Mei, J.A. Kelber. *J. Phys. Cond. Matt.* 22 (2010) 302002.
18. L. Kong, C. Bjelkevig, S. Gaddam, M. Zhou, Y.H. Lee, G.H. Han, H.K. Jeong, N. Wu, Z. Zhang, J. Xiao, P.A. Dowben, J.A. Kelber. *J. Phys. Chem. C* 114 (2010) 21618.
19. S. Gaddam, C. Bjelkevig, S. Ge, K. Fukutani, P.A. Dowben, J.A. Kelber. *J. Phys. Cond. Matt.* 23 (2011) 072204.
20. G. Lippert, J. Dabrowski, M. Lemme, C. Marcus, O. Seifarth, G. Lupina. *Phys. Stat. Sol. B* 248 (2011) 2619.
21. O. Seifarth, G. Lippert, J. Dabrowski, G. Lupina, W. Mehr. *Proceedings of IEEE Conference on Semiconductors* (Dresden, September 27–28, 2011), p. 1.
22. M. Zhou, F.L. Pasquale, P.A. Dowben, A. Boosalis, M. Schubert, V. Darakchieva, R. Yakimova, J.A. Kelber. *J. Phys. Cond. Matt.* 24 (2012) 072201.
23. M.A. Fanton, J.A. Robinson, B.E. Weiland, M. LaBella, K. Trumbell, R. Kasarda, C. Howsare, M. Hollander, D.W. Snyder. Abstract for the Graphene 2011 Conference (Bilbao, Spain, 2011).
24. R.S. Shishir, D.K. Ferry. *J. Phys. Cond. Matt.* 21 (2009) 232204.
25. C.R. Dean, A.F. Young, I. Meric, C. Lee, L. Wang, S. Sorgenfrei, K. Watanabe, T. Taniguchi, P. Kim, K.L. Shepard, J. Hone. *Nature Nanotechnol.* 5 (2010) 722.
26. J.A. Kelber, S. Gaddam, C. Vamala, S. Eswaran, P.A. Dowben. *Proc. SPIE* 8100 (2011) 81000Y-1
27. J. Hass, F. Varchon, J.E. Millan-Otoya, M. Sprinkle, N. Sharma, W.A. de Heer, C. Berger, P.N. First, L. Magaud, E.H. Conrad. *Phys. Rev. Lett.* 100 (2008) 125504.

28. Z. Peng, Z. Yan, Z. Sun, J.M. Tour. *ACS Nano* 5 (2011) 8241.
29. E. Kim, H. An, H. Jang, W. Cho, N. Lee, W. Lee, J. Jung. *Chem. Vap. Dep.* 17 (2011) 9.
30. W. Gannett, W. Regan, K. Watanabe, T. Taniguchi, M.F. Crommie, A. Zettl. *Appl. Phys. Lett.* 98 (2011) 242105.
31. A. Reina, X. Jia, J. Ho, D. Nezich, H. Son, V. Bulovic, M.S. Dresselhaus, J. Kong. *Nano Lett.* 9 (2009) 30.
32. J. Zhou, L. Wang, R. Qin, J. Zheng, W.N. Mei, P.A. Dowben, S. Nagase, Z. Gao, J. Lu. *J. Phys. Chem. C* 115 (2011) 25280.
33. P.A. Cox. *The Electronic Structure and Chemistry of Solids*. Oxford University Press, Oxford, 1987.
34. X. Li, X. Wu, M. Sprinkle, F. Ming, M. Ruan, Y. Hu, C. Berger, W.A. de Heer. *Phys. Stat. Sol.i A* 207 (2010) 286.
35. S.Y. Zhou, G.-H. Gweon, A.V. Federov, P.N. First, W.A. de Heer, D.-H. Lee, F. Guinea, A.H. Castro Neto, A. Lanzara. *Nature Materials* 6 (2007) 770.
36. M. Cheli, P. Michetti, G. Iannaccone. *IEEE Trans. Electron Dev.* 57 (2010) 1936.
37. P. Michetti, M. Cheli, G. Iannaccone. *Appl. Phys. Lett.* 96 (2010) 133508.
38. Y. Zhang, T. Tang, C. Cirit, Z. Hao, M.C. Martin, A. Zettl, M.F. Crommie, R.Y. Shen, F. Wang. *Nature* 459 (2009) 820.
39. M.Y. Han, B. Ozyilmaz, Y. Zhang, P. Kim. *Phys. Rev. Lett.* 98 (2007) 206805.
40. Y. Son, M.L. Cohen, S.G. Louie. *Phys. Rev. Lett.* 97 (2006) 216803.
41. A. Nagashima, N. Tejima, Y. Gamou, T. Kawai, C. Oshima. *Phys. Rev. Lett.* 75 (1995) 3918.
42. C. Oshima, A. Nagashima. *J. Phys. Cond. Matt.* 9 (1997) 1.
43. A. Goriachko, Y. He, M. Knapp, H. Over, M. Corso, T. Brugger, S. Berner, J. Osterwalder, T. Greber. *Langmuir* 23 (2007) 2928.
44. K. Watanabe, T. Taniguchi, H. Kanda. *Nature Mater.* 3 (2004) 404.
45. A.B. Preobrajenski, A.S. Vinogradov, N. Martensson. *Surf. Sci.* 582 (2005) 21.
46. C. Oshima, A. Itoh, E. Rokuta, T. Tanaka, K. Yamashita, T. Sakuri. *Sol. St. Commun.* 116 (2000) 37.
47. T. Lin, Y. Wang, H. Bi, D. Wan, F. Huang, X. Xie, M. Jiang. *J. Mater. Chem.* 22 (2012) 2859.
48. A.C. Ferrari, J.C. Meyer, V. Scardaci, C. Casiraghi, M. Lazzeri, F. Mauri, S. Piscanec, D. Jiang, K.S. Novoselov, S. Roth, A.K. Geim. *Phys. Rev. Lett.* 97 (2006) 187401.
49. U. Gelius. *Phys. Scripta* 9 (1974) 133.
50. L. Ci, L. Song, C. Jin, D. Jariwala, D. Wu, Y. Li, A. Srivastava, Z.F. Wang, K. Storr, L. Balicas, F. Liu, P.M. Ajayan. *Nature Mater.* 9 (2010) 430.
51. I. Forbeaux, J.-M. Themlin, J.-M. Debever. *Phys. Rev. B* 58 (1998) 16396.
52. X. Hong, A. Posadas, K. Zou, C.H. Ahn, J. Zhu. *Phys. Rev. Lett.* 102 (2009) 136808.
53. Y. Zhang, Y. Tan, H.L. Stormer, P. Kim. *Nature* 438 (2005) 201.
54. J.D. Ferguson, A.W. Weimar, S.M. George. *Thin Sol. Films* 413 (2002) 16.
55. G. Giovannetti, P.A. Khomyakov, G. Brocks, P.J. Kelly, J. van den Brink. *Phys. Rev. B* 76 (2007) 073103.
56. B. Lee, L. Wang, C.D. Spartary, S.G. Louie. *Phys. Rev. B* 76 (2007) 245114.
57. F. Rohr, K. Wirth, J. Libuda, D. Cappus, M. Baumer, H.-J. Freund. *Surf. Sci.* 315 (1994) L977.
58. D. Cappus, C. Xu, D. Ehrlich, B. Dillmann, C.A. Ventrice Jr., K. Al Shamery, H. Kuhlenbeck, H.-J. Freund. *Chem. Phys.* 177 (1993) 533.
59. M. Hassel, H.-J. Freund. *Surf. Sci.* 325 (1995) 163.
60. V.K. Lazarov, R. Plass, H.-C. Poon, D.K. Saldin, M. Weinert, S.A. Chambers, M. Gajdardziska-Josifovska. *Phys. Rev. B* 71 (2005) 115434.
61. J. Goniakowski, C. Noguera. *Phys. Rev. B* 66 (2002) 085417.
62. Y.S. Dedkov, M. Fonin, U. Rudiger, C. Laubschat. *Phys. Rev. Lett.* 100 (2008) 107602.
63. J.A. Kelber, M. Zhou, S. Gaddam, F. Pasquale, L. Kong, P.A. Dowben. *Proc. ECS* (2012) in press.

64. K.V. Emtsev, F. Speck, T. Seyller, L. Ley. *Phys. Rev. B.* 77 (2008) 155303.
65. K.V. Emtsev, A. Bostwick, K. Horn, J. Jobst, G.L. Kellogg, L. Ley, J.L. McChesney, T. Ohta, S.A. Reshanov, J. Rohrl, E. Rotenberg, A.K. Schmid, D. Waldmann, H.B. Weber, T. Seyller. *Nature Mater.* 8 (2009) 203.
66. X.Y. Chen, K.H. Wong, C.L. Mak, X.B. Yin, M. Wang, J.M. Liu, Z.G. Liu. *J. Appl. Phys.* 91 (2002) 5728.
67. C. Argile, G.E. Rhead. *Surf. Sci. Rep.* 10 (1989) 277.
68. S.C. Petito, M.A. Langell. *J. Vac. Sci. Technol. A* 22 (2004) 1690.
69. Briggs and M. P. Seah (eds.), *Practical Surface Analysis 2nd Edition, Vol. I: Auger and Photoelectron Spectroscopy,* Wiley Interscience, NY (1990) 657 pages.
70. H. Ago, Y. Ito, N. Mizuta, K. Yoshida, B. Hu, C.M. Orofeo, M. Tsuji, K. Ikeda, S. Mizuno. *ACS Nano* 4 (2010) 7407.
71. Z. Ni, Y. Wang, T. Yu, Z. Shen. *Nano Res.* 1 (2008) 273.
72. S. Franzen. *J. Phys. Chem. C* 113 (2009) 5912.
73. J. Park, W.C. Mitchel, L. Grazulis, H.E. Smith, K.G. Eyink, J.J. Boeckl, D.H. Tomich, S.D. Pacley, J.E. Hoelscher. *Adv. Mater.* 22 (2010) 4140.
74. E. Moreau, F.J. Ferrer, D. Vignaud, S. Godey, X. Wallart. *Phys. Stat. Sol. A* 207 (2010) 300.
75. Y.G. Semenov, K.W. Kim, J.M. Zavada. *Appl. Phys. Lett.* 91 (2007) 153105.
76. S. Cho, Y. Chen, M.S. Fuhrer. *Appl. Phys. Lett.* 91 (2007) 123105.
77. W. Han, K. Pi, K. Wang, H. Wang, M. McCreary, Y. Li, W. Bao, P. Wei, J. Shi, C.N. Laun, R.K. Kawakami. *Proc. SPIE* 7398 (2009) 739819-1.
78. B. Dlubak, P. Seneor, A. Anane, C. Barraud, C. Deranlot, D. Deneuve, B. Servet, R. Mattana, F. Petroff, A. Fert. *Appl. Phys. Lett.* 97 (2010) 092502.

6 Aligned Carbon Nanotubes for Interconnect Application

Yang Chai, Minghui Sun, Zhiyong Xiao, Yuan Li, Min Zhang, and Philip C.H. Chan

CONTENTS

6.1 INTRODUCTION

The complex interconnects link the electronic components in the very large-scale integration (VLSI) and provide signal and power to them. Nowadays, the speed of the VLSI is more and more dominated by the signal transmission in the interconnects instead of the transistor switching, because most of the on-chip capacitances of the VLSI are associated with the interconnects [1]. State-of-the-art copper interconnect technology is expected to run into its physical limit as the feature size of the integrated circuit (IC) technology continues to scale down to nanometer size [2]. The increasing copper resistivity due to surface and grain boundary scattering results in the increasing resistance-capacitance delay, and it has become a near-term challenge for the copper interconnects according to the International Technology Roadmap for Semiconductors (ITRS). The current density through the copper interconnects also increases as the feature size scales down, which aggravates the electromigration (EM) reliability of the copper interconnect.

To enable the continuing scaling of interconnects, researchers must develop a more conductive and robust interconnect conductor [2]. A variety of methods have been investigated by scientists and engineers, including the replacement of copper with new materials (metal silicides, carbon nanotube, and graphene nano-ribbon), optimization of metal interconnect, e.g., metallic phonon engineering of silver [3] and metallic geometrical resonance [4], and other interconnect schemes, e.g., optical interconnect [5] and wireless interconnect [6].

Among these approaches, carbon nanotube (CNT) has emerged as a promising candidate for next-generation interconnect conductors because of its ballistic transport characteristics, high thermal conductivity, high current carrying capacity, and high aspect ratio. Theoretical works have predicted that the CNT interconnect can outperform the copper interconnect at most levels of interconnection hierarchy [1,7–9]. The CNT interconnect can be made of either densely packed single-walled CNTs (SWNTs) or larger-diameter multiwalled CNTs (MWNTs) [7,8,10]. In this chapter, we shall present the recent experimental advancements in the field of CNT interconnects, mainly focusing on the works conducted in our group at the Hong Kong University of Science and Technology.

6.2 CONTROLLED GROWTH OF ALIGNED CNTS

The electrical and thermal conduction in the CNT is highly anisotropic due to its quasi-one-dimensional (1D) feature. Straight and aligned CNTs are preferred for the interconnect application, as this configuration provides the shortest carrier conduction path compared to the random entangled CNTs. Since each conductive channel in CNT has a quantum limit (12.9 kΩ), we desire the high-density aligned CNTs in parallel to minimize the overall interconnect resistance. In the VLSI interconnect hierarchy, different locations in the same level are connected by the isolated interconnect lines, and different levels of the interconnections are connected by the contacts/vias. To use CNT for VLSI interconnections, this requires precise control on the direction of the aligned CNTs. The chemical vapor deposition (CVD) method has been demonstrated to be deterministic for controlled growing carbon nanostructures [11].

6.2.1 VERTICALLY ALIGNED CNTS

The contact and via are usually the smallest and most abundant features in VLSI [12]. It is challenging to seamlessly fill the high aspect ratio via with Cu, W, or Ru. The voiding in the contact/via will lead to significant reliability problems. The 1D feature of the CNT enables us to fill the high aspect ratio contact/via hole from the bottom with the CNTs. The CNT itself also does not suffer from the EM-like failure because of the strongest carbon-carbon chemical bonding.

The thermal randomization during the typical CNT growth process results in the entangled CNTs without any alignment. To align the CNT in the preferred direction, it is desirable to provide an external force to overcome the thermal randomization

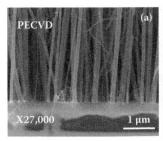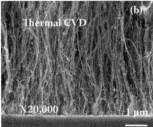

FIGURE 6.1 SEM images of the vertically aligned CNT. (a) Vertically aligned CNTs grown by PECVD method. (b) Vertically aligned CNTs grown by thermal CVD from the decomposition of ferrocene. The CNTs are entangled due to the thermal randomization.

during the CNT growth process. A usual method is the so-called crowding effect. The CNTs are grown from high-density catalyst particles. The growth direction of CNTs is restricted by the neighboring CNTs, only allowing them to grow in the vertical direction [13–15]. In our work, we used ferrocene as an in situ high-density catalyst to grow the vertically aligned CNT by the thermal CVD method [13–15]. In the second example, the electric field has been shown to guide the growth direction of CNT [16]. In the plasma-enhanced chemical vapor deposition (PECVD) chamber, a vertical built-in electric field can be generated on the blanket substrate due to the electron accumulation on the substrate surface, which interacts with the induced dipole in the CNT, and align it in the vertical direction [17]. Figure 6.1 shows the representative scanning electron microscopy (SEM) images of the vertically aligned CNTs in our experiments, grown with the two methods we mentioned above. We clearly observe that the CNT grown by PECVD is much straighter than those grown by thermal CVD.

Figure 6.2(a) schematically illustrates the mechanism of the alignment control of the CNT by electric field. An electric dipole is induced by the external electric field inside the CNT. This dipole is along the axis of the 1D CNT. The interaction between the dipole and the external electric field steers the direction of the CNT at every stage of the growth process, resulting in the perfect alignment of the CNT in the direction of the electric field. Figure 6.2(b) and (c) shows the vertically aligned CNT film and a single vertical and freestanding CNT grown by PECVD. This suggests the electric field is an effective tool for growing the aligned CNTs. However, the defects introduced in the plasma growth environment still remain an open question.

6.2.2 HORIZONTALLY ALIGNED CNTs

The controlled growth of the horizontally aligned and high-density CNTs on Si substrate is still challenging. The aligned SWNTs grown on quartz or sapphire substrate have low density (<20/μm) and require the complex transfer process [18]. Our group has developed two methods to fabricate the high-density horizontally aligned CNTs directly on Si substrate. By designing and fabricating the microstructures, we modified the built-in electric field near the sidewall of the microstructures in the horizontal direction [19]. The interaction between the electric field and the induced dipole results in the growth of horizontally aligned CNTs [19]. An alternative approach is to grow

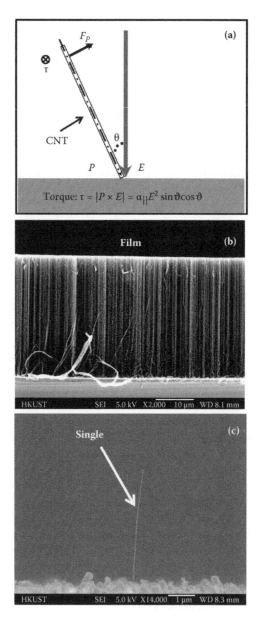

$$\text{Torque: } \tau = |P \times E| = \alpha_{||} E^2 \sin\vartheta\cos\vartheta$$

FIGURE 6.2 Alignment control of the CNT by electric field. (a) Schematic of the interaction between the induced dipole in CNT and the external electric field. SEM images of (b) the vertically aligned CNT film and (c) a single vertical and freestanding CNT.

the vertically aligned CNTs from the catalyst stripes first, then immerse the sample into isopropyl alcohol to orientate them to the desired direction. The capillary force between the CNT and the liquid guides the CNT to align horizontally on the substrate surface [20]. Figure 6.3 shows the horizontally aligned CNTs by the two methods.

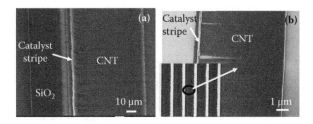

FIGURE 6.3 SEM images of the horizontally aligned CNTs. (a) Horizontally aligned CNTs directly grown by electron-shading effect. (b) Horizontally aligned CNTs leveled down from vertically aligned CNTs by liquid treatment.

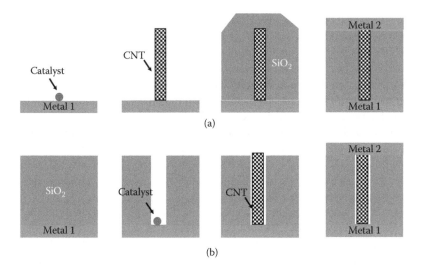

FIGURE 6.4 Schematic of the process flow (a) bottom-up approach without the via hole opening process, and (b) buried catalyst approach, which requires via hole opening, similar to the conventional Cu via process.

6.3 INTEGRATION SCHEMES WITH IC TECHNOLOGY

Integrating the CNTs in ICs as interconnects requires researchers to develop a cost-effective integration scheme without scarifying the overall performance and reliability of the interconnect. The CNT interconnects need to interface with other materials, including metal, dielectrics, barrier layer, and catalyst. This makes the integration of the CNTs challenging. We shall describe some of the attempts to integrate the CNT into the VLSI system.

6.3.1 INTEGRATION SCHEMES

Two approaches have been developed for the integration of the vertical CNT via: bottom-up approach and buried catalyst. They are schematically illustrated in Figure 6.4 [21]. In one example, the CNT is first grown on the substrate. The interlayer dielectrics

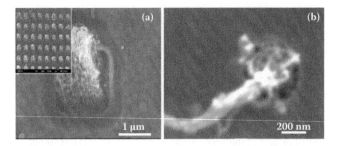

FIGURE 6.5 SEM images of the CNT via with different feature size. (a) 1 μm and (b) 300 nm diameter.

are then deposited. The excess dielectric is removed by chemical-mechanical polish process to expose the CNT tip. The top metal is then in contact with the CNT tip, as shown in Figure 6.4(a). This bottom-up approach avoids the via hole opening process, and offers the ultra-scaling potential to a few nanometers (around the diameter of CNT). However, this method is only applicable to the large-diameter MWNT or carbon nanofiber with strong mechanical strength, which allows it to withstand the harsh dielectric layer deposition process [22].

In the second example, the so-called buried catalyst approach, the via hole is etched first and the CNTs are grown from the bottom catalyst. The order of the process flow is similar to the copper interconnect process. In the via hole opening process, the harsh etching environment may affect the catalyst surface and cause the catalyst particles not to function properly.

In our work, we used a process flow similar to that in the buried catalyst approach. To avoid the adverse influence of the etching process on the catalyst surface, we optimized the process flow by depositing catalyst after the via hole etching. The photoresist pattern for the via hole etching also served as the mask of the catalyst liftoff, as shown in Figure 6.4(b). Unlike the Cu via process, the CNT via does not require a high-resistance and thick liner to interface between the conductor and the dielectric. Figure 6.5 shows the SEM images of the vertical CNT via with different feature sizes. The integration of CNT into the nanoscale via hole using this method still has poor density CNTs, lower than that from the blanket catalyst film.

The integration of the horizontally aligned CNTs as the interconnect line has been demonstrated by several methods, including the direct growth [23], transfer [18], and dielectrophoresis [24]. The key challenges are the fabrication of the densely packed CNTs and the formation of good electrical contact to the metal in the meantime.

6.3.2 DENSITY ENHANCEMENT

Theoretical calculations suggest that the density of the multiple CNT conduction channels has to be larger than $10^{13}/cm^2$ to outperform the copper interconnect [25]. In the CVD fabrication method, the CNT is grown from the catalyst particle. In order to reach the projected density of the CNTs for the interconnect application, we need to prepare monodispersed catalyst particles on the substrate. Currently, the catalyst for the CNT growth by CVD method is mostly prepared by physical vapor

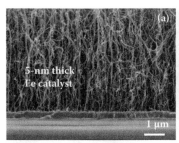 

FIGURE 6.6 SEM images of the CNTs from different catalyst thicknesses. (a) 5 nm and (b) 10 nm thick Fe thin film.

deposition. The effect of catalyst thickness is a key factor to determine the density and diameter of the CNTs. The density of particles as a function of the catalyst thickness can be described according to: $\rho_s = n/S = 6t/(\pi d^3) \propto t/d^3$, where n is the number of the catalyst particles, d is the average diameter of the CNT, and t is the thickness of catalyst film [23]. The ultra-thin catalyst film helps to form the dense catalyst nanoparticles. Figure 6.6 shows the CNTs grown from different catalyst thicknesses. The thinner catalyst produces the smaller-diameter and higher-density CNTs.

The catalyst "poisoning" is a key factor for reducing the density of CNTs, where part of the catalyst particles is deactivated during the CNT growth process. This catalyst poisoning has been proved to be caused by the coating of amorphous carbon on the surface of catalyst particles [26]. To achieve the high-density catalyst particles with activation, the catalyst particles should be matched with an appropriate feeding rate of carbon source. Oxygen and plasma have been used to remove the amorphous carbon coating on the catalyst, and keep the activity of the catalyst during the CNT growth process [27].

6.3.3 INTERFACIAL CONTACT

The small contact area to the CNT makes the electrical coupling between the CNT and the surrounding conductor extremely difficult [28]. To use the CNT for VLSI interconnection, it is necessary to establish a low-resistance electrical contact to the CNT. Palladium (Pd), a noble metal with good wetting interactions with the CNT, was found to have good electrical contact to the CNT [29]. Recent works show that the metal wettability to the CNT is a key factor for the low-resistance electrical contact [30]. Our results showed the contact resistance to the CNT is diameter dependent [23], which indicates that the contact resistance of the metal/CNT interface depends on the contact area at the interface, which is closely related to the wettability of the metal to the CNT.

The metal/CNT contact can be classified into the end-contact (metal/CNT tip) and side-contact (metal/CNT sidewall) configurations. Depending on whether the pentagon/heptagon cap exists on the end of the CNT, the CNT tip can be classified into the open-end and the close-end. For the open-end multiwalled CNT without the cap, it provides contacts to not only the outer shell but also the inner walls of the

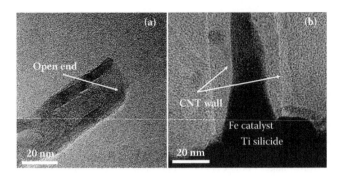

FIGURE 6.7   TEM images of (a) the open end of the CNT by the abrupt termination growth and (b) the root of the CNT on Ti silicide.

CNT, contributing more electrical conduction channels [31]. We have developed an abrupt termination process of the CNT growth using the PECVD method to control the end structure of the CNT. The plasma was abruptly switched off to end the CNT growth process, and a large hydrogen stream was introduced to dilute the carbon reaction [32]. Figure 6.7(a) shows a transmission electron microscopy (TEM) image of the open-end CNT by the abrupt termination process. To realize this open-end CNT tip configuration for better electrical contact, some research groups have demonstrated chemical-mechanical polish to remove the cap of the close end of the CNT [25,33].

The electrical contact between the CNT and the metal is very difficult. To enlarge the contact area to the CNT and form good electrical contact between the CNT and other conductors, we have used an interfacial layer between the CNT and the conductor. The contact between CNT and doped Si typically shows Schottky behavior [34]. We used Ti silicide as the interfacial layer, which has metal-like resistivity, and its Fermi level is close to that of Si. Figure 6.7(b) shows a TEM image of the interface between the CNT and the doped Si. We can clearly see that the Fe catalyst particle is encapsulated by the CNT shells, and penetrates into the Ti silicide. This enables formation of good electrical contact between the root of the CNT and the metal elecrode. We have also used a graphitic interfacial layer between the metal and the CNT, showing a substantial contact improvement [28]. The graphic interfacial layer improves the wettability of the metal to the CNT, and probably forms chemical bonding with the CNT. Although this method is demonstrated with the horizontal and side-contact configuration, it can be also extended to the vertical CNT via.

6.4   ELECTRICAL CHARACTERIZATION

Owing to the high electrical conductivity, CNT has been proposed for carrying the signal and power at the DC and RF region. The DC resistances of the CNT interconnect consist of three parts: the quantum resistance, the diffusive resistance, and the imperfect contact resistance between CNT and metal [23]. For the impedance of the CNT in the high-frequency region, the researchers need to investigate the parasitic effect.

6.4.1 DC CHARACTERIZATION

Table 6.1 lists the resistance comparisons of the CNT via fabricated by different methods. The low-resistance CNT via requires both high density and good electrical contact. One single CNT with 100 nm diameter and 25 μm length has been reported with the resistance of 34.4 Ω [31]. Dijon and coworkers reported the density of 8 × 10^{12}/cm² in December 2010, the world record for the highest density of vertical aligned CNTs until now [25]. In their works, the resistance for a 1 μm diameter CNT via is 10 kΩ [25]. Awano et al. developed a process flow similar to the damascene copper interconnects, and realized open-end CNTs by chemical-mechanical polish process [33]. A 2 μm via filled with a vertical CNT array has been shown with the resistance of 0.6 Ω [33]. Our experimental results have shown 20–108 Ω for a via (Φ = 1.2 μm) filled with the vertically aligned CNTs, which is still higher than the theoretical value [35]. This is believed to result from the high metal/CNT contact resistance, the low site density of CNTs, and the low ratio of metallic CNT in the aligned CNTs [36].

From the comparisons, we can see that the resistance of the CNT via at micrometer scale is still much larger than the resistance of copper via because of the low density of the CNTs and the high contact resistance. To overcome these engineering hurdles, we have demonstrated the concept of the copper/CNT composite, in which the gaps between the CNTs are filled by electroplated copper [36]. Table 6.2 shows

TABLE 6.1
Resistance Comparisons of the CNT Vias Fabricated by Different Methods

	Via Diameter	Density	Growth Method	Resistance
Dijon	1 μm	10^{12}/cm²	Base growth	10 kΩ
Awano	2 μm	10^{11}/cm²	PECVD	0.6 Ω
Li	100 nm	Single	Arc discharge	34.4 Ω
Ours	1.2 μm	10^9/cm²	PECVD	20–108 Ω

TABLE 6.2
Resistivity Comparisons of the Copper/CNT Composite Fabricated by Different Methods

	Process	ρ_{cu} (μΩ·cm)	$\rho_{cu/CNT}$ (μΩ·cm)	$\rho_{cu/CNT}/\rho_{cu}$
Liu	Electrophoretic	2.345	2.048	67.32%
Yang	Ultrasonic ECP	1.67	1.65	98%
Baik	Molecule mixing	N/A	N/A	90%
Ours	ECP	1.89	2.22	117%

Note: ECP = electrochemical plating.

the comparisons of the related works. Researchers have demonstrated lower resistivity of the composite than of copper only, or comparable to copper [37–39]. This CNT-based composite could be used for the multilevel interconnects.

6.4.2 RF CHARACTERIZATION

Researchers have also investigated the signal transmission in the CNT interconnect at the radio frequency (RF) region. The experimental results have demonstrated that CNT can carry high-frequency currents at least as well as DC currents operating up to 20 GHz [40]. We carried out the RF characterizations on both low-density single-walled CNTs and high-density aligned multiwalled CNTs. The high density of the CNTs enables us to study the high-frequency transport of the CNT at 40 GHz, as shown in Figure 6.8(a) [41]. The metal/CNT shows capacitive contact characteristics. The transport property of the CNT exhibits induction over 10 GHz because of the presence of the kinetic inductance (Figure 6.8b). We have constructed a lumped resistance-inductance-capacitance (RLC) model to de-embed the inductance, capacitance, and resistance (Figure 6.8c). The obtained kinetic inductance is close to the theoretical value [40,41]. We also investigated the kinetic inductance of the CNT with different channel lengths, experimentally, validating the existence of kinetic inductance of CNT in the diffusive transport region [41].

6.5 ELECTROMIGRATION CHARACTERIZATION

The state-of-the-art copper interconnect suffers from electromigration (EM) as a result of the high current density ($>10^6$ A/cm^2). This will cause structural damage by metal ion transport, resulting in a short circuit or open circuit in the integrated circuit system. As the feature size of the IC technology scales down to nanometer size, the current density through the interconnect increases dramatically. From the long-term perspective, the reliability of the copper interconnect caused by the EM becomes inevitable at nanoscale. The strongest carbon-carbon covalent bonding in the CNT enables it to carry very high current density.

6.5.1 ELECTRICAL BREAKDOWN OF CNT

The high current carrying capacity ($>10^9$ A/cm^2) and the electrical breakdown have both been demonstrated in the CNT devices. To understand these two contradictory phenomena, we need to investigate the effects of the testing conditions on the reliability of the CNT interconnects. The electrical breakdown of the CNT has been reported in air by the simple DC sweep. In this electrical breakdown process, heat was generated by the high current density through the CNT, and raised to a certain temperature. The defective sites in the CNT react with the oxygen in air. The sp^2 chain of the CNT is unzipped with the chemical reaction of -O-C-O- at the elevated temperature [42,43].

However, for the high-quality CNTs with few defects, or the CNT device testing in an oxygen-free environment, or isolated from air by passivation [35], the CNT

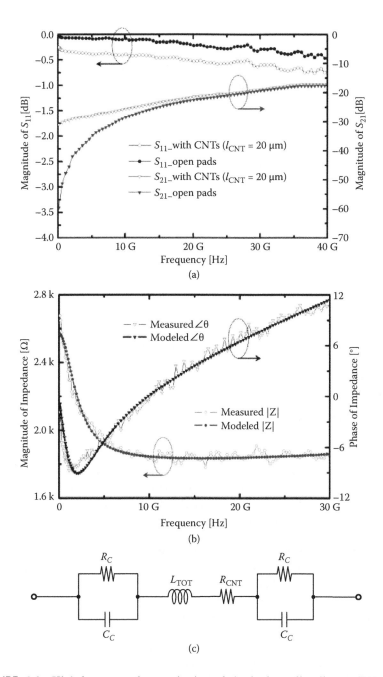

FIGURE 6.8 High-frequency characterization of the horizontally aligned CNT array. (a) Measured S_{11} and S_{21} magnitudes of a device structure with 20 μm long CNT array and the corresponding open pad structure. (b) Measured CNT impedance and the fitting results of RLC model. (c) Lumped RLC model for aligned CNTs.

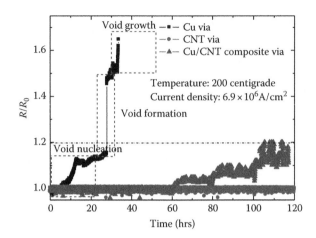

FIGURE 6.9   Resistance evolution of Cu, CNT, and Cu/CNT composite via stressing at the same current density and temperature.

shows high current carrying capacity without any EM-like degradation. Figure 6.9 shows the resistance evolution of the CNT via as a function of the constant current stressing time. We can clearly observe that the CNT via passivated with oxide carries a high current density without any failure over 100 h. The electrical breakdown is not observed in this oxygen-free condition even at high temperature.

6.5.2   ELECTROMIGRATION OF CARBON-BASED INTERCONNECT

As we have discussed in Section 6.4, the copper/CNT composite exhibits comparable or even smaller resistivity than copper. This composite is promising for the interconnect application. The CNT has been widely used in composite to enhance the mechanical strength. The EM of metal is caused by the microscopic "electron wind" force, which is a result of the high-density electron bombardment, and exchange of the momentum with metal atoms. Similar to enhancing the mechanical strength of the metal by the CNT reinforcement, the EM resistance of copper interconnects can be also improved by the CNT loading [44,45]. We have used two test structures to study the EM properties of copper/CNT composite: Kelvin structure for via EM and Blech structure for line EM [44]. Figure 6.9 shows the resistance evolution of copper and copper/CNT composite via as a function of constant current density stressing. The copper/CNT composite via shows much better EM resistance than copper via only.

We also used the Blech structure to study the EM property, which allows us to directly observe the atomic migration of the metal interconnect. Figure 6.10 shows a typical EM of metal migration near the cathode side. Obviously, the copper/CNT composite line shows a much slower void growth rate. The average critical current density length threshold products of pure copper and copper/CNT composite were estimated to be 1800 and 5400 A/cm, respectively [45].

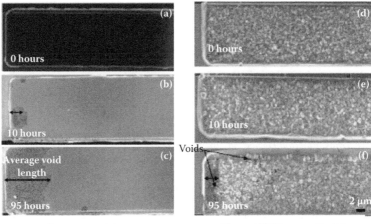

Pure copper stripes   Copper/CNT composite stripes

FIGURE 6.10 Electromigration testing using Blech structure. (a–c) Void growth in the cathode side of the Cu stripe. (d–f) Void growth in the cathode side of the Cu/CNT composite stripe.

6.6 CONCLUSIONS

We have shown our experimental results on the controlled CNT growth, the integration of the CNT via with IC technology, and the electrical and electromigration characterizations. These results suggested that the carbon-based interconnect is a promising candidate for future VLSI interconnect, and may provide a long-term solution for interconnects as the feature size scales down to nanometer size.

To realize the practical application of the CNT interconnect, we still need to overcome many engineering hurdles, including the enhancement of CNT density to the densely packed aligned CNTs, the optimization of the CNT interface to reduce the imperfect contact resistance, the growth of high-quality CNTs with less defects, and the development of cost-effective integration schemes [46].

REFERENCES

1. A. Naeemi and J.D. Meindl. Carbon nanotube interconnects. *Annu. Rev. Mater. Res.*, 39, 255–275, 2009.
2. Y. Chai, M.H. Sun, Z.Y. Xiao, Y. Li, M. Zhang, and P.C.H. Chan. Pursuit of future interconnect technology with aligned carbon nanotube array. *IEEE Nanotechnol. Mag.*, 5(1), 22–26, 2011.
3. M. Kralj, A. Siber, P. Pervan, M. Milun, T. Valla, P.D. Johnson, and D.P. Woodruff. Moving surface-vacuum barrier effects temperature dependence of photoemission from quantum-well states in Ag/V(100): Moving surface-vacuum barrier effects. *Phys. Rev. B*, 64, 085411–085419, 2001.
4. N. Trivedi and N.W. Ashcroft. Quantum size effects in transport properties of metallic films. *Phys. Rev. B*, 38, 12298–12309, 1988.
5. K. Cadien, M. Reshotko, B. Block, A. Bowen, D. Kencke, and P. Davids. Challenges for on-chip optical interconnects. *Proc. SPIE*, 5730, 133–143, 2005.

6. M.F. Chang, V.P. Roychowdhury, L. Zhang, H. Shin, and Y.X. Qian. RF/wireless interconnect for inter- and intra-chip communications. *Proc. IEEE*, 89, 456–466.

7. A. Naeemi, R. Sarvari, and J.D. Meindl. Performance comparison between carbon nanotube and copper interconnects for gigascale integration (GSI). *IEEE Electron Device Lett.*, 26, 84–86, 2005.

8. A. Naeemi and J.D. Meindl. Performance modeling for single- and multiwall carbon nanotubes as signal and power interconnects in gigascale systems. *IEEE Trans. Electron Device*, 55, 2574–2582, 2008.

9. A. Naeemi and J.D. Meindl. Design and performance modeling for single-walled carbon nanotubes as local, semilocal, and global interconnects in gigascale integrated systems. *IEEE Trans. Electron Device*, 54, 26–37, 2007.

10. A. Naeemi and J.D. Meindl. Compact physical model for multi-walled carbon nanotubes. *IEEE Electron Device Lett.*, 27, 338–340, 2006.

11. A.V. Melechko, V.I. Merkulov, T.E. McKnight, M.A. Guillorn, K.L. Klein, D.H. Lowndes, and M.L. Simpson. Vertically aligned carbon nanofibers and related structures: Controlled synthesis and directed assembly. *J. Appl. Phys.*, 97, 2005.

12. P.C.H. Chan, Y. Chai, M. Zhang, and Y.Y. Fu. The application of carbon nanotubes in CMOS integrated circuits. In *IEEE International Conference on Solid-State and Integrated Circuit Technology (ICSICT)*, 2008, pp. 534–536.

13. Y. Chai, X.L. Zhou, P.J. Li, W.J. Zhang, Q.F. Zhang, and J.L. Wu. Nanodiode based on a multiwall CN_x/carbon nanotube intramolecular junction. *Nanotechnology*, 16, 2134–2137, 2005.

14. Y. Chai, Q.F. Zhang, and J.L. Wu. A simple way to CN_x/carbon nanotube intramolecular junctions and branches. *Carbon*, 44, 687–691, 2006.

15. Y. Chai, L.G. Yu, M.S. Wang, Q.F. Zhang, and J.L. Wu. Low-field emission from iron oxide-filled carbon nanotube arrays. *Chin. Phys. Lett.*, 22, 911–914, 2005.

16. Y.G. Zhang, A.L. Chang, J. Cao, Q. Wang, W. Kim, Y.M. Li, N. Morris, E. Yenilmez, J. Kong, and H.J. Dai. Electric-field-directed growth of aligned single-walled carbon nanotubes. *Appl. Phys. Lett.*, 79, 3155–3157, 2001.

17. C. Bower, W. Zhu, S. Jin, and O. Zhou. Plasma-induced alignment of carbon nanotubes. *Appl. Phys. Lett.*, 77, 830–832, 2000.

18. N. Patil, A. Lin, E.R. Myers, K. Ryu, A. Badmave, C.W. Zhou, H.-S.P. Wong, and S. Mitra. Wafer-scale growth and transfer of aligned single-walled carbon nanotubes. *IEEE Trans. Nanotechnol.*, 8(4), 498–504, 2009.

19. Y. Chai, Z.Y. Xiao, and P.C.H. Chan. Electron-shading effect on the horizontal growth of carbon nanotubes using plasma-enhanced chemical vapor deposition. *Appl. Phys. Lett.*, 94(4), 043116-1–043116-3, 2009.

20. Z.Y. Xiao, Y. Chai, Y. Li, M.H. Sun, and P.C.H. Chan. Integration of horizontal carbon nanotube device on silicon substrate using liquid evaporation. In *Proceedings of 60th IEEE Electronic and Component Technology Conference*, 2010, pp. 943–947.

21. F. Kreupl, A.P. Graham, G.S. Duesberg, W. Steinhogl, M. Liebau, E. Unger, and W. Honlein. Carbon nanotubes in interconnect applications. *Microelectronic Eng.*, 64, 399–408, 2002.

22. J. Li, Q. Ye, A. Cassell, H.T. Ng, R. Stevens, J. Han, and M. Meyyappan. Bottom-up approach for carbon nanotube interconnects. *Appl. Phys. Lett.*, 82, 2491–2493, 2003.

23. Y. Chai, Z.Y. Xiao, and P.C.H. Chan. Horizontally aligned carbon nanotubes for interconnect applications: Diameter-dependent contact resistance and mean free path. *Nanotechnology*, 21, 235705-1–235705-5, 2010.

24. G.F. Close, S. Yasuda, B. Paul, S. Fujita, and H.-S.P. Wong. A 1 GHz integrated circuit with carbon nanotube interconnect and silicon transistor. *Nano Lett.*, 8(2), 706–709, 2007.

25. J. Dijon, H. Okuno, M. Fayolle, T. Vo, J. Pontcharra, D. Acquaviva, D. Bouvet, A.M. Ionescu, C.S. Esconjauregui, B. Capraro, E. Quesnel, J. Robertson. Ultra-high density carbon nanotubes on Al-Cu for advanced vias. *Int. Electron Device Meeting Tech. Dig.*, 33.4.1–33.4.4, 2010.

26. T. Yamada, A. Maigne, M. Yudasaka, K. Mizuno, D.N. Futaba, M. Yumura, S. Iijima, K. Hata. Revealing the secret of water-assisted carbon nanotube syntheis by microcopic observation on the interaction of water on the catalyst. *Nano Lett.*, 8, 4288–4292, 2008.

27. G. Zhong, T. Iwasaki, J. Roberston, and H. Kawarada. Growth kinetics of 0.5 cm vertically aligned single-walled carbon nanotubes. *J. Phys. Chem. B*, 111, 1907–1910, 2007.

28. Y. Chai, A. Hazeghi, K. Takei, H.Y. Chen, P.C.H. Chan, A. Javey, and H.-S. Philip Wong. Graphitic interfacial layer to carbon nanotube for low electrical contact resistance. *Int. Electron Device Meeting Tech. Dig.*, 210–213, 2010.

29. A. Javey, J. Guo, Q. Wang, M. Lundstrom, and H.J. Dai. Ballistic carbon nanotube field-effect transistors. *Nature*, 424(7), 654–657, 2003.

30. S.C. Lim, J.H. Jang, D.J. Bae, G.H. Han, S. Lee, I.-S. Yeo, and Y.H. Lee. Contact resistance between metal and carbon nanotube interconnects: Effect of work function and wettability. *Appl. Phys. Lett.*, 95, 264103-1–264103-3, 2009.

31. H.J. Li, W.G. Lu, J.J. Li, X.D. Bai, and C.Z. Gu. Multichannel ballistic transport in multiwall carbon nanotubes. *Phys. Rev. Lett.*, 95(8), 086601, 2005.

32. Y. Chai, Z.Y. Xiao, and P.C.H. Chan. Low-resistance of carbon nanotube contact plug to silicon. *IEEE Electron Device Lett.*, 30(8), 811–813, 2009.

33. D. Yokoyama, T. Iwasaki, T. Yoshida, H. Kawarada, S. Sato, T. Hyakushima, M. Neihi, and Y. Awano. Low temperature grown carbon nanotube interconnect using inner shells by chemical mechanical polishing. *Appl. Phys. Lett.*, 91(26), 263101-1–263101-3, 2007.

34. X.J. Yang, M.A. Guillorn, D. Austin, A.V. Melechko, H.T. Cui, H.M. Meyer, V.I. Merkulov, J.B.O. Caughman, D.H. Lowndes, and M.L. Simpson. Fabrication and characterization of carbon nanofiber-based vertically integrated Schottky barrier junction diodes. *Nano Lett.*, 3(12), 1751–1755, 2003.

35. Y. Chai and P.C.H. Chan. High electromigration-resistant copper/CNT composite for interconnect applications. *Int. Electron Device Meeting Tech. Dig.*, 607–610, 2008.

36. Y. Chai, K. Zhang, M. Zhang, P.C.H. Chan, and M.M.F. Yuen. Carbon nanotube/copper composites for via filling and thermal management. In *Proceedings of 57th IEEE Electronic and Component Technology Conference*, 2007, pp. 1224–1229.

37. P. Liu, J.H. Wu, D. Xu, Y.Z. Pan, C. You and Y.F. Zhang. CNTs/Cu composite thin films fabricated by electrophoresis and electroplating techniques. In *Proceedings of 2nd Nanoelectronic Conference*, Shanghai, China, 2008, pp. 975–978.

38. Y.L. Yang, Y.D. Wang, Y. Ren, C.S. He, J.N. Deng, J. Nan, J.G. Chen, and L. Zuo. Single-walled carbon nanotube-reinforced copper composite coatings prepared by electrodeposition under ultrasonic field. *Mater. Lett.*, 62, 47–50, 2008.

39. S. Baik, B. Lim, R.S. Ryu, D. Choi, B. Kim, S. Oh, B.H. Sung, J.H. Choi, and C.J. Kim. Mechanical and electrical properties of carbon nanotubes in copper-matrix nanocomposites. *Solid State Phenomena*, 120, 285–288, 2007.

40. M. Zhang, X. Huo, P.C.H. Chan, Q. Liang, and Z.K. Tang. Radio-frequency characterization for the single-walled carbon nanotubes. *Appl. Phys. Lett.*, 88(16), 163109-1–163109-3, 2006.

41. M.H. Sun, Z.Y. Xiao, Y. Chai, Y. Li, and P.C.H. Chan. Inductance properties of in-situ grown horizontally aligned carbon nanotubes. *IEEE Trans. Electron Device*, 58(1), 229–235, 2011.

42. P.M. Ajayan and B.I. Yakobson. Materials science—Oxygen breaks into carbon world. *Nature*, 441, 818–819, 2006.

43. Y. Chai, M. Zhang, J.F. Gong, and P.C.H. Chan. Reliability evaluation of carbon nanotube interconnect in a silicon CMOS environment. In *Proceedings of 8th IEEE International Conference on Electronic Materials and Packaging*, Hong Kong, December 2006, pp. 343–347.
44. Y. Chai, P.C.H. Chan, Y.Y. Fu, Y.C. Chuang, and C.Y. Liu. Copper/carbon nanotube composite interconnect for enhanced electromigration resistance. In *Proceedings of 58th IEEE Electronic and Component Technology Conference*, 2008, pp. 412–420.
45. Y. Chai, P.C.H. Chan, Y.Y. Fu, Y.C. Chuang, and C.Y. Liu. Electromigration studies of Cu/CNT composite using Blech structure. *IEEE Electron Device Lett.*, 29(9), 1001–1003, 2008.
46. Y. Chai, M. H. Sun, Z. Y. Xiao, Y. Li, M. Zhang and Philip C. H. Chan. Towards future VLSI interconnects using aligned carbon nanotubes. *IEEE International Conference on Very Large Scale Integration*, 2011, pp. 248–253.

7 Monolithic Integration of Carbon Nanotubes and CMOS

Huikai Xie, Ying Zhou, Jason Johnson, and Ant Ural

CONTENTS

7.1   INTRODUCTION

As discussed in the previous chapter, carbon nanotubes (CNTs) have been explored for various applications with great success. With the extraordinary and unique electrical, mechanical, and chemical properties of CNTs, it is also highly desired to develop complementary metal-oxide-semiconductor (CMOS)–CNT hybrid integration technology that can take advantage of the powerful signal conditioning and processing capability of state-of-the-art CMOS technology. CNTs may be used either as an integral part of CMOS circuits or as sensing elements to form functional nanoelectromechanical systems (NEMSs).

In CNT-CMOS circuits, or nanoelectronics applications, nanoscale CNT–field effect transistors (FETs) can be integrated with conventional CMOS circuits and form various functional blocks, e.g., CNT memory devices. Since the further scaling of silicon-based memories is expected to approach its limits in the near future, as

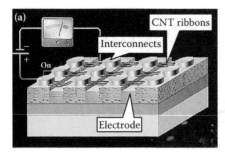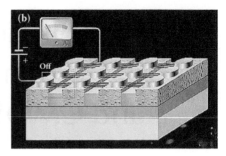

FIGURE 7.1　The structure of NRAM at (a) on and (b) off states. (From I. Nantero, NRAM®, 2000–2009, http://www.nantero.com/mission.html.)

the capacitor charge and writing/erasing voltages cannot be scaled down [1], a highly scalable hybrid CMOS-molecular nonvolatile memory has been widely explored, where CNT-based memories are connected to the readout CMOS logic circuitry [2,3]. The nanotube random access memory (NRAM) used in [3] is shown in Figure 7.1. It is a nonvolatile, high-density, high-speed, and low-power nanomemory developed by Woburn, Massachusetts-based Nantero, and it is claimed to be a universal memory chip that can replace DRAM, SRAM, flash memory, and ultimately hard-disk storage [3,4].

For sensing applications, CNTs can be used as mechanical, chemical, and radiation sensors. The integration of CMOS circuits with CNT sensors can provide high-performance interfacing, advanced system control, and powerful signal processing to achieve so-called smart sensors. Such CMOS circuits for CNT sensors can regulate the sensing temperature [5], increase the dynamic range [6], improve the measurement accuracy [7], and provide multiple readout channels to realize electronically addressable nanotube chemical sensor arrays [8]. Various sensors based on such CMOS-CNT hybrid systems have been demonstrated, including integrated thermal and chemical sensors [9–11].

Other efforts have been made to use multiwall carbon nanotubes (MWNTs) as the CMOS interconnect for high-frequency applications [12], or to apply CNT-based nanoelectromechanical switches for leakage reduction in CMOS logic and memory circuits [13].

Currently, monolithic integration of CMOS and CNT is still very challenging. Most CMOS-CNT systems have been realized by either a two-chip solution or complicated CNT manipulations. In this chapter, we will review various CNT synthesis technologies and CMOS-CNT integration approaches. Particularly, we will focus on the localized heating CNT synthesis method, based on which the integration of CNT on foundry CMOS has been demonstrated.

7.1.1　CARBON NANOTUBE SYNTHESIS

Great efforts have been made to investigate the CNT growth mechanism, but it is still not completely understood. Based on transmission electron microscope (TEM) observation, Yasuda et al. proposed that CNT growth starts with rapid formation of rod-like carbons, followed by slow graphitization of walls and formation of hollow structures inside the rods [14].

There are three main methods for carbon nanotube synthesis: arc discharge [15–17], laser ablation [18–20], and chemical vapor deposition (CVD) [21–24]. The first two methods involve evaporation of solid-state carbon precursors and condensation of carbon atoms to form nanotubes. The high temperature (thousands of degrees Celsius) ensures perfect annealing of defects, and hence these two methods produce high-quality nanotubes. However, they tend to produce a mixture of nanotubes and other by-products, such as catalytic metals, so the nanotubes must be selectively separated from the by-products. This requires quite challenging post-growth purification and manipulation.

In contrast, the CVD method employs a hydrocarbon gas as the carbon source and metal catalysts heated in a tube furnace to synthesize nanotubes. It is commonly accepted that the synthesis process starts with hydrocarbon molecules adsorbed on the catalyst surface. Then the carbon is decomposed from the hydrocarbon and diffuses into the catalytic particles. Once the supersaturation is reached, carbons start to precipitate onto the particles to form carbon nanotubes. After that, nanotubes can grow by adding carbons at the top of the tubes if the particles are weakly adhered to the substrate surface. Nanotubes can also grow from the bottom if the particles are strongly adhered [25]. The former is called the tip-growth model and the latter is called the base-growth model. Compared to the arc discharge and laser ablation methods, CVD uses much lower synthesis temperature, but it is still too high to directly grow on CMOS substrates. In addition, CVD growth provides an opportunity to directly manufacture substantial quantities of individual carbon nanotubes. The diameter and location of the grown CNTs can be controlled via catalyst size [26] and catalyst patterning [27], and the orientation can be guided by an external electric field [28]. Suitable catalysts that have been reported include Fe, Co, Mo, and Ni [29].

Besides the CNT synthesis, electrical contacts need to be created for functional CNT-based devices. It is reported that molybdenum (Mo) electrodes form good ohmic contacts with nanotubes and show excellent conductivity after growth, with resistance ranging from 20 kΩ to 1 MΩ per tube [30]. The electrical properties can be measured without any post-growth metallization processing. The resistance, however, tends to increase over time, which might be due to the slight oxidation of the Mo in air. Other than that, electron-beam lithography is generally used to place post-growth electrodes. Several metals, such as gold, titanium, tantalum, and tungsten, have been investigated as possible electrode materials, and palladium top contacts are believed to be most promising [31].

7.1.2 CMOS-CNT INTEGRATION CHALLENGES AND DISCUSSION

As mentioned previously, a complete system with carbon nanotubes and microelectronic circuitry integrated on a single chip is needed to fully utilize the potentials of nanotubes for emerging nanotechnology applications. This monolithic integration requires not only high-quality nanotubes but also a robust fabrication process that is simple, reliable, and compatible with standard foundry CMOS processes. Such a CMOS-compatible integration process up to now remains a challenge, primarily due to the material and temperature limitations posted by CMOS technology [32,33].

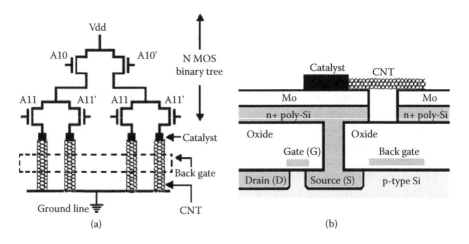

FIGURE 7.2 (a) Circuit schematic of the decoder consisting of NMOS and single-walled carbon nanotubes. (b) Schematic of the cross section of the decoder chip, interconnected by phosphorus-doped n+ polysilicon and molybdenum. (From Y.-C. Tseng et al., *Nano Letters*, 4, 123–127, 2004. With permission.)

As discussed in Section 7.1.1, CVD has been widely used to synthesize nanotubes. Researchers have been working on growing CNTs directly on CMOS substrate using thermal CVD methods. For example, Tseng et al. demonstrated, for the first time, a process that monolithically integrates SWNTs with N-type metal-oxide-semiconductor (NMOS) FETs in a CVD furnace at 875°C [34]. However, the high synthesis temperature (typically 800–1000°C for SWNT growth [35]) would damage aluminum metallization layers and change the characteristics of the on-chip transistors. Ghavanini et al. have assessed the deterioration level of CMOS transistors that were applied with a CVD synthesis condition, and reported that one P-type metal-oxide-semiconductor (PMOS) transistor lost its function after the thermal CVD treatment (610°C, 22 min) [32]. As a result, the integrated circuits in Tseng's thermal CVD CNT synthesis, as shown in Figure 7.2, can only consist of NMOS and use n+ polysilicon and molybdenum as interconnects, making it incompatible with foundry CMOS processes.

To address this problem, one possible solution is to grow nanotubes at high temperature first, and then transfer them to the desired locations on another substrate at low temperature. However, handling, maneuvering, and integrating these nanostructures with CMOS chips/wafers to form a complete system are very difficult. In the early stage, atomic force microscope (AFM) tips were used to manipulate and position nanotubes into predetermined locations under the guide of scanning electron microscope (SEM) imaging [36,37]. Although this nanorobotic manipulation realized precise control over both the type and location of CNTs, its low throughput becomes the bottleneck for large-scale assembly. Other post-growth CNT assembly methods that have been demonstrated so far include surface functionalization [38], liquid-crystalline processing [39], dielectrophoresis (DEP) [40–43], and large-scale transfer of aligned nanotubes grown on quartz [44,45].

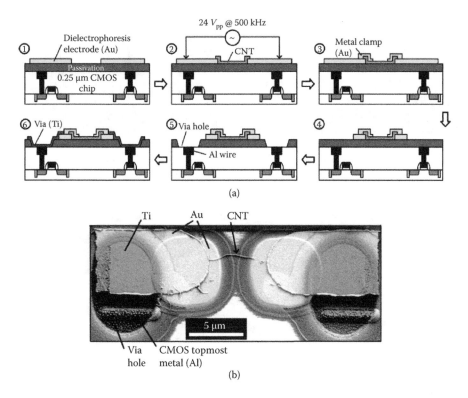

(a)

(b)

FIGURE 7.3 (a) Process flow to integrate MWNT interconnects on CMOS substrate. (b) SEM image of one MWNT interconnect (wire and via). (From G.F. Close et al., *Nano Letters*, 8, 706–709, 2008. With permission.)

Among these methods, CMOS-CNT integration based on the DEP-assisted assembly technique has been reported, and a 1 GHz integrated circuit with CNT interconnects and silicon CMOS transistors has been demonstrated by Close et al. [12]. The fabrication process flow and the assembled MWNT interconnect are shown in Figure 7.3. The DEP process provides the capability of positioning the nanotubes precisely in a noncontact manner, which minimizes the parasitic capacitances and allows the circuits to operate above 1 GHz. However, to immobilize the DEP-trapped CNTs in place and to improve the contact resistances between CNTs and the electrodes, metal clamps have to be selectively deposited at both ends of the CNTs (Figure 7.3a, step 3). The process complexity and low yield (~8%, due to the MWNT DEP assembly limitation) are still the major concerns.

Alternatively, some other attempts have been made to develop low-temperature growth using various CVD methods [46–48]. Hofmann et al. reported vertically aligned carbon nanotubes grown at temperatures as low as 120°C by plasma-enhanced chemical vapor deposition (PECVD) [47]. However, the decrease in growth temperature jeopardizes both the quality and yield of the CNTs, as evident from their published results (shown in Figure 7.4). The synthesized products are actually defect-rich, less crystalline, bamboo-like structured carbon nanofibers rather than MWNTs or SWNTs.

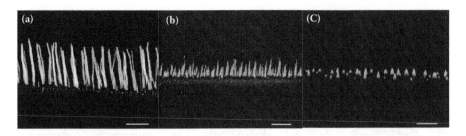

FIGURE 7.4 SEM images of vertically aligned CNFs grown by PECVD deposition at (a) 500°C, (b) 270°C, and (c) 120°C. Scale bars: a and b, 1 μm; c, 500 nm. (From S. Hofmann et al., *Applied Physics Letters*, 83, 135–137, 2003. With permission.)

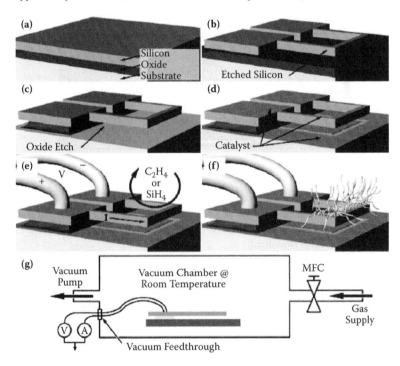

FIGURE 7.5 Fabrication process and localized heating concept. (From O. Englander et al., *Applied Physics Letters*, 82, 4797–4799, 2003. With permission.)

To accommodate both the high-temperature requirement (800–1000°C) for high-quality SWNT synthesis and the temperature limitation of CMOS process-ing (<450°C), CNT synthesis based on localized heating has drawn great interest. Englander et al. demonstrated, for the first time, the localized synthesis of silicon nanowires and carbon nanotubes based on resistive heating using microheaters [49]. The fabrication processes and concepts are shown in Figure 7.5. Operated inside a room temperature chamber, the suspended microelectromechanical system (MEMS) structures serve as microheaters to provide high temperature at predefined regions

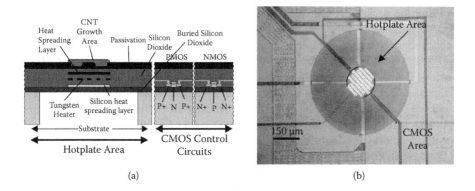

FIGURE 7.6 (a) Schematic of the cross-sectional layout of the chip. (b) Optical image of the device top view, showing the tungsten interdigitated electrodes on top of the membranes. Heater radius = 75 μm, membrane radius = 280 μm. (From M.S. Haque et al., *Nanotechnology*, 19, 025607, 2008. With permission.)

for optimal nanotube growth, leaving the rest of the chip area at low temperature. Using the localized heating concept, direct integration of nanotubes at specific areas can be potentially achieved in a CMOS-compatible manner, and there is no need for additional assembly steps. Attracted by this promising technique, several research groups have followed up, and localized CNT growth on various MEMS structures has been demonstrated [50–52]. However, the devices typically have large sizes, and their fabrication processes are not fully compatible with the standard foundry CMOS processes. Although this concept has solved the temperature incompatibility problem between CNT synthesis and circuit protection, the fabrication processes of microheater structures still have to be well designed to fit into standard CMOS foundry processes, and the materials of microheaters have to be selected to meet the CMOS compatibility criteria.

Progress toward complete CMOS-CNT systems has been made. On-chip CNT growth using CMOS micro-hotplates was later demonstrated by Haque et al. [53]. As shown in Figure 7.6, tungsten was used to fabricate both the micro-hotplates (as the thermal source) and interdigitated electrodes for nanotube contacts. MWNTs have been successfully synthesized on the membrane, and simultaneously connected to circuits through tungsten metallization. Although tungsten can survive the high-temperature growth process, and has high connectivity and conductivity, Franklin et al. reported that no SWMTs were found to grow from catalyst particles on the tungsten electrodes, presumably due to the high catalytic activity of tungsten toward hydrocarbons [30]. Further, although monolithic integration has been achieved, the utilization of tungsten, a refractory metal, as an interconnect metal is limited in CMOS foundry, especially for mixed-signal CMOS processes. From the CMOS point of view, tungsten is not a good candidate as an interconnect metal compared with aluminum and copper. Other than the material, this approach is limited to silicone on insulator (SOI) CMOS substrates, requires a backside bulk micromachining process, and has low integration density.

From the reviews above, we can see that monolithic CMOS-CNT integration is desirable to utilize the full potential of nanotubes for emerging nanotechnology applications. However, the existing approaches, although each has its own merits, still cannot meet all the requirements and realize complete compatibility with CMOS processes. To solve this problem, a simple and scalable monolithic CMOS-CNT integration technique using a novel maskless post-CMOS surface micromachining processing has been developed and will be presented in the following sections. This approach is fully compatible with commercial foundry CMOS processes and has no specific requirements on the type of metallization layers and substrates.

Since CMOS fabrication is costly and time-consuming even through the multi-project wafer (MPW) service provided by MOSIS, mock-CMOS substrates will first be used for the process development and basic conceptual verification. A mock-CMOS substrate is a silicon substrate with multiple layers of metals and dielectrics but without any diffusion layers.

7.2 CNT SYNTHESIS BY LOCALIZED RESISTIVE HEATING ON MOCK-CMOS

The localized heating concept is illustrated in Figure 7.7. A microstructure is thermally isolated by suspending it over a micromachined cavity. The microstructure has a heater embedded. When a current is injected into the heater, the temperature of the microstructure will rise. Because of the thermal isolation provided by the cavity, the temperature of the substrate outside the cavity will not change much. The key is the heater design.

7.2.1 MICROHEATER DESIGN

The local temperature distribution and the maximum temperature are the key parameters of the microheater structures. Since SWNT growth requires temperature as high as 800°C or above, the resistivity of the heater, thermal isolation, thermal stresses, and structural stiffness must be taken into consideration when designing the microheaters. A 3D model of a mock-CMOS microheater is shown in Figure 7.7. Suspended microstructures are created over a microcavity for good thermal isolation. Resistors are integrated as the heating source, and the local electrical field (E-field) and growth temperature can be controlled by the microheater geometry and power supply.

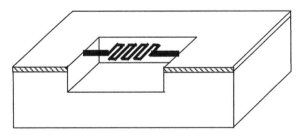

FIGURE 7.7   Structural demonstration of the mock-CMOS platinum microheater.

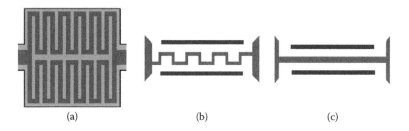

(a)   (b)   (c)

FIGURE 7.8 (a) Large hotplate with a meander Pt microheater embedded. (b) Serpentine microheater design. (c) Straight-line microheater design.

Three types of microheaters have been designed. The first microheater design (Figure 7.8a) is a micro-hotplate with a meander-shaped microheater embedded. The hotplate is supported by two anchored short beams. This design has a large growth area with relatively uniform temperature distribution, and potential as a gas sensor. CNTs are expected to grow on the hotplate surface. The second design (Figure 7.8b) has a serpentine shape. This design introduces trenches between the microheater (the central part) and the silicon dioxide secondary wall (the two straight lines on the side) for suspended CNT growth. With the secondary wall grounded, this configuration is also designed to study the local E-field distribution and the impact of the E-field on the CNT alignment. The third design (Figure 7.8c) is further simplified into one straight line for studying temperature and CNT density distribution along the microheater. The distribution information will facilitate the further microheater scaling. The minimization of heating elements offers more accurate local control of E-field, temperature, and growth rate. Thus the position, quantity, length, direction, and properties of CNTs can be better controlled. Moreover, smaller heating elements require less power, and lead to a higher CMOS-CNT integration density.

Several different materials have been investigated as the electrode material for CVD growth of CNTs in the literature. Metals with relatively low melting temperature, such as gold, become discontinuous as they are balling up during the high-temperature growth process, while some other candidates, such as Ti and Ta, tend to react with hydrogen and form volatile metal hydride at high temperature [30]. In our experiment, platinum (Pt) has been chosen as the microheater material due to its compatibility with CNT growth. Pt is a refractory metal with very high melting temperature, is widely used as a contact electrode in traditional CVD CNT synthesis procedures, and has demonstrated good contacts with CNTs, with resistance ranging from 10 to 50 kΩ [12].

7.2.2 Device Fabrication and Microheater Characterization

7.2.2.1 Device Fabrication

Based on the above microheater designs, a process flow is proposed to fabricate the devices. Figure 7.9 shows the cross-sectional view of the process flow. The fabrication process starts from the deposition of a 0.5 μm thick SiO_2 (Figure 7.9a). A Cr/Pt/Cr heater film is then sputtered and patterned using a liftoff process, in which the 200 nm

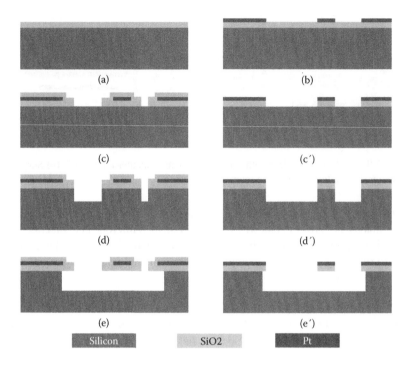

FIGURE 7.9 Cross-sectional view of the proposed process flow. (a) PECVD SiO$_2$ deposition. (b) Pt sputtering and liftoff to form heater and pads. (c, c′) Top PECVD SiO$_2$ deposition and patterning. (d, d′) Anisotropic Si dry etch. (e, e′) Isotropic Si dry etch and heater release.

thick Pt is the heater and the 30 nm thick Cr is the adhesive layer for Pt (Figure 7.9b; the Cr layers are not shown). Next another 0.5 μm thick top SiO$_2$ layer is deposited and patterned. Depending on the mask design, the top SiO$_2$ layer can remain to form an oxide/Pt/oxide sandwich structure (Figure 7.9c), or the SiO$_2$ over the microheater can be etched away to form a Pt/oxide bimorph structure (Figure 7.9c′). In the latter case, direct contact between Pt electrodes and CNTs will be formed during the synthesis process. Next using the patterned SiO$_2$ layer (Figure 7.9d) or the Pt heater itself (Figure 7.9d′) as the etching mask, an anisotropic deep-reactive-ion etching (DRIE) of silicon is performed to create trenches around the heater. Finally, isotropic silicon dry etching is performed to undercut the silicon underneath to release the microheater hotplates (or bridges) suspended over the cavity (Figure 7.9e and e′). The localized heating is realized by using a DRIE silicon dry-etching process to form a cavity to obtain a good thermal isolation. In Section 7.3, we will see that the process proposed for mock-CMOS microheater fabrication is fully transferable for releasing microheaters integrated in CMOS substrates.

Three types of microheaters that are introduced in Section 7.2.1 have been fabricated and characterized. Figure 7.10 shows some SEM pictures of fabricated microheaters. The first design (Figure 7.10a) is an 87 × 87 μm^2 micro-hotplate with a symmetric meander Pt heater embedded.

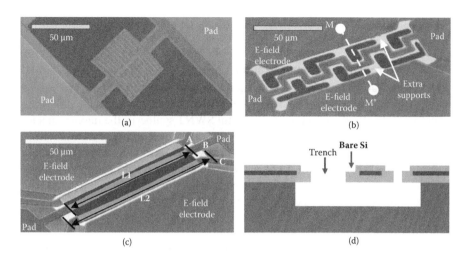

FIGURE 7.10 SEM pictures of fabricated microheaters. (a) Design 1: Pt heater embedded in a micro-hotplate. (b) Design 2: Pt heater in a curved shape. (c) Design 3: Pt heater with two straight lines in parallel, labeled A, B, and C, respectively. L_1 and L_2 represent the effective length of the heater (80 μm) and the length of the cavity (about 95 μm), respectively. (d) Cross-sectional view of the device (line MM" in b).

The second (Figure 7.10b) is a serpentine microheater design. The dark region is the etch-through openings that are patterned during the step shown in Figure 7.9c. It corresponds to the trench shown in Figure 7.10d. The white region surrounding the center Pt heater in the SEM picture in Figure 7.10b is pure silicon oxide with no silicon underneath, as illustrated in Figure 7.10d. Therefore the white region outlines the size of the microcavity below the microheater. Its bright color is an artifact due to the charging effect of the sample during SEM imaging. Due to its serpentine shape, some pure silicon oxide was left between etch openings as the extra mechanical supports. This was found to be necessary. A test structure with a similar serpentine shape but no extra mechanical supports sagged down obviously after the release. Although this suspended microstructure still can function as a microheater, one can imagine that the mismatch of the thermal coefficient of expansion between CNT and Pt heaters may induce thermal stress that will increase the probability of detachment of the synthesized CNTs from the microheaters.

The third design (Figure 7.10c) is a straight-line Pt microheater. It is 5 μm wide and 120 μm long. The top SiO_2 layer is etched during the step shown in Figure 7.9c′, and thus the Pt is exposed to facilitate the electrical contact with CNTs. The exposed Pt is 80 μm long, defined as the effective length of the microheater L_1. Similarly, the white region is pure SiO_2, representing the cavity boundary, and the effective length is labeled L_2. In this design, two extra parallel Pt lines are placed on the left and right sides of the heater and labeled A and C, respectively. They are used as the second walls for CNT landing during the growth, and also as the second electrodes for extracting electrical signals after CNT growth. The trenches are 3 μm (left) and 6 μm (right) wide, respectively. For each microheater, there are two big pads for applying electrical voltage to generate heating. Another pair of electrodes is designed to apply a proper E-field if necessary (as shown in Figure 7.10, labeled "E-field electrode").

7.2.2.2 Microheater Characterization

After device fabrication, the microheaters are characterized before performing growth experiments. The characterization is designed to measure the following properties. First, for the purpose of successful SWNT synthesis, the maximum temperature that can be reached by resistive heating and the reliability of the microheaters under such high-temperature conditions are two key factors. Then for the purpose of integrating SWNTs on CMOS chips, the local temperature distribution is a major concern, i.e., a sharp temperature gradient is desired so that the temperature can decrease rapidly from the heater center toward the substrate.

7.2.2.2.1 Maximum Temperature Estimation

The mechanical robustness at high temperature was first tested by baking microheaters in an oven at 900°C for 20 min. Both the oxide/Pt/oxide and Pt/oxide bimorph structures were expected to undergo strain incompatibility primarily due to the thermal expansion mismatch of different materials. Although slight sags were observed, all the microheaters survived the high-temperature treatments with no rupture occurring.

Then the working temperature of microheaters was evaluated using the micro-hotplate design. It has been reported that platinum has the optimum thermoresistive characteristics, and Pt resistance thermometers have served as the international standard for temperature measurements between –259.34 and 630.75°C [54]. In a linear approximation, the relation between resistance and temperature is given as follows:

$$R_T = R_0 \left(1 + \alpha \cdot \Delta T\right) \qquad (7.1)$$

in which R_0 is the initial resistance, R_T is the resistance dependent on temperature, α is the temperature coefficient of resistance (TCR), and ΔT is the temperature change. The de facto industrial standard value of TCR is 0.00385/°C, but thin-film platinum exhibits a coefficient that tends to decrease with the thickness. The value of TCR used for our 200 nm thick Pt microheater is about 0.00373/°C, which is obtained from Clayton's report [54]. The microheater was electrically connected using silver epoxy and then characterized. The experimental current-voltage relation was plotted in Figure 7.11a. The resistance under each power supply can be calculated, and the corresponding temperature can be estimated based on the change of the Pt heater resistance and the TCR of Pt. The resistance of the micro-hotplate was about 46 Ω at room temperature. It increased to 212.5 Ω when the applied current was increased to 14 mA. Therefore the maximum temperature was estimated to be above 900°C based on Equation 7.1 (Figure 7.11b).

For the temperature obtained by this method, it should be noted that the calculated temperature is only an approximation and subject to the following assumptions. First, the resistance measurement is simplified. The resistance measured is actually the sum of the contact resistance and the heater's resistance: $R_{Total} = R_{contact} + R_{resistor}$. Further, the overall heater's resistance, $R_{resistor}$, should consider both the 200 nm Pt

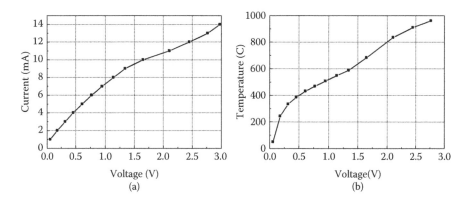

FIGURE 7.11 (a) I-V characterization of the microheater. (b) Temperature estimated based on resistance change.

FIGURE 7.12 Microscopic images under applied voltages of (a) 2.28 V, (b) 2.62 V, and (c) 3.00 V, respectively.

layer and the 30 nm Cr adhesive layers. At 20°C, the electrical resistivity of Cr is 125 n$\Omega \cdot$m, while the electrical resistivity of Pt is 105 n$\Omega \cdot$m [55,56]. Second, under Joule heating, the microheater actually exhibits varying temperature along the heater rather than one uniform temperature. Thus the calculated temperature is an approximation of the average temperature of the entire heater.

7.2.2.2.2 Heating Experiment

During the I-V characterization experiment, red glowing of the microheaters under different voltages was clearly observed under an optical microscope, as shown in Figure 7.12. This red glowing can be switched between "on" and "off" instantaneously by controlling the power supply, indicating much shorter response times than traditional CVD processes. The localized microheating combined with this fast response greatly reduces the total power consumption and improves the temperature budget of post-CMOS processing. In addition, we need an indicator to determine when the microheater has reached the required high temperature so that we can stop increasing the power and start the CNT synthesis process. Due to different structure designs or fabrication variations between devices, individual microheaters require different electrical powers to reach the same temperature. Thus neither current nor voltage is a good indicative parameter unless each microheater is characterized under the growth condition and the temperature-power relation is established for each device before the real growth. Instead, the incandescence, the emission of visible light from a hot body due to its temperature, of platinum microheaters observed as red glowing indicates that the heater reaches the required high temperature. In practice, we start flowing in synthesis gases once the red glowing is observed.

7.2.2.2.3 High-Temperature Reliability

After assessing the maximum temperature, the reliability of the microheaters at high temperature was then evaluated. Although bulk platinum has a high melting point (1768°C [56]) and is extraordinarily stable at high temperature, the degradation of platinum thin films at high temperature has been reported by several groups [57–60]. As shown in Figure 7.13a, we observed a kink point in the curve at higher voltage, and a negative slope between 1.2 and 1.4 V before the microheater was broken. To study the negative slope region, another newly released platinum thin-film

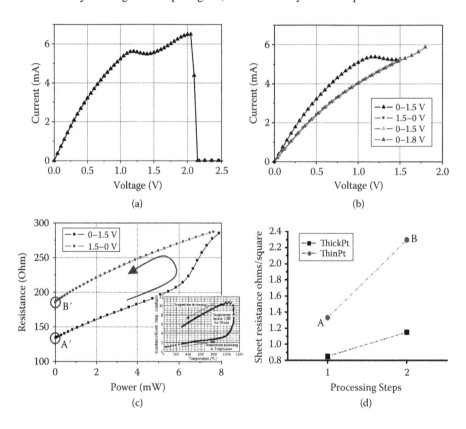

(a)

(b)

(c)

(d)

FIGURE 7.13 (a) Embedded straight-line platinum microheater I-V characteristics. Negative slope was observed between 1.2 and 1.4 V. (b) Heater characteristics under repeated sweepings. (c) Microheater resistance under repeated sweepings. Inset: resistance characterization of the 100 nm Pt/10 nm Ti film, reported in literature [57]. (d) Sheet resistance for 205 nm Pt/15 nm Ta and 145 nm Pt/15 nm Ta thin film, (1) before and (2) after 830°C heat treatment, as reported in literature [61].

microheater with the same design was tested. The voltage was gradually increased from 0 to 1.5 V and then swept back to 0 V. The sweepings were repeated from 0 to 1.5 V and from 0 to 1.8 V. As shown in Figure 7.13b, the negative slope only occurred during the first 0–1.5 V voltage sweeping. After that, the electrical characteristics changed and became repeatable as a result of a permanent change in the platinum resistance.

Briand et al. reported that the resistivity of the Pt/Ta thin film increased after a heat treatment of 95 min at 830°C, regardless of the film thickness (Figure 7.13d) [61]. Replotting the electrical characteristics we obtained (Figure 7.13b) as resistance versus power in Figure 7.13c, we found that the resistance increased from 134 Ω to 185 Ω after the voltage sweeping. This resistance increase is explained as the result of self-heating treatment through resistive Joule heating, with points A′ and B′ corresponding to points A and B in Briand's paper, respectively. There might be a critical point that is associated with this high-temperature degradation. This critical temperature might be different for Pt thin films with different configurations, such as thickness, structure, type of adhesion layer, etc. Firebaugh et al. reported that the resistance of their 100 nm Pt/10 nm Ti film was well behaved up to ~900°C, beyond which holes started to form in the film and resistance increased rapidly [57] (Figure 7.13c, inset).

For our device, the critical temperature was believed to be reached at the center of the microheater when the power was increased to around 6 mW, and the rest of the microheater later underwent similar temperature treatment when the input power was continuously increased to about 8 mW.

Several factors attribute to the degradation of platinum thin films, including interdiffusion and reaction between the Pt layer and the adhesive layer, stress, and platinum silicide formation [60], but the agglomeration of continuous thin films into islands of material is believed to be the dominant mechanism [57]. Agglomeration involves the nucleation and growth of holes in the film, and it is driven by the high surface-to-volume ratio of the thin films. The surface diffusivity is dependent on the temperature, and the diffusion of metal atoms on surface tends to reduce the surface-to-volume ratio through capillarity [62]. The sizes for the initial holes must be larger than the thickness-dependent thermodynamic critical radius, and the holes will then grow under the surface diffusion-driven capillarity [63]. As reported by Firebaugh et al. [57], holes started to form in the platinum thin film at around 900°C, and gradually increased in size with time, resulting in the discontinuity (Figure 7.14). Note that the hole growth rate is slow and the film lifetime is sufficiently long, compared with the time required for CNT growth. To avoid agglomeration, thin-film thickness greater than 1 μm was recommended [57]. For our 200 nm platinum microheaters, we noticed that although the film is not sufficiently thick, all the microheaters are capable of withstanding a sufficient amount of time required for the CNT growth process.

7.2.2.2.4 Local Temperature Distribution

In addition to the maximum temperature and the heater reliability, the local temperature distribution is also of vital importance. It was investigated using a QFI InfraScope. The thermal image of the straight-line microheater is shown in

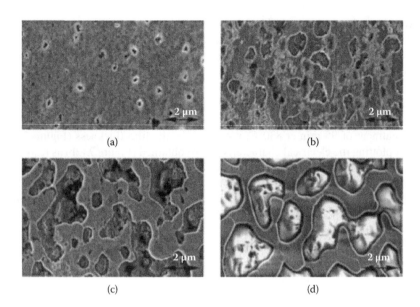

FIGURE 7.14   The formation and growth of holes in the 100 nm Pt/10 nm Ti thin film. (a) After 0 h at 900°C. (b) After 2 h at 900°C. (c) After 6 h at 900°C. (d) After 9 h at 900°C. Reported in literature [57].

Figure 7.15a. Since the maximum working temperature of this infrared imager is 400°C, only 0.6 V was applied to the microheater. Figure 7.15d and e shows the temperature distributions along and transverse to the platinum heater line, respectively. Note that the InfraScope stage was heated to 60°C to facilitate accurate temperature measurement for this image. It is clearly shown that the temperature is near uniform at the region ±30 μm from the microheater center. This is the optimal region for CNT growth. Meanwhile, even though the temperature is as high as 400°C around the heater center, the temperature outside the microcavity drops quickly to about 100°C. Therefore the temperature distribution is compatible with the post-CMOS processing even when the supply power must be increased to provide the desired growth temperature of ~900°C. By accurately choosing the spacing between CMOS circuits and microheaters, this localized microheating method can integrate CNTs at a close proximity of the CMOS devices on the same chip. For designs 1 and 2, the corners of the microheaters have slightly higher temperatures due to the current crowding effect, as evident from their thermal images (as shown in Figure 7.15).

7.2.3   ROOM TEMPERATURE CARBON NANOTUBE SYNTHESIS

After device fabrication and characterization, the samples were coated with alumina-supported iron catalyst by drop-drying. Two contact pads were connected to a voltage-controlled power supply by clamps. And then the sample was placed into a quartz tube. After 5 min of argon purging, the microheater was heated up to the state that red glowing could be observed. For example, design 2 needs a supply voltage of 3–3.5 V. Then a mix of 1000 sccm CH_4, 20 sccm C_2H_4, and 500 sccm H_2 was flowed

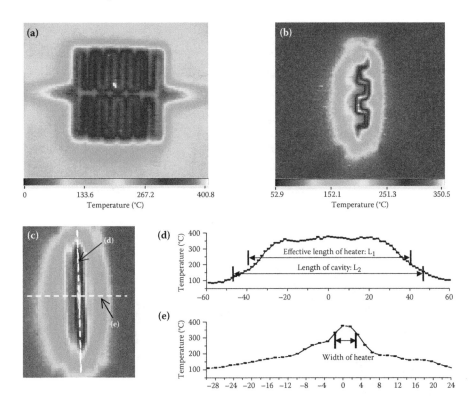

FIGURE 7.15 Thermal imaging. (a) Thermal image of design 1 by applying 0.93 V DC voltage. (b) Thermal image of design 2 by applying 1.2 V DC voltage. (c) Thermal image of design 3 by applying 0.6 V DC voltage. (d, e) Corresponding temperature distribution along and transverse to the microheater, indicating good thermal isolation. (The substrate temperature is 60°C.)

into the quartz tube for CNT growth. After a 15 min growth, SWNTs and MWNTs with diameters ranging from 1–10 nm were successfully synthesized on all three types of suspended microstructures.

Figure 7.16 shows a dense film of CNTs grown on the micro-hotplate surface. Some interesting coiled nanostructures, as shown in the insets in Figure 7.16, are observed on many samples. The orientations of these CNTs are random since there was no guiding electrical field during growth.

On the other hand, for the other two designs with trenches and secondary landing walls, the supplied voltage simultaneously introduces an E-field (about 0.1 ~ 1.0 V/μm) between the microheater and a nearby ground electrode/oxide wall. As a result, most of the suspended CNTs grown on these two types of microheaters exhibit a significant alignment along the E-field perpendicular to the cold wall, as shown in Figure 7.17. As demonstrated in our experiments, with proper designed microheaters and trench widths, the desired temperature profile and E-field distribution can be obtained by the same power supply. The extra pair of backup electrodes designed to enhance the E-field was not used, and thus can be removed to simplify the design in

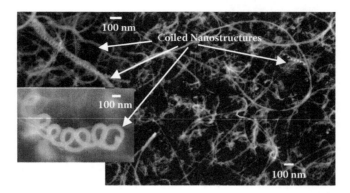

FIGURE 7.16 Dense film of CNTs over micro-hotplate surface (design 1). Insets: Coiled nanostructures observed in the growth.

FIGURE 7.17 Localized synthesis of CNTs suspended across the trench, showing good CNT alignment. (a, b) Zoom-in SEM of CNTs grown on design 2. (c) Zoom-in SEM of CNTs grown on design 3 (from second batch, with no Cr on top). Insets: SEMs of overall microheaters.

the future. Although the microheater corners have a higher temperature and stronger E-field, comparing the SEM images (shown in Figure 7.17a and b) reveals that there is no significant difference between the growth around the corners and the rest of the microheater. When the microheater is further simplified into one straight line (design 3), we find that the alignment is further improved, as shown in Figure 7.17c, and the CNT growth is uniform along the length of the entire microheater, except on the small regions next to the anchors. These results are in good agreement with the measured temperature distribution. Hence the microheater geometry with either a relatively larger hotplate or a small hotspot can be customized to control the temperature distribution for regulating the CNT growth for various applications.

7.3   MASKLESS POST-CMOS-CNT SYNTHESIS ON FOUNDRY CMOS

In the previous section, localized CNT synthesis on mock-CMOS substrates has been demonstrated. We have proved that, based on the voltage-controlled localized heating, suspended on-chip microheaters can provide both uniform high temperature for high-quality CNT growth and good thermal isolation for the CMOS compatibility requirement. Repeatable and well-aligned CNT growth can be realized, and the simple straight-line microheater is promising for further scaling down to a hotspot. However, a number of vital questions have yet to be settled. First, platinum may not be available in foundry CMOS processes, and thus some other materials need to be used to form microheaters. Second, the process flow described in Section 7.2.2 requires two lithography steps for patterning microheaters and release openings, which is still too complicated for post-CMOS processes. Third, the integration presented in Section 7.2 did not involve the interconnection of CNTs with CMOS circuits in the monolithic integration. To address these issues, we present a simple and scalable CMOS-CNT integration approach in this section. CNTs are selectively synthesized on polysilicon microheaters embedded aside the CMOS circuits using localized heating and maskless post-CMOS surface micromachining techniques. There is no need of any photomasks, shadow masks, or metal deposition to achieve the localized synthesis and the CNT-polysilicon electrical contact. Successful monolithic CMOS-CNT integration has been demonstrated [64]. And it is verified that the electrical characteristics of the neighboring NMOS and PMOS transistors are unchanged after CNT growth [64].

7.3.1   INTEGRATION PRINCIPLES AND DEVICE DESIGN

As illustrated in Figure 7.18, the basic idea of the monolithic integration approach is to use maskless post-CMOS MEMS processing to form microcavities for thermal isolation and use the gate polysilicon to form the heaters for localized heating as well as the nanotube-to-CMOS interconnect. The microheaters, made of the gate polysilicon, are deposited and patterned along with the gates of the transistors in the standard CMOS foundry processes. Except the shape and dimensions, the polysilicon microheaters are equivalent to the transistor gate, and thus they share exactly the same subsequent processes, i.e., the vias, interconnects, passivation, and I/O pads. One of the top metal layers (i.e., the metal 3 layer, as shown in Figure 7.18b) is also patterned during the CMOS fabrication. It is used as an etching mask in the following post-CMOS microfabrication process for creating the microcavities. Finally, the polysilicon microheaters are exposed and suspended in a microcavity on a CMOS substrate, while the circuits are covered under the metallization and passivation layers, as illustrated in Figure 7.18b. Unlike the traditional thermal CVD synthesis that heats up the whole chamber to above 800°C, the device with embedded microheaters

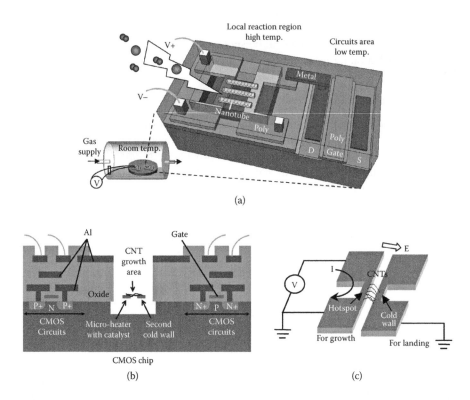

FIGURE 7.18 (a) The 3D schematic showing the concept of the CMOS-integrated CNTs. The CVD chamber is kept at room temperature all the time. (b) Cross-sectional view of the device. (c) The schematic 3D microheater showing the local synthesis from the hotspot and self-assembly on the cold landing wall under the local electric field.

works like a miniaturized CVD array: The CVD chamber is kept at room temperature all the time, with only the microheaters activated to provide the local high temperature for CNT growth.

The top view of a microheater design is shown in Figure 7.18c. The configuration is similar to that of the platinum microheaters. There are two polysilicon bridges: one as the microheater for generating high temperature to initiate CNT growth and the other for CNT landing. With the cold wall grounded, an E-field perpendicular to the surface of the two bridges will be induced during CNT growth. Activated by localized heating, the nanotubes will start to grow from the hotspot (i.e., the center of the microheater) and will eventually reach the secondary cold bridge under the influence of the local E-field. Since both the microheater bridge and the landing bridge are made of a gate polysilicon layer and have been interconnected with the metal layers in the CMOS foundry process, the as-grown CNTs can be electrically connected to CMOS circuitry on the same chip without any post-growth clamping or connection steps.

As discussed in Section 7.2.1, the microheater design is of vital importance. The gate polysilicon layer is thin and its thickness is determined by the foundry CMOS

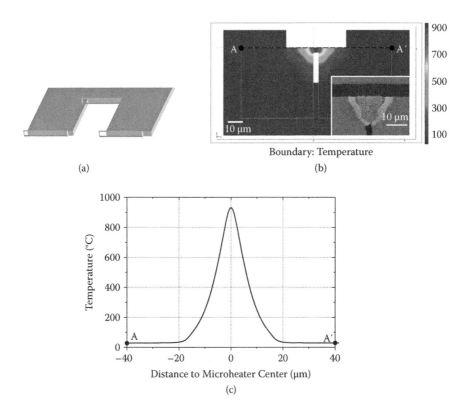

FIGURE 7.19 (a) A typical heater design with stripe-shaped resistor. (b) Simulated temperature distribution along the surface of the microheater at an applied voltage of 2.5 V through the pads. The area of polysilicon microheater is 3 × 3 μm, and a thickness of 0.35 μm is chosen according to the CMOS foundry process. Inset: An SEM image of a microheater after CNT growth. (c) The line plot of the temperature along the heater (line AA′ in b).

processes. For the AMI 0.5 μm CMOS process [65] used in this work, the polysilicon thickness is 0.35 μm. Since the serpentine design showed no meaningful advantages of the corner effect but required extra mechanical supports in the previous mock-CMOS synthesis experiments, a straight-line-shaped microheater is adopted for its superior mechanical robustness and simple design. A typical heater design is shown in Figure 7.19a, which is basically a polysilicon resistor. Since the temperature has to reach at least as high as 800°C for SWNT growth and drop quickly to avoid deteriorating the surrounding CMOS circuits, the thermal isolation, thermal stresses, and structural stiffness must be carefully considered during microheater design. Since thermal resistance is proportional to the length of a resistor, short resistors tend to dissipate heat faster, so that reaching high temperature requires more power, while long resistors tend to have mechanical stiffness issues. In addition, the current density limitation of the polysilicon resistors and the limitation of the release process do not allow us to design microheaters with too narrow width. As a result of considering all these factors, the heating unit first investigated is a 3 μm long and 3 μm wide polysilicon resistor.

TABLE 7.1
MOSIS AMI C5 Technology

Structure	Min	Typ	Max	Units
N+ poly sheet res	23	30	37	Ohms/sq
CMP M3 thickness	7000	7700	8400	Å
CMP M3 to M2 dielectric	10,000	11,000	12,000	Å
CMP M2 thickness	5000	5700	6400	Å
CMP M2 to M1 dielectric	10,000	11,000	12,000	Å
CMP M1 thickness	5700	6400	7100	Å
Poly thickness	3000	3500	4000	Å
Field ox under poly	3500	4000	4500	Å
Via allowed current density	1.6 (85°C), 0.6 (125°C)			mA/cnt
Metal allowed current density per width	2.2 (85°C); 0.85 (125°C)			mA/μm

Electrothermal modeling in a multiphysics finite element method tool, COMSOL [66], has been used to simulate and optimize the microheater design. The simulation results are shown in Figure 7.19b and c, where a typical polysilicon sheet resistance of 30 Ω/□ and other material properties are chosen according to the employed foundry process [65] (see Table 7.1). For the simulation, the convection and radiation heating losses are neglected as the heating area is small. The substrate bottom surface is assumed to stay at room temperature. Figure 7.19b shows the temperature distribution when a 2.5 V activation voltage is applied to the heater. It shows good agreement with the post-growth surface pattern (Figure 7.19b, inset SEM image), which is believed to reflect the temperature distribution during the growth. Figure 7.19c plots the temperature distribution along the microheater. It shows an ideal condition for CMOS-CNT integration: a very small, localized high-temperature region for CNT growth and a sharp temperature decrease toward the substrate.

Based on this 3×3 μm^2 heating unit, a series of heater design variations have been investigated. First, to exploit the geometry limitations (mainly the mechanical robustness and electrothermal properties), six line-shaped microheaters are designed with dimension variations. The width ranges from 1.2 to 6 μm, and the length ranges from 1.2 to 40 μm. Second, two types of secondary walls are designed: One is a bridge parallel to the microheater for uniform E-field formation (Figure 7.20a), and the other is a sharp tip (or multiple tips) to form a converged E-field (Figure 7.20b). The gap between the two polysilicon microstructures is typically 3–6 μm in order to obtain the proper electric field and facilitate the CNT landing. Third, opposing sharp tips (Figure 7.20c) are also designed to facilitate the CNT bridge formation since CNTs tend to attach to the nearest support boundary [67]. Finally, instead of grounding the secondary wall, some designs have five configurable inputs. As illustrated in Figure 7.20a, four pads can input different voltages to tune the electric field, and the fifth input (not shown) is connected to the substrate as a global back gate for studying the electrical gating effect.

The final CMOS chip includes test circuits and 13 embedded microheaters. The schematic layout is shown in Figure 7.21. Microheaters are placed around the center,

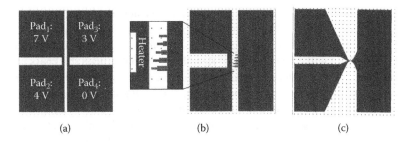

(a)   (b)   (c)

FIGURE 7.20 Schematic layouts showing (a) a 3 × 10 μm microheater with a paralleled bridge as the secondary wall and four configurable input pads, (b) a 6 × 20 μm microheater with multiple tips as landing walls, and (c) a microheater with opposing sharp tips.

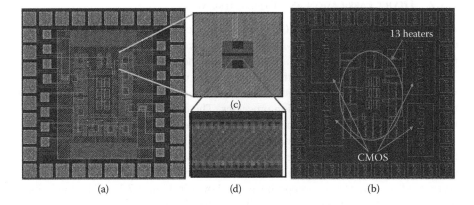

(a)   (d)   (b)

FIGURE 7.21 (a, b) Schematic layouts of the chip, including test circuits and 13 embedded microheaters. The spacings between the microheaters and circuits vary from 36 to 60 μm. (c, d) Close-up views of a microheater with metal 3 as etching opening mask and metal 1 as etching protection.

and they are independent from each other. Four test circuits are placed close to the microheaters with spacings ranging from 36 to 60 μm. The spacings are chosen based on the temperature distribution investigation shown in Figure 7.15. The sizes of the NMOS and PMOS transistors, which are the subject of our investigation, are W_n/L_n = 3.6 μm/0.6 μm and W_p/L_p = 7.2 μm/0.6 μm, respectively. All the areas outside the central square in Figure 7.21c are covered with the metal 3 layer. It is patterned as the etching mask that covers all the circuit area beneath but has 13 etching openings, with one opening for each individual microheater. The etching opening as shown in Figure 7.21c determines the size of the microcavity. The microcavities will be formed when the top silicon dioxide and bottom silicon are all etched away during the post-CMOS MEMS processes, leaving only the suspended microheaters, as shown in Figure 7.18a and b.

However, there is a selective etching issue. In the cases where the microheaters are made of platinum, specific etch chemistry can be used to etch away silicon dioxide or silicon completely with little etching to platinum. Thus etching protection for the heaters is not necessary, and the platinum microheaters themselves can be used as masks once the heaters' shapes have been patterned. When switching the heater material to polysilicon, this is no longer the case. Both the dioxide etchant (dry etch) and silicon etchant will etch polysilicon. Recall that the polysilicon thickness is only 0.35 μm. Thus etching sequence needs to be carefully designed and etching protection is essential. In our design, the metal 1 layer above the microheater is patterned with the same shape as the microheater, but with a slightly greater width, as shown in Figure 7.21d. This patterned metal 1 layer will protect the microheaters during the first few steps when the etchants can react with polysilicon, and then it will be removed using polysilicon-safe etching recipes. Details will be presented in the device fabrication section.

7.3.2 DEVICE FABRICATION AND CHARACTERIZATION

After the layout design, the CMOS chips are fabricated through MOSIS using the commercial AMI 0.5 μm three-metal CMOS process [65]. The gate oxide thickness is 13.5 nm. The estimated layer thicknesses and other parameters are listed in Table 7.1. Several parameters in the table have been used in the modeling in the previous section. In addition, the thickness of each layer is used to estimate the etching time in the post-CMOS fabrication processes. The current density limits have also been taken into consideration when designing the metallization.

Figure 7.22a shows the schematic cross-sectional view of the CMOS chip after the foundry processes but before any post-CMOS MEMS fabrication. The metal 1 and metal 3 layers have been patterned as described in the previous section. The corresponding optical microscope image is shown in Figure 7.23a. The total chip area is 1.5×1.5 mm^2. Due to this small size, individual chips are mounted on top of a 4-inch carrier wafer for easy handling and processing. The center big square is the metal 3 layer. In addition to the 13 etching openings to expose the microheaters, there are 6 extra opening windows in the center. These are dummy structures for etching rate control. There are 32 outer pads for microheaters and 16 inner pads for circuits. These pads, as well as the area uncovered by metal 3, need to be covered with photoresist for protection before any release processes.

The maskless post-CMOS MEMS fabrication process flow that used to release the polysilicon microheaters is shown in Figure 7.22. It starts with the reactive ion etching (RIE) of silicon dioxide (Figure 7.22b). The metal 3 layer of the CMOS substrate is used as etching mask to protect the CMOS circuit area, and it also defines the cavity opening size. Within the cavity, the anisotropic etching, which uses a mixture of CHF_3 and O_2 as the etch chemistry, is mainly in the vertical direction. As a result, three steep trenches are formed, but the oxide under metal 1 is almost unattacked. Thus the polysilicon microheaters are entirely wrapped inside silicon dioxide. Next the exposed aluminum is etched by RIE (Figure 7.22c), using BCl_3, Cl_2, and Ar. At this point, the photoresist should be removed since the pad protection is no longer needed in the following steps, and removing it after the release step may damage the suspended microstructures. Then an anisotropic deep-reactive-ion etching (DRIE) of silicon is performed (Figure 7.22d) using SF_6 and C_4F_8 as the etching and passivation chemistry, respectively, to create a roughly 6 μm deep trench

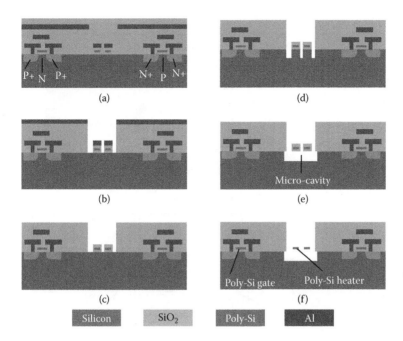

FIGURE 7.22 Mastless post-CMOS MEMS fabrication process flow: (a) CMOS chip from foundry. (b) SiO$_2$ dry etch. (c) Al etch. (d) Anisotropic Si dry etch. (e) Isotropic Si dry etch and heater release. (f) SiO$_2$ wet etch.

around the heater. Next an isotropic silicon etching using only SF$_6$ is performed to undercut the silicon under the microheaters (Figure 7.22e), resulting in suspended microheaters in microcavities. In these two steps of silicon etching, polysilicon will be quickly etched away if exposed. The silicon dioxide will also be etched, but at a much slower rate. As mentioned previously, metal 1 is designed slightly wider than the polysilicon microheaters. The width difference determines the thickness of the sidewall protective silicon dioxide. The oxide sidewall must withstand the etching during the two silicon etching steps so that the polysilicon microheaters will remain unattacked. The final step is to etch away the thin oxide protective layer surrounding the microheaters using a 6:1 buffered oxide etchant (BOE) at room temperature for approximately 5 min to expose the polysilicon for electrical contact with CNTs (Figure 7.22f). This BOE wet etch has a very high etching selectivity between silicon and silicon dioxide. Overall, since all the required etching masks have been patterned in the foundry processes, the post-CMOS MEMS fabrication is simple and easy to control, requiring only RIE, DRIE, and BOE etching.

Optical microscope images of the CMOS chip before and after the post-CMOS processing are shown in Figure 7.23a and b, respectively. A close-up optical image of one microheater is shown in Figure 7.23c. The nearby circuit, although visible, is protected under a silicon dioxide layer. Only the microheater and cold wall within the microcavity are exposed. To satisfy the spacing, the microstructures are rearranged such that the microheater is closest to the circuits and the secondary wall is on the opposite side. The polysilicon heater is connected to the metal interconnect by a

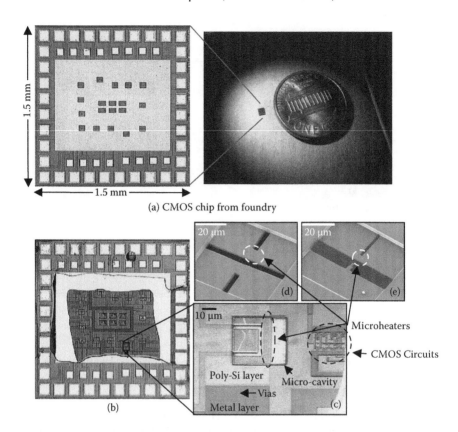

FIGURE 7.23 (a) The CMOS chip photograph (1.5 × 1.5 mm²) after foundry process. (b) The CMOS chip photograph after post-CMOS process (before final DRIE step). (c) Close-up optical image of one microheater and nearby circuit. CMOS circuit area, although visible, is protected under silicon dioxide layer. Only the microheater and cold wall within the micro-cavity are exposed to synthesis gases. Polysilicon heater and metal wire are connected by vias. (d, e) Close-up SEM images of two microheaters.

number of vias. The quantity of the vias is determined by two factors: the current that is required for Joule heating and the maximum current that one single via can withstand (see Table 7.1). SEM images of two microheaters are shown in Figure 7.23d and e, with the resistances of 97 and 117 Ω, respectively. The design is simple and the released microstructures are mechanically robust. All 13 microheaters, including the sharp-tip design and the 40 µm long design, survived the post-CMOS processes. Five chips have been fabricated with a yield of 100%, indicating the robustness of the maskless post-CMOS MEMS processes.

Similar to the mock-CMOS samples, polysilicon microheaters were also characterized before the CNT growth experiments. The measured current-voltage relation is plotted in Figure 7.24a. However, extracting the temperature-dependent resistivity from the I-V characterization and then using the resistivity to estimate the temperature becomes difficult due to the following reasons. First, grain boundaries in polysilicon exhibit charge carrier trapping that contributes to the temperature-dependent

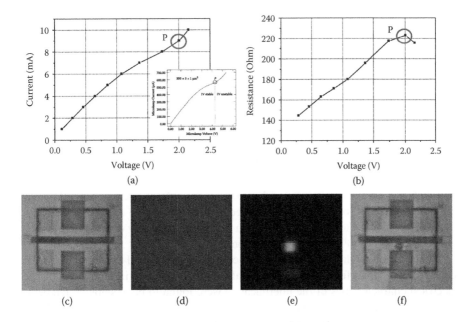

FIGURE 7.24 (a) I-V characterization of a 6 × 20 μm microheater. (b) Resistance extracted from the I-V characterization. (c–f) Microscopic images of a microheater: (c) before applying power, (d) starting to glow at 2.20 V, (e) showing bright glowing at 2.38 V, and (f) burnt after applying 2.47 V. Inset: Filament I-V characteristics showing the kink point. (From C.H. Mastrangelo et al., *IEEE Transactions on Electron Devices*, 39, 1363–1375, 1992. With permission.)

behavior [68]. Second, dopant concentrations have a significant impact on the temperature-dependent resistivity of polysilicon. For low to moderate dopant levels ($\leq 10^{18}$ cm^{-3}), increased temperature offers higher thermal energy that excites more dopant electrons to the conduction band [69], resulting in a negative TCR. Such temperature-dependent behavior is undesirable since it will cause eletrothermal instability at high temperature. For heavily doped polysilicon ($\geq 10^{19}$ cm^{-3}), if neglecting the grain boundary effects, the temperature-dependent resistivity can be defined as

$$\frac{1}{\rho} = p\,|e|\,(\mu_e + \mu_h) \tag{7.2}$$

in which ρ is the resistivity, p is the free electron concentration, e is the magnitude of an electron charge, and μ_e and μ_h are the electron and hole mobility, respectively [70]. Since the majority of the dopants' outermost electrons in heavily doped N+ silicon are already in the conduction band, the free electrons that are thermally activated from the donor no longer dominate the resistivity. Instead, the temperature-dependent behavior is dominated by the carrier mobilities at high doping levels, resulting in a positive TCR and a linearly increased resistance within the temperature range from 300 to 800 K. Lattice and impurity scattering mobility, μ_L and μ_I, respectively, are found to have significant contributions to the charge carrier mobility at high temperature [70]. The mobility terms can be expressed in the following forms:

$$\begin{cases} \mu_L \propto T^a & -2.7 < a < -1.5 \\ \mu_I \propto T^b & 1.5 < b < 2 \end{cases} \tag{7.3}$$

and the temperature-dependent resistivity can be expressed by a general equation [70]:

$$\rho = \alpha_1 + \alpha_2 T^{\alpha_3} \tag{7.4}$$

in which the parameters α_1, α_2, and α_3 have to be extracted empirically. For our microheaters, based on the polysilicon sheet resistance and the thickness, the resistivity ρ is estimated to be 1.05×10^{-5} ohm·m, and the impurity doping level is estimated to be above 10^{19} cm^{-3} using Equation 7.2.

Although a high doping level requirement has been satisfied and a positive TCR has been confirmed from the resistance calculation (as plotted in Figure 7.24b), the electrical characteristics of the microheaters were found to be unstable at temperatures beyond the polysilicon recrystallization temperature T_{cr} (at roughly 870 K [71]), and the linear increase in resistance no longer exists. Mastrangelo et al. reported that the resistance of a heavily doped polysilicon filament decreases when the bias is beyond a kink point (Figure 7.24 inset, point P) [72]. For the microheater we investigated, the kink point P happens around a bias voltage of 2 V (see Figure 7.23a and Figure 7.24b). Possible mechanisms reported include current-induced resistance decrease [73,74], filamentation [75], and the polysilicon thermal breakdown [76,77].

Continuing to increase the bias voltage, red glowing was first observed in the dark at 2.20 V (Figure 7.24d), and it quickly became much brighter at 2.38 V. The heater was burnt at the center under a bias voltage of 2.47 V, as evident from the comparison between Figure 7.24c and Figure 7.24f. As previously discussed, the extraction of temperature above the polysilicon recrystallization temperature is quite difficult due to its unstable electrical characteristics. Again, the incandescence of the microheater becomes a good indication of high temperature. Ehmann et al. carefully calibrated a microheater, which is also made of n-doped CMOS gate polysilicon, and estimated that the average heater temperature was about 1200 K when incandescence was observed in the dark [78]. In addition, Englander et al., who first locally synthesized CNTs on suspended polysilicon MEMS structures, reported that based on their growth results, the barely glowing condition, which is the condition when glowing is just started and still weak, offered the proper temperature (850 to 1000°c) for maximum CNT growth (prior glowing was too cold while bright glowing was too hot) [49]. Therefore the microheaters we designed and fabricated are capable of providing the high temperature required for CNT growth, and it can withstand a sufficient amount of time under the barely glowing state. The local temperature distribution is not characterized. Instead, the performances of test circuits and individual transistors are recorded, and will be compared with their performances after CNT growth to assess the effectiveness of the thermal isolation.

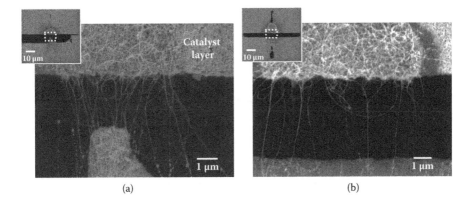

(a)   (b)

FIGURE 7.25 Localized synthesis of carbon nanotubes grown from (a) the 3 × 3 μm micro-heater and (b) the 6 × 6 μm microheater, suspended across the trench and connecting to the polysilicon tip/wall. Insets: SEM images of the overall microheater.

7.3.3 ON-CHIP SYNTHESIS OF CARBON NANOTUBES

After the microheater release, the CMOS chips were carefully taken off from the carrier wafer and then wire bonded to DIP packages. Next the chips were coated with alumina-supported iron catalyst by drop-drying on the surface [28]. The whole package was then placed into a quartz chamber. The on-chip microheaters were turned on by applying appropriate voltages such that bare growing was observed. The supplied voltage also introduced a local E-field of about 0.1 ~ 1.0 V/μm. The CNT growth was carried out using 1000 sccm CH_4, 15 sccm C_2H_4, and 500 sccm H_2 for 15 min, while the chamber remained at room temperature all the time.

After 15 min of growth, CNTs were successfully synthesized. Two SEM images with locally synthesized CNTs are shown in Figure 7.25. The individual suspended CNTs in Figure 7.25a were grown from the 3 × 3 μm microheater, as shown in Figure 7.23e, and landed on the near polysilicon tip. In the configuration with a parallel bridge as the secondary landing wall, the growth exhibits less convergence due to the less converging E-field. Similar growth occurred on 11 out of 13 microheaters. The 1.2 × 1.2 μm microheater and the one with opposing sharp tips were broken during the growth. The reason might be the presence of super-local heating that caused the failure at grain boundaries.

7.3.4 CHARACTERIZATION OF CARBON NANOTUBES AND CIRCUIT EVALUATIONS

After the growth, the as-grown CNTs were characterized by measuring the resistance between two polysilicon microstructures (the microheater and the landing wall) at room temperature and at atmosphere. The I-V characteristic is shown in Figure 7.26. The typical resistances of in situ-grown CNTs are in the range of several MΩ. This large resistance is primarily attributed to the contact resistance between the polysilicon and the CNTs. Recall that the final step of the post-CMOS fabrication is meant to completely remove the silicon dioxide, and thus fully expose the polysilicon surface. However, this step has been done long before the growth process

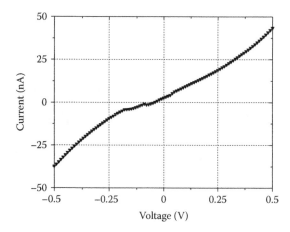

FIGURE 7.26　I-V characteristics of the as-grown CNTs. The I-V curve is measured between two polysilicon microstructures contacting the CNTs.

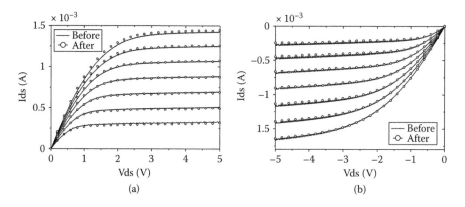

FIGURE 7.27　DC electrical characteristics of single transistors before and after CNT growth. (a) Drain current (Ids) versus grain voltage (Vds) for NMOS transistors under seven different gate voltages. (b) Drain current (Ids) versus grain voltage (Vds) for PMOS transistors under seven different gate voltages.

due to the subsequent steps, such as wire bonding, package, and catalyst deposition. A thin layer of native oxide is expected to naturally grow on the polysilicon electrode surface with a typical thickness of about 2 nm [79]. Other factors, such as defects along the CNTs, might also contribute to the large resistance.

After the successful on-chip synthesis of CNTs, the impact of the localized heating on the nearby circuits was assessed. As mentioned in Section 7.3.1, the spacings range from 36 to 60 μm. Simple circuits, such as inverters, were first tested and their proper functions were verified after synthesis. Then more accurate electrical characteristics were measured at the transistor level. Figure 7.27 shows the DC electrical characteristics of individual NMOS and PMOS transistors before and after the CNT growth. The tests were performed at room temperature using a Keithley

4200 semiconductor characterization system. The drain current versus drain-source voltage (Ids-Vds) plots show no significant change after the synthesis, demonstrating the CMOS compatibility of this integration approach.

It should be noted that after a short period (approximately 1 month), this $M\Omega$ resistance became infinite. But the synthesis on platinum microheaters exhibited $k\Omega$ resistances that stayed almost the same after 1 year. These two synthesis experiments used the same catalyst recipe and same growth procedure. The differences come from the electrode material, which results in two types of contacts: one is poly-silicon/CNTs versus Pt/CNTs, and the other is polysilicon/catalyst versus Pt/catalyst. There might be two reasons that cause the failure in the polysilicon heater case. First, it might be the result of the continuous oxidation of the polysilicon in air. Another possible reason might be due to the weaker adhesion of the CNTs and catalyst particles to the polysilicon. In our practice, the nanometer-scale iron catalyst particles are separated by and supported on alumina, a thin nonconductive layer. As shown in Figure 7.28, the catalyst on the top of the silicon dioxide surface consists of discrete particles, with the heater outline clearly visible (Figure 7.28a); the catalyst on the top of the platinum surface has a fluffy appearance, with the underneath platinum electrode partially exposed (Figure 7.28b).

However, the catalyst layer appears much thicker on CMOS chips, as evident from Figure 7.28c–f. The catalyst layer extends over trenches (Figure 7.28c). A thick catalyst layer on one of the heaters even cracks and peels off (Figure 7.28d). Some designed shapes (e.g., the multi-tip landing wall) that have successfully survived the post-CMOS fabrication (Figure 7.28e) are unable to be preserved after catalyst deposition (Figure 7.28f). The mock-CMOS devices have platinum microheaters on the surface (Figure 7.28g), while the CMOS chips have polysilicon microheaters hidden inside the microcavities (Figure 7.28h). The topographic profiles, the surface conditions of different materials, and the further miniaturized feature size might all contribute to the thicker catalyst layer, which in return might block the nanotube/polysilicon contacts or result in weak adherence when CNTs reach the secondary electrode.

To improve the chemical stability and mechanical robustness, several possible solutions are recommended for future work. First, instead of using aluminum-supported iron catalyst particles and the drop-drying method, thin-film metal catalysts can be deposited using various techniques, such as evaporation, sputtering, or molecular beam epitaxy techniques [80]. These thin films will "ball up" and break up into particles during the growth as long as the film thickness does not exceed the critical thickness [81]. In this way, the structure shape can be preserved, and the thickness of the catalyst layer can be controlled. Second, contact activation through electrical breakdown in the inert gas environment has been proven to be effective for healing the 2 nm native oxide [82]. Third, polymer deposition after the CNT growth has been reported to be capable of stabilizing the nanotube/electrode contact resistances [83]. This coating might be helpful for preventing the polysilicon from oxidation over time.

7.4 CONCLUSION

Monolithic CNT-CMOS integration can be realized using custom CMOS processes by employing refractory metals for interconnect and SOI CMOS processes.

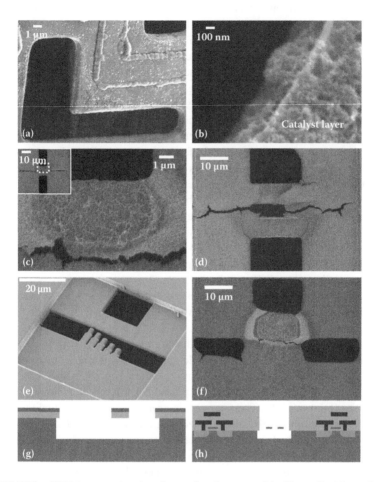

FIGURE 7.28 SEM images showing the catalyst layers on (a) silicon dioxide surface, (b) platinum microheater surface, and (c) polysilicon microheater surface. (d) An SEM image of one polysilicon microheater with the thick catalyst layer peeled off. (e, f) SEM images of one polysilicon microheater before and after catalyst deposition. Multifigure structure was completely covered by the catalyst layer. (g, h) Schematic cross-sectional views of mock-CMOS platinum heater and polysilicon heater embedded in CMOS chips.

Localized heating is promising for CNT-CMOS integration. Monolithic CNT-CMOS integration has been demonstrated on foundry CMOS substrate by combining MEMS microfabrication and localized heating. The post-CMOS microfabrication is maskless. The CNT growth does not affect the characteristics of the transistors on the same chip. As a hotplate is formed from creating the localized heating, integrated CNT-based gas sensors can be made. This CNT-CMOS integration technique has a wide range of applications in chemical, temperature, stress, and radiation sensing.

REFERENCES

1. International Technology Roadmap for Semiconductors. SEMATECH, Austin, TX, 2001. http://public.itrs.net/.
2. R.J. Luyken and F. Hofmann. Concepts for hybrid CMOS-molecular non-volatile memories. *Nanotechnology*, 14, 273–276, 2003.
3. W. Zhang, N.K. Jha, and L. Shang. Nature: A hybrid nanotube/CMOS dynamically reconfigurable architecture. In *Design Automation Conference*, 2006 43rd ACM/IEEE, 2006, pp. 711–716.
4. I. Nantero. NRAM®. 2000–2009. http://www.nantero.com/mission.html.
5. U. Frey, M. Graf, S. Taschini, K.-U. Kirstein, and A. Hierlemann. A digital CMOS architecture for a micro-hotplate array. *IEEE Journal of Solid-State Circuits*, 42, 441–450, 2007.
6. M. Grassi, P. Malcovati, and A. Baschirotto. A 0.1% accuracy 100Ω-$20M\Omega$ dynamic range integrated gas sensor interface circuit with 13+4 bit digital output. *Proceedings of ESSCIRC*, 351–354, 2005.
7. M. Malfatti, D. Stoppa, A. Simoni, L. Lorenzelli, A. Adami, and A. Baschirotto. A CMOS interface for a gas-sensor array with a 0.5%-linearity over $500k\Omega$-to-$1G\Omega$ range and ±2.5°C temperature control accuracy. In *ISSCC Dig. Tech. Papers*, 2006, pp. 294–295.
8. M. Malfatti, M. Perenzoni, D. Stoppa, A. Simoni, and A. Adami. A high dynamic range CMOS interface for resistive gas sensor array with gradient temperature control. In *IMTC 2006—Instrumentation and Measurement Technology Conference*, 2006, pp. 2013–2016.
9. V. Agarwal, C.-L. Chen, M.R. Dokmeci, and S. Sonkusale. A CMOS integrated thermal sensor based on single-walled carbon nanotubes. In *IEEE Sensors 2008 Conference*, 2008, pp. 748–751.
10. T.S. Cho, K.-J. Lee, J. Kong, and A.P. Chandrakasan. A 32-uW 1.83-kS/s carbon nanotube chemical sensor system. *IEEE Journal of Solid-State Circuits*, 44, 659–669, 2009.
11. F. Udreal, S. Maeng, J.W. Gardner, J. Park, M.S. Haquel, S.Z. Ali, Y. Choi, P.K. Guhal, S.M.C. Vieiral, H.Y. Kim, S.H. Kim, K.C. Kim, S.E. Moon, K.H. Park, W.I. Milne, and S.Y. Oh. Three technologies for a smart miniaturized gas-sensor: SOI CMOS, micromachining, and CNTs—Challenges and performance. In *IEEE International Electron Devices Meeting, IEDM 2007*, 2007, pp. 831–834.
12. G.F. Close, S. Yasuda, B. Paul, S. Fujita, and H.-S.P. Wong. A 1 GHz integrated circuit with carbon nanotube interconnects and silicon transistors. *Nano Letters*, 8, 706–709, 2008.
13. R.S. Chakraborty, S. Narasimhan, and S. Bhunia. Hybridization of CMOS with CNT-based nano-electromechanical switch for low leakage and robust circuit design. *IEEE Transactions on Circuits and Systems*, 54, 2480–2488, 2007.
14. A. Yasuda, N. Kawase, and W. Mizutani. Carbon-nanotube formation mechanism based on in situ TEM observations. *J. Phys. Chem. B*, 106, 13294–13298, 2002.
15. D.S. Bethune, C.H. Kiang, M.S. DeVries, G. Gorman, R. Savoy, J. Vazquez, and R. Beyers. Cobalt-catalysed growth of carbon nanotubes with single-atomic-layer walls. *Nature*, 363, 605–607, 1993.
16. C. Journet, W.K. Maser, P. Bernier, A. Loiseau, M.L.D.L. Chapelle, S. Lefrant, P. Deniard, R. Lee, and J.E. Fischerk. Large-scale production of single-walled carbon nanotubes by the electric-arc technique. *Nature*, 388, 756–758, 1997.
17. T.W. Ebbesen and P.M. Ajayan. Large-scale synthesis of carbon nanotubes. *Nature*, 358, 220–222, 1992.
18. A. Thess, R. Lee, P. Nikolaev, H. Dai, P. Petit, J. Robert, C. Xu, Y.H. Lee, S.G. Kim, A.G. Rinzler, D.T. Colbert, G.E. Scuseria, D. Tománek, J.E. Fischer, and R.E. Smalley. Crystalline ropes of metallic carbon nanotubes. *Science*, 273, 483–487, 1996.
19. M. Yudasaka, T. Komatsu, T. Ichihashi, and S. Iijima. Single-wall carbon nanotube formation by laser ablation using double-targets of carbon and metal. *Chemical Physics Letters*, 278, 102–106, 1997.

20. T. Guo, P. Nikolaev, A. Thess, D.T. Colbert, and R.E. Smalley. Catalytic growth of single-walled nanotubes by laser vaporization. *Chemical Physics Letters*, 243, 49–54, 1995.
21. M. Endo, K. Takeuchi, K. Kobori, K. Takahashi, H. Kroto, and A. Sarkar. Pyrolytic carbon nanotubes from vapor-grown carbon fibers. *Carbon*, 33, 873–881, 1995.
22. J. Kong, A.M. Cassell, and H. Dai. Chemical vapor deposition of methane for single-walled carbon nanotubes. *Chemical Physics Letters*, 292, 567–574, 1998.
23. A.M. Cassell, J.A. Raymakers, J. Kong, and H. Dai. Large scale CVD synthesis of single-walled carbon nanotubes. *J. Phys. Chem. B*, 103, 6484–6492, 1999.
24. J.H. Hafner, M.J. Bronikowski, B.R. Azamian, P. Nikolaev, A.G. Rinzler, D.T. Colbert, K.A. Smith, and R.E. Smalley. Catalytic growth of single-wall carbon nanotubes from metal particles. *Chemical Physics Letters*, 296, 195–202, 1998.
25. S.B. Sinnott, R. Andrews, D. Qian, A.M. Rao, Z. Mao, E.C. Dickey, and F. Derbyshire. Model of carbon nanotube growth through chemical vapor deposition. *Chemical Physics Letters*, 315, 25–30, 1999.
26. C.L. Cheung, A. Kurtz, H. Park, and C.M. Lieber. Diameter-controlled synthesis of carbon nanotubes. *J. Phys. Chem. B*, 106, 2429–2433, 2002.
27. Y. Choi, J. Sippel-Oakley, and A. Ural. Single-walled carbon nanotube growth from ion implanted Fe catalyst. *Applied Physics Letters*, 89, 153130, 2006.
28. A. Ural, Y. Li, and H. Dai. Electric-field-aligned growth of single-walled carbon nanotubes on surfaces. *Applied Physics Letters*, 81, 3464–3466, 2002.
29. M. Meyyappan. *Carbon nanotubes: Science and applications.* New York: CRC, 2005.
30. N.R. Franklin, Q. Wang, T.W. Tombler, A. Javey, M. Shim, and H. Dai. Integration of suspended carbon nanotube arrays into electronic devices and electromechanical systems. *Applied Physics Letters*, 81, 913–915, 2002.
31. Z. Chen, J. Appenzeller, J. Knoch, Y.-M. Lin, and P. Avouris. The role of metal-nanotube contact in the performance of carbon nanotube field-effect transistors. *Nano Letters*, 5, 1497–1502, 2005.
32. F.A. Ghavanini, H.L. Poche, J. Berg, A.M. Saleem, M.S. Kabir, P. Lundgren, and P. Enoksson. Compatibility assessment of CVD growth of carbon nanofibers on bulk CMOS devices. *Nano Letters*, 8, 2437–2441, 2008.
33. G.K. Fedder. CMOS-based sensors. Presented at 5th International Conference on Sensor (IEEE-Sensors), Irvine, CA, 2005.
34. Y.-C. Tseng, P. Xuan, A. Javey, R. Malloy, Q. Wang, J. Bokor, and H. Dai. Monolithic integration of carbon nanotube devices with silicon MOS technology. *Nano Letters*, 4, 123–127, 2004.
35. J. Kong, H.T. Soh, A.M. Cassell, C.F. Quate, and H. Dai. Synthesis of individual single-walled carbon nanotubes on patterned silicon wafers. *Nature*, 395, 878–881, 1998.
36. X.M.H. Huang, R. Caldwell, L. Huang, S.C. Jun, M. Huang, M.Y. Sfeir, S.P. O'Brien, and J. Hone. Controlled placement of individual carbon nanotubes. *Nano Letters*, 5, 1515–1518, 2005.
37. P.A. Williams, S.J. Papadakis, M.R. Falvo, A.M. Patel, M. Sinclair, A. Seeger, A. Helser, R.M. Taylor II, S. Washburn, and R. Superfine. Controlled placement of an individual carbon nanotube onto a microelectromechanical structure. *Applied Physics Letters*, 80, 2574–2576, 2002.
38. J. Liu, M.J. Casavant, M. Cox, D.A. Walters, P. Boul, W. Lu, A.J. Rimberg, K.A. Smith, D.T. Colbert, and R.E. Smalley. Controlled deposition of individual single-walled carbon nanotubes on chemically functionalized templates. *Chemical Physics Letters*, 303, 125–129, 1999.
39. H. Ko and V.V. Tsukruk. Liquid-crystalline processing of highly oriented carbon nanotube arrays for thin-film transistors. *Nano Letters*, 6, 1443–1448, 2006.

40. T. Schwamb, N.C. Schirmer, B.R. Burg, and D. Poulikakos. Fountain-pen controlled dielectrophoresis for carbon nanotube-integration in device assembly. *Applied Physics Letters*, 93, 193104, 2008.
41. J. Chung, K.-H. Lee, J. Lee, and R.S. Ruoff. Toward large-scale integration of carbon nanotubes. *Langmuir*, 20, 3011–3017, 2004.
42. A. Vijayaraghavan, S. Blatt, D. Weissenberger, M. Oron-Carl, F. Hennrich, D. Gerthsen, H. Hahn, and R. Krupke. Ultra-large-scale directed assembly of single-walled carbon nanotube devices. *Nano Letters*, 7, 1556–1560, 2007.
43. P. Makaram, S. Selvarasah, X. Xiong, C.-L. Chen, A. Busnaina, N. Khanduja, and M.R. Dokmeci. Three-dimensional assembly of single-walled carbon nanotube interconnects using dielectrophoresis. *Nanotechnology*, 18, 395204, 2004.
44. S.J. Kang, C. Kocabas, H.-S. Kim, Q. Cao, M.A. Meitl, D.-Y. Khang, and J.A. Rogers. Printed multilayer superstructures of aligned single-walled carbon nanotubes for electronic applications. *Nano Letters*, 7, 3343–3348, 2007.
45. K. Ryu, A. Badmaev, C. Wang, A. Lin, N. Patil, L. Gomez, A. Kumar, S. Mitra, H.-S.P. Wong, and C. Zhou. CMOS-analogous wafer-scale nanotube-on-insulator approach for submicrometer devices and integrated circuits using aligned nanotubes. *Nano Letters*, 9, 189–197, 2009.
46. H.E. Unalan and M. Chhowalla. Investigation of single-walled carbon nanotube growth parameters using alcohol catalytic chemical vapour deposition. *Nanotechnology*, 16, 2153–2163, 2005.
47. S. Hofmann, C. Ducati, J. Robertson, and B. Kleinsorge. Low-temperature growth of carbon nanotubes by plasma-enhanced chemical vapor deposition. *Applied Physics Letters*, 83, 135–137, 2003.
48. K. Ryu, M. Kang, Y. Kim, and H. Jeon. Low-temperature growth of carbon nanotube by plasma-enhanced chemical vapor deposition using nickel catalyst. *Japanese Journal of Applied Physics*, 42, 3578–3581, 2003.
49. O. Englander, D. Christensen, and L. Lin. Local synthesis of silicon nanowires and carbon nanotubes on microbridges. *Applied Physics Letters*, 82, 4797–4799, 2003.
50. Y. Zhou, J. Johnson, L. Wu, S. Maley, A. Ural, and H. Xie. Design and fabrication of microheaters for localized carbon nanotube growth. Presented at the 8th IEEE Conference on Nanotechnology, Dallas, TX, 2008.
51. S. Dittmer, O.A. Nerushev, and E.E.B. Campbell. Low ambient temperature CVD growth of carbon nanotubes. *Applied Physics A*, 84, 243–246, 2006.
52. A. Jungen, C. Stampfer, M. Tonteling, S. Schiesser, D. Sarangi, and C. Wierold. Localized and CMOS compatible growth of carbon nanotubes on a 3×3 μm^2 microheater spot. Presented at the 13th International Conference on Solid-State Sensors, Actuators and Microsystems, Seoul, Korea, 2005.
53. M.S. Haque, K.B.K. Teo, N.L. Rupensinghe, S.Z. Ali, I. Haneef, S. Maeng, J. Park, F. Udrea, and W.I. Milne. On-chip deposition of carbon nanotubes using CMOS microhotplates. *Nanotechnology*, 19, 025607, 2008.
54. W.A. Clayton. Thin-film platinum for appliance temperature control. *IEEE Transactions on Industry Applications*, 24, 332–336, 1988.
55. http://en.wikipedia.org/wiki.
56. http://en.wikipedia.org/wiki/Platinum.
57. S.L. Firebaugh, K.F. Jensen, and M.A. Schmidt. Investigation of high-temperature degradation of platinum thin films with an in situ resistance measurement apparatus. *Journal of Microelectromechanical Systems*, 7, 128–135, 1998.
58. J.O. Olowolafe, R.E. Jones, A.C.C. Jr., R.I. Hetide, C.J. Mogab, and R.B. Gregory. Effects of anneal ambients and Pt thickness on Pt/Ti and Pt/Ti/TiN interfacial reactions. *Journal of Applied Physics*, 73, 1764–1772, 1992.

59. G.R. Fox, S. Trolier-McKinstry, S.B. Krupanidhi, and L.M. Casas. Pt/Ti/SiO$_2$/Si substrates. *J. Mater. Res.*, 10, 1508–1515, 1995.

60. K.H. Park, C.Y. Kim, Y.W. Jeong, H.J. Kwon, K.Y. Kim, J.S. Lee, and S.T. Kim. Microstructures and interdiffusions of Pt/Ti electrodes with respect to annealing in the oxygen ambient. *Journal of Materials Research*, 10, 1790–1794, 1995.

61. D. Briand, S. Heimgartner, M. Dadras, and N.F.D. Rooij. On the reliability of a platinum heater for micro-hotplates. In *The 16th European Conference on Solid-State Transducers*, 2002. pp. 474–477.

62. E. Jiran and C.V. Thompson. Capillary instabilities in thin, continuous films. *Thin Solid Films*, 208, 23–28, 1991.

63. D.J. Srolovitz and M.G. Goldiner. The thermodynamics and kinetics of film agglomeration. *Journal of Minerals, Metals, and Materials Society*, 31–36, 1995.

64. Y. Zhou, J.L. Johnson, A. Ural, and H. Xie. Localized growth of carbon nanotubes on CMOS substrate at room temperature using maskless post-CMOS processing. *IEEE Transactions on Nanotechnology*, 2009.

65. AMI semiconductor 0.50μm C5 process. http://www.mosis.com/products/fab/vendors/amis/c5/.

66. COMSOL. http://www.comsol.com/.

67. A. Jungen, S. Hofmann, J.C. Meyer, C. Stampfer, S. Roth, J. Robertson, and C. Hierold. Synthesis of individual single-walled carbon nanotube bridges controlled by support micromachining. *Journal of Micromechanics and Microengineering*, 17, 603–608, 2007.

68. R.P. Manginell. Physical properties of polysilicon. PhD thesis, University of New Mexico, 1997.

69. J.W.D. Callister. *Material Science and Engineering*. New York: Wiley, 1994.

70. A.A. Geisberger, N. Sarkar, M. Ellis, and G.D. Skidmore. Electrothermal properties and modeling of polysilicon microthermal actuators. *Journal of Microelectromechanical Systems*, 12, 513–523, 2003.

71. E. Kinsbron, M. Sternheim, and R. Knoell. Crystallization of amorphous silicon films during low pressure chemical vapor deposition. *Applied Physics Letters*, 42, 835–837, 1983.

72. C.H. Mastrangelo, J.H.-J. Yeh, and R.S. Muller. Electrical and optical characteristics of vacuum-sealed poly silicon microlamps. *IEEE Transactions on Electron Devices*, 39, 1363–1375, 1992.

73. Y. Amemiya, T. Ono, and K. Kato. Electrical trimming of heavily doped polycrystalline silicon I resistors. *IEEE Transactions on Electron Devices*, 26, 1738–1742, 1979.

74. K. Kato, T. Ono, and Y. Amemiya. A physical mechanism of current-induced resistance decrease in heavily doped polysilicon resistors. *IEEE Transactions on Electron Devices*, 29, 1156–1161, 1982.

75. C.N. Berglund. Thermal filaments in vanadium dioxide. *IEEE Transactions on Electron Devices*, 16, 432–437, 1969.

76. H.A. Schafft. Second breakdown—Comprehensive review. *Proceedings of the IEEE*, 55, 1272–1288, 1967.

77. K. Ramkumar and M. Satyam. Negative-resistance characteristics of polycrystalline silicon resistors. *Journal of Applied Physics*, 62, 174–176, 1987.

78. M. Ehmann, P. Ruther, M.V. Arx, and O. Paul. Operation and short-term drift of polysilicon-heated CMOS microstructures at temperatures up to 1200 K. *Journal of Micromechanics and Microengineering*, 11, 397–401, 2001.

79. M.J. Madou. *Fundamentals of Microfabrication*. 2nd ed. Boca Raton, FL: CRC Press, 2002.

80. R.-M. Liu, J.-M. Ting, J.-C.A. Huang, and C.-P. Liu. Growth of carbon nanotubes and nanowires using selected catalysts. *Thin Solid Films*, 420–421, 2002.

81. Y.Y. Wei, G. Eres, V.I. Merkulov, and D.H. Lowndes. Effect of catalyst film thickness on carbon nanotube growth by selective area chemical vapor deposition. *Applied Physics Letters*, 78, 1394–1396, 2001.

82. Y. Jiang, M.Q.H. Zhang, T. Kawano, C.Y. Cho, and L. Lin. Activation of CNT nano-to-micro contact via electrical breakdown. Presented at MEMS 2008, Tucson, AZ, 2008.

83. C.-L. Chen, V. Agarwal, S. Sonkusale, and M.R. Dokmeci. Integration of single-walled carbon nanotubes on to CMOS circuitry with parylene-C encapsulation. Presented at the 8th IEEE Conference on Nanotechnology, Dallas, TX, 2008.

8 Applications of Carbon Nanotubes in Biosensing and Nanomedicine

Aihua Liu and Huajun Qiu

CONTENTS

8.1 INTRODUCTION

Since the discovery of carbon nanotubes (CNTs) by Iijima [1], CNTs have aroused enormous interest because of their unique properties, which involve high specific surface area, high electronic conductivity, outstanding chemical stability, unique optical and electrochemical properties, etc. [2–7].

Depending on the number of graphene layers from which a single CNT is composed, CNTs are classified as single-walled carbon nanotubes (SWNTs) and multiwalled carbon nanotubes (MWNTs). As a 1D carbon nanostructure, the lengths of CNTs vary from several hundred nanometers to several millimeters, and their diameters depend on their classes: SWNTs are 0.4–2 nm in diameter and MWNTs are 2–100 nm in diameter.

CNTs have been studied for applications in a wide variety of areas, including composite materials [8], nanoelectronics [9,10], field-effect emitters [11], hydrogen storage [12], etc. In recent years, increasing efforts have been devoted to exploring the potential applications of CNTs in biosensing and drug delivery in biological systems [13–22]. For example, due to the combination of excellent conductivity, good electrochemical properties, and nanometer dimensions, CNTs can be plugged directly into individual redox enzymes for better transduction in electrochemical third-generation enzyme biosensors [23–34]. Moreover, alignment of CNTs has created the potential for an electrode that resists nonspecific adsorption of proteins, but that can interact with individual biomolecules [35,36]. The sensitivities of the optical properties of CNTs to binding events have also been exploited to make entirely nanoscale, highly sensitive, and multiplexed optical biosensors that could be used inside cells or dispersed through a system to capture the small amount of analyte in a sample [37]. However, the real application of CNTs is usually inhibited by their poor solubility in aqueous solutions. To enhance the solubility of CNTs, both covalent and noncovalent functionalizations have been used [5,38–43]. Chemical covalent functionalization of CNTs usually destroys the sp^2 structure of CNTs, and therefore damages their intrinsic properties. Thus noncovalent modification of CNTs is of great significance. Various noncovalent modification strategies have been used for different applications [44].

In this review, we first summarize the various strategies used for the dispersion/functionalization of CNTs, including covalent and noncovalent methods. CNT-based electrochemical and optical biosensors are then discussed. We also briefly overview the recent advances in developing CNT-based drug delivery and tumor therapy. Finally, we discuss the outlook for future development.

8.2 DISPERSION AND FUNCTIONALIZATION OF CNTS

Raw CNTs have highly hydrophobic surfaces, and are not soluble in aqueous solutions. For real applications, surface chemistry (functionalization) is required to solubilize CNTs. Surface functionalization of CNTs can be covalent or noncovalent. Chemical reactions forming bonds with CNT sidewalls are carried out by covalent functionalization, while noncovalent functionalization exploits favorable interactions such as hydrophobic interaction and π-π stacking between molecules and the CNT surface, producing aqueous stable CNTs wrapped by different molecules. Different functionalization strategies for CNTs have been reviewed recently by Karousis et al. [44].

8.2.1 COVALENT FUNCTIONALIZATION OF CNTS

Various covalent reactions have been developed to functionalize CNTs. Oxidation is one of the most common ones. CNT oxidation has been carried out with oxidizing agents such as nitric acid [5,38]. During the process, carboxyl groups are formed at the ends of CNTs as well as the defects on the CNT sidewalls. Zeng et al. observed sp^3 carbon atoms on SWNTs after CNT oxidation, and the oxidized SWNTs were further covalently conjugated with amino acids [45]. Although oxidized CNTs are soluble in water, they aggregate in the presence of salts due to charge screening effects, and thus may not be quite suitable for biological applications due to the high salt content of most biological solutions. Another widely used covalent reaction to functionalize CNTs is the cylcoaddition reaction, which occurs on the aromatic sidewalls instead of the ends and defects of CNTs, as in the oxidation case. [2+1] Cycloadditions can be conducted by photochemical reaction of CNTs with azides (Figure 8.1A) [46,47] or carbene generating compounds via the Bingel reaction (Figure 8.1B) [48,49]. Prato and coworkers developed a 1,3-dipolar cycloaddition reaction on CNTs, which is now a commonly used strategy (Figure 8.1C) [50,51]. Despite the robustness of the covalent functionalization, the intrinsic physical properties of CNTs, such as electron conductivity, photoluminescence, and Raman scattering, are often damaged after chemical reactions due to the disrupted CNT structure. For example, the intensities of Raman scattering and photoluminescence of SWNTs are dramatically decreased after covalent modification, reducing the potential of optical applications of CNTs. Covalent functionalization is useful in some areas, such as drug delivery, where the optical properties of CNTs are not applied.

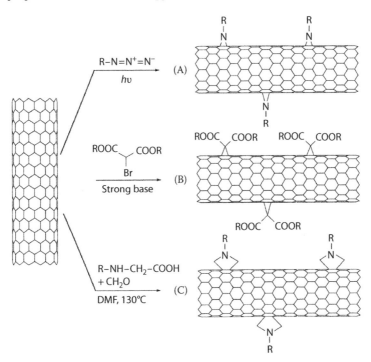

FIGURE 8.1 Schemes of covalent functionalization of CNTs: (A) photoinduced (1, 2) addition of azide compounds with CNTs; (B) Bingel reaction on CNTs; (C) 1,3-dipolar cylcoaddition on CNTs. "R" in the figure is normally a hydrophilic domain that would make CNTs water soluble.

8.2.2 NONCOVALENT FUNCTIONALIZATION OF CNTS

Noncovalent functionalization of CNTs holds great promise because it does not disrupt the large π-electronic surface. Over the past decade, it has been reported that noncovalent engineering of CNT surfaces with amphiphilic surfactants and polymers is remarkable for dispersing CNTs [52–55]. Various amphiphiles have been used to suspend CNTs in aqueous solutions, with hydrophobic domains attached to the CNT surface via van der Waals forces/hydrophobic interaction and polar heads for water solubility [56].

Traditional surfactants, including sodium dodecyl sulfate, Triton X-100, etc., have been used to suspend CNTs in water [57]. CNTs solubilized by those amphiphiles with relatively high critical micelle concentrations (CMCs) are typically not stable without an excess of surfactant molecules in the solution. For biological applications, large amounts of surfactants may lyse cell membranes and denature proteins, which means that they may not be suitable in biological environments. The polyaromatic graphitic surfaces of CNTs are accessible to the binding of aromatic molecules via π-π stacking [53,58]. Taking advantage of the π-π interaction between pyrene and CNT surfaces, pyrene derivatives have been used for the noncovalent dispersion and functionalization of CNTs [53,59,60]. The pyrene derivative-modified CNT surface can also be used for further functionalization. For example, Chen and coworkers showed that proteins can be immobilized on SWNTs functionalized by an amine-reactive pyrene derivative (Figure 8.2A) [53]. Recently, Wu and coworkers reported the use of pyrene-conjugated glycodendrimers for the dispersion of CNTs [59]. Other aromatic molecules such as porphyrin derivatives have also been used [61]. Another interesting work was carried out by Kam and coworkers [62,63]. They studied the dispersion and functionalization of CNTs by PEGylated phospholipids (PL-PEGs) (Figure 8.2B). The two hydrocarbon chains of the lipid are strongly anchored onto the CNT surface, with the hydrophilic polyethylene glycol (PEG) chain extending into the aqueous phase, imparting CNTs water solubility and biocompatibility. Further conjugation of biological molecules on PL-PEG-CNTs can be achieved by using the functional group at the PEG terminal [62,63].

We developed poly(acrylic acid) (PAA) for the dispersion of CNTs [40,64,65]. Due to the presence of carboxyl groups on the backbone of PAA, the PAA-CNTs complex is readily soluble in water. By studying the effect of solution pH on the PAA-CNT dispersion, it was observed that the complex was stable in the pH range of 3–8 (Figure 8.3A). It was also observed that the PAA-CNT complex was more tolerant to ionic strength than the poly(4-styrenesulfonic acid)-CNT complex. The PAA-CNT composites are stable in the solution whose ionic strengths are lower than 40 mM NaCl and 2 mM $MgCl_2$ (Figure 8.3B), indicating its potential application in biological conditions with high ionic strength [64]. In another work, we studied the spectroscopic properties of PAA-CNTs. From Fourier transform infrared (FT-IR) spectra, the characteristic peaks for CNTs are unchanged due to no new chemical bonds formed in PAA-CNTs, indicating that the electronic structures of the CNTs are intact [40]. The obvious blue-shift of the peak at 266 nm for the C=C double bonds of raw CNTs upon polymer wrapping, the systematic upshift in peak position, the enhancement in the band intensities of characteristic Raman bands of CNTs, and the disappearance of ¹H

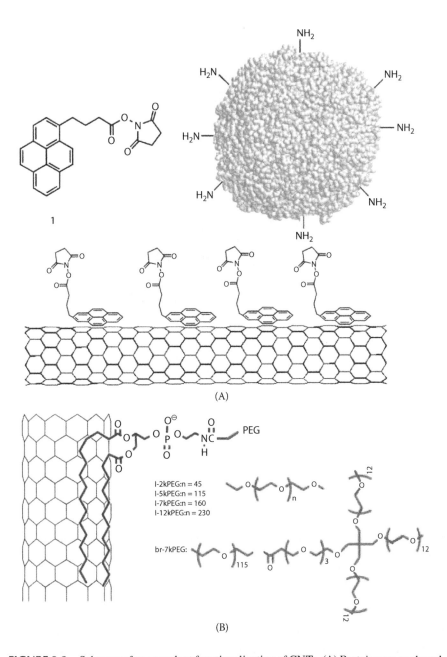

(A)

(B)

FIGURE 8.2 Schemes of noncovalent functionalization of CNTs. (A) Proteins are anchored on the SWNT surface via pyrene π-π stacked on a CNT surface. (B) A SWNT functionalized with PEGylated phospholipids. l-PEG means linear PEG and br-PEG means branched PEG. (Reproduced from R.J. Chen et al., *Journal of the American Chemical Society*, 123, 3838–3839, 2001. Copyright © 2001 American Chemical Society. With permission.)

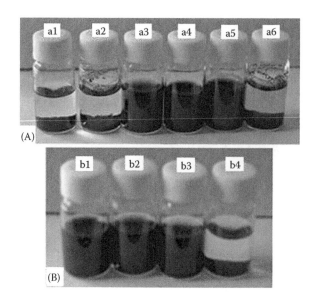

FIGURE 8.3 Photographs of vials containing suspensions of MWNTs (1 mg ml⁻¹) in different solutions: (A) absolute ethanol (a1), pure water (a2), 1 mg ml⁻¹ aqueous PAA solution (pH 3) (a3), 1 mg ml⁻¹ aqueous PAA solution (pH 4) (a4), 1 mg mL⁻¹ aqueous PAA solution (pH 7) (a5), and 1 mg ml⁻¹ aqueous PAA solution (pH 10) (a6); (B) 1 mg ml⁻¹ PAA-MWNTs complex aqueous solution (pH 4) containing different ionic strengths: 0 mM (b1), 20 mM NaCl + 2 mM MgCl₂ (b2), 40 mM NaCl + 2 mM MgCl₂ (b3), and 200 mM NaCl (b4). (Reproduced from A.H. Liu et al., *Biosensors and Bioelectronics*, 22, 694–699, 2006. Copyright © 2006 Elsevier B.V. With permission.)

NMR spectra for the PAA-CNT complex were observed, which indicate the strong binding of PAA to the CNT surface via the hydrophobic interaction [40].

Imidazolium-based room temperature ionic liquids have also been used to disperse CNTs in large amounts [66]. It was reported that the van der Waals interaction between ionic liquids and CNTs other than the previous assumed cation-π interaction contributed most for the dispersion [67]. Therefore the electronic structure of CNTs in the dispersions can be kept intrinsically. These ionic liquids, which possess very high dielectric constants, can effectively shield the strong π-π stacking interaction among CNTs, and thus evidently disperse CNTs [67].

Biomacromolecules have also been studied for CNT dispersion. Single-stranded DNA (ssDNA) molecules have been widely used to solubilize SWNTs by the π-π stacking between aromatic DNA base units and CNT surfaces [68,69]. However, a recent report by Moon et al. showed that DNA molecules coated on SWNTs could be cleaved by nucleases in the serum, suggesting that DNA-functionalized SWNTs might not be stable in biological environments containing nucleases [70]. Karajanagi et al.'s group reported the solubilization of CNTs in water by different proteins [39]. Proteins will bind CNTs by hydrophobic interaction, and the dispersion of CNTs is at the individual level [39].

For potential biological applications, an ideal noncovalent functionalization of CNTs should have the following characteristics. First, the coating molecules should be biocompatible. Second, the coating should be very stable to resist detachment from the CNT surface in biological environments, especially in serum with high salt and protein contents. The amphiphilic coating molecules should have very low CMC values to ensure that the modified CNTs are stable after the removal of excess coating molecules from the CNT suspension. Last but not least, the coating molecules should have functional groups that are available for bioconjugation with other functional molecules, such as drugs, antibodies, etc., to create CNT conjugates with various functions for different applications.

8.3   ELECTROCHEMICAL BIOSENSORS

Electrochemical detection offers several advantages over conventional optical measurements, such as portability, less expensive equipment, higher performance with lower background, and the ability to carry out measurements in turbid samples. During the past few years there have been a large amount of reports about CNT-based electrochemical biosensors for the detection of diverse biological molecules, such as DNA, viruses, antigens, disease markers, and whole cells. An important part of the success of CNTs for these applications is their large specific surface area and ability to promote electron transfer in electrochemical reactions [71–74]. Moreover, CNTs can be readily modified with other functional materials, such as metal/metal oxide nanoparticles, protein, DNA, aptamer, etc., to improve their electrocatalytic activities and selectivities.

8.3.1   Biosensing Based on the Electrocatalytic Activities of CNTs and Metal/Metal Oxide Nanoparticle-Modified CNTs

Well-dispersed CNT-modified electrodes have exhibited substantially improved electrocatalytic activities for the oxidation of H_2O_2 and NADH, which can be used for sensitive H_2O_2 and NADH amperometric biosensing [64,75,76]. Since H_2O_2 and NADH are two very important enzyme mediators, the CNT-based electrodes can also be used for the detection of various other molecules by the incorporation of enzymes onto the electrodes. For example, Wang and coworkers used Nafion (a sulfonated tetrafluoroethylene-based polymer) to incorporate glucose oxidase (GOx) into a MWNT-based composite electrode for the detection of glucose. The GOx-modified electrode will detect the H_2O_2 concentration produced by the oxidation of glucose by GOx [75]. The composite electrode shows substantially greater sensitivity to glucose, in particular at low potentials (0.05 V vs. Ag/AgCl electrode), with negligible interference from dopamine (DA), uric acid (UA), or ascorbic acid (AA), which are biological molecules that commonly interfere with electrochemical detection of glucose. It is also found that CNT-based electrodes can accelerate electron transfer from NADH molecules, decreasing the overpotential and minimizing surface fouling, which are properties that are particularly useful for addressing the limitations of NADH oxidation at ordinary electrodes [64,77,78]. Based on the electrocatalytic activity of CNTs for NADH oxidation and the biocatalytic activity of alcohol dehydrogenase (ADH), the enzyme-modified CNT electrodes can also be used for alcohol

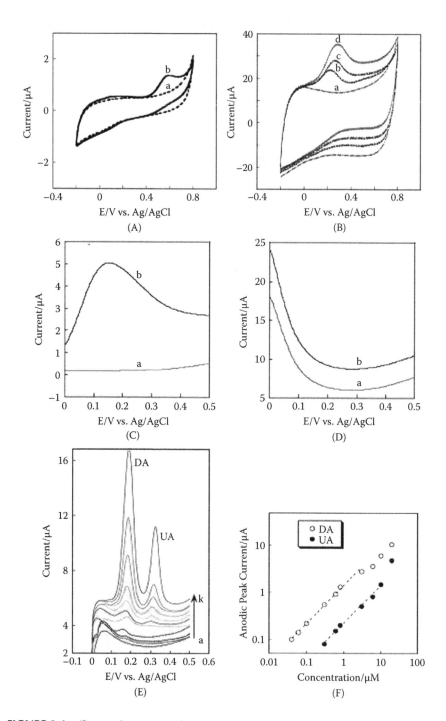

FIGURE 8.4 (See caption on opposite page).

detection [78]. This detection strategy, i.e., electrochemical detection of products produced by enzyme-catalyzed reaction, has been widely used in CNT-based electrochemical biosensors. This strategy, which exploits both the catalytic activities of enzymes and the electrocatalytic activity of CNTs, has also been used for electrochemical detection of environmental pollutants (e.g., organophosphate pesticides) by the combination of cholinesterase or acetylcholinesterase with CNTs [79–81].

We studied the electrocatalytic activity of PAA-MWNT complex-modified glassy carbon electrode (GCE) toward NADH oxidation [64]. The overpotential for the oxidation peak of NADH on a PAA-MWNT-modified electrode is significantly reduced (at 0.315 V vs. Ag/AgCl electrode) with enhanced current density compared with that on bare GCE (Figure 8.4A and B). The PAA-MWNT-modified electrode was also very stable and could be used for the detection of NADH with a linear range of 4–100 μM and detection limit of 1 μM. In another study, the PAA-MWNT-modified GCE was used for simultaneous detection of DA and UA at their physiological levels in the presence of AA [65]. It was found that the oxidation of AA on the PAA-MWNT electrode was completely suppressed (Figure 8.4C and D), which might be due to the electrostatic repulsion interaction between AA and the modified electrode. However, the PAA-MWNT electrode exhibited enhanced electrocatalytic activities toward DA and UA oxidation. The oxidation peaks were at ca. 0.18 V for DA and 0.35 V for UA, indicating that the CNT-modified electrode can be used for simultaneous detection of DA and UA in the presence of excess AA (Figure 8.4E and F). The detection limits were 20 nM for DA and 110 nM for UA. Other CNT-based electrodes have also been studied for the detection of DA, UA, or AA. A detailed comparison of the analytical performances of various CNT-based electrodes for DA detection can be found in a recent review by Jacobs and coworkers [18].

In order to enhance their electrocatalytic activities and sensing performances, the properties of CNTs can also be capitalized on by combining them with other functional materials, such as metal and metal oxide nanoparticles [82–84]. Electroless reduction of adsorbed metal salts on CNTs and direct electrodeposition are usually used to decorate metal nanoparticles on CNTs. Metal nanoparticles such as Au, Ag, Pt, Pd, and Cu have been successfully decorated on CNTs. However, metal

---

FIGURE 8.4   (See figure on opposite page). Cyclic voltammograms of PAA/GCE (A) in the absence of NADH (a) and in the presence of 0.1 mM NADH (b). Cyclic voltammograms of PAAMWNTs/GCE (B) in the presence of 0 mM (a), 0.1 mM (b), 0.2 mM (c), and 0.3 mM NADH (d). Scan rate: 0.1 V s^{-1}. Differential pulse voltammograms of PAA/GCE (C) and PAA-MWNTs/GCE (D) in the presence of 0 μM AA (a) and 300 μM AA (b). Differential pulse voltammometry conditions: equilibration time, 5 s; potential amplitude, 75 mV; step height, 4 mV; frequency, 50 Hz. Linear sweep voltammograms of PAA-MWNTs/GCE (E) in the presence of a mixture of 300 μM AA with varying concentrations of DA and UA: DA (0) + UA (0) (a), DA (40 nM) + UA (40 nM) (b), DA (60 nM) + UA (60 nM) (c), DA (0.1 μM) + UA (0.1 μM) (d), DA (0.3 μM) + UA (0.3 μM) (e), DA (0.6 μM) + UA (0.6 μM) (f), DA (0.8 μM) + UA (0.8 μM) (g), DA (3 μM) + UA (3 μM) (h), DA (6 μM) + UA (6 μM) (i), DA (10 μM) + UA (10 μM) (j), DA (20 μM) + UA (20 μM) (k). The electrolyte is 0.1 M phosphate buffer (pH 7.4). Scan rate: 0.1 V/s. Calibration curves (F) for DA and UA. The anodic peak currents at <0.18 V (DA) and <0.35 V (UA) were collected. (Reproduced from A.H. Liu et al., *Biosensors and Bioelectronics*, 23, 74–80, 2007. Copyright © 2007 Elsevier B.V. With permission.)

nanoparticles prepared by simple deposition on raw or oxidized CNTs are usually not very stable and not uniform in particle size and coverage. To solve these problems, Wu and coworkers used 3,4,9,10-perylene tetracarboxylic acid (PTCA) to modify CNTs first [85], making the CNT surface covered with a large number of carboxyl groups (one PTCA molecule contains four carboxyl groups). These carboxyl groups not only obviously improve the hydrophilicity of CNTs, but also efficiently anchor the precursors of noble metal ions by coordination or electrostatic interaction and disperse noble metal NPs on the CNT surface in the subsequent reduction process (Figure 8.5A). After reduction, the Pt nanoparticles (PtNPs) are more uniformly distributed on the CNT surface than conventional acid-oxidized CNTs (Figure 8.5B) [85]. They also reported the dispersion and modification of CNTs with an ionic liquid polymer, which then serves as the medium to stabilize and anchor metal nanoparticles [86]. The metal nanoparticles (Pt and PtRu alloy) decorated on CNTs showed smaller particle size, better stability and dispersion, as well as enhanced electrocatalytic activities [86]. Similarly, Guo's group [87] and Wang's group [88] also reported ionic-liquid-modified CNTs for metal nanoparticles decoration. In their work, ionic liquid was covalently modified on the CNT surface; nevertheless, these strategies involved acid oxidation pretreatment of CNTs, which might cause some structural damage to the CNTs. Dai and coworkers developed a versatile and effective approach for decorating CNTs with metallic nanoparticles through substrate-enhanced electroless deposition [89,90]; these modified CNTs had enhanced electrochemical activity when incorporated into working electrodes. Ritz and coworkers reported the reversible attachment of Pt alloy nanoparticles (PtCo, PtNi, and PtFe) to nonfunctionalized CNTs by their simple integration in the organometallic synthesis [91]. This procedure involves only a single synthetic step, whereby the crucial parameters for the particle size, shape, and attachment are found in the correct balance of the ligands oleylamine and oleic acid [91]. Wei and coworkers also developed an effective strategy for the fabrication of PtNP-modified CNTs [92]. In their work, CNTs were first modified by a bifunctional linker (Z-glycine N-succinimidyl ester), and then a protein (hemoglobin) was immobilized on CNTs by the linker. $PtCl_6^{2-}$ ions were adsorbed on the protein-CNT surface by electrostatic interaction between protein and $PtCl_6^{2-}$. A CNT-PtNP hybrid with uniform PtNP size, shape, and dispersion was obtained by chemically reducing $PtCl_6^{2-}$ with $NaBH_4$. Also, by drawing on a composite system, Claussen and coworkers fabricated Au-coated Pd (Au/Pd) nanocube-modified SWNTs for electrochemical biosensing [93]. First, SWNTs grew vertically in the pores of porous anodic alumina (PAA) until they protruded from the pore and extended laterally along the PAA surface (Figure 8.6A). Then, by electrodeposition, Pd nanowires were formed within the pore and Pd nanocubes were formed on the SWNTs (Figure 8.6B). Finally, a thin layer of Au was coated on the Pd nanocube surface by electrodeposition (Figure 8.6C). The Au/Pd nanocubes, which were of homogeneous size and shape, were integrated within an electrically contacted network of SWNTs. The Pd provided a low-resistance contact between the SWNTs and Au interfaces, while the Au offered the biocompatibility necessary for further biofunctionalization via thiol linking. After immobilization of GOx on the nanocomposite, the bioelectrode can be used for the detection of glucose with a linear range from 10 μM to 50 mM

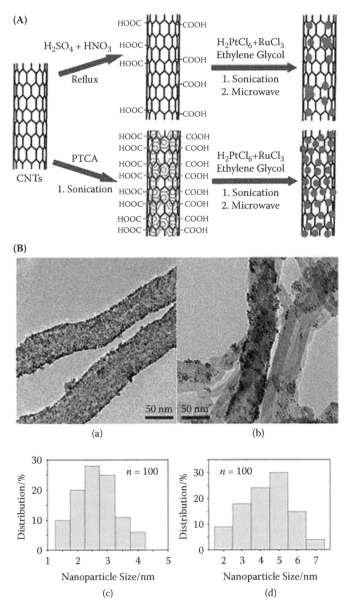

FIGURE 8.5 Schematic diagram for dispersion of PtRu NPs on acid-oxidized CNTs and PTCA-functionalized CNTs (A). TEM images and size distribution of PtRu NPs of PtRu/CNTs-PTCA (a, c) and PtRu/CNTs–acid-oxidized (b, d) nanohybrids (B). (Reproduced from B.H. Wu et al., *Nano Today*, 6, 75–90, 2011. Copyright © 2011 Elsevier B.V. With permission.)

and detection limit of 1.3 μM [93]. Very recently, Sahoo et al. developed a novel method for the fabrication of silver nanoparticle (AgNP)-modified CNTs [94]. In this method, Ag is dissolved from a Ag anode and then electrodeposited on a CNT-based cathode (Figure 8.7A). The density of AgNPs on CNTs can be controlled by

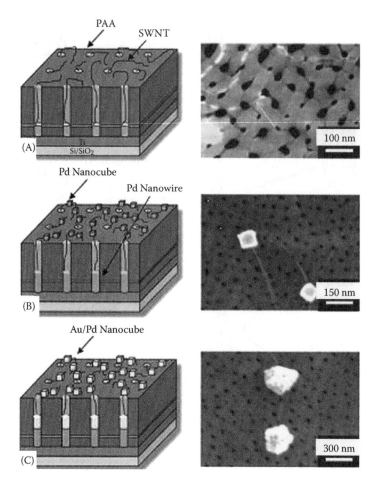

FIGURE 8.6 Tilted cross-sectional schematics with corresponding top view field emission scanning electron microscopy (FESEM) micrographs portraying sequential fabrication process steps: (A) SWNTs grown from the pores of the PAA via microwave plasma enhanced chemical vapor deposition (MPCVD) (FESEM shows a SWNT protruding from a pore and extending along the PAA surface), (B) electrodeposition of Pd to form Pd nanowires in pores and Pd nanocubes on SWNTs (two such nanocubes are shown in corresponding FESEM), and (C) electrodeposition to coat the existing Pd nanocubes with a thin layer of Au. (Reproduced from J.C. Claussen et al., *ACS Nano*, 3, 37–44, 2009. Copyright © 2009 American Chemical Society. With permission.)

changing the deposition time, applied potential, and location of CNTs with respect to the anode. At low potential, single AgNP is attached at the open ends of CNTs (Figure 8.7B), whereas at high potential, intermediate and full coverage of AgNPs is observed (Figure 8.7C and D). As the potential is further increased, fractals of AgNPs along CNTs are formed (Figure 8.7E). The AgNP-CNTs can be used for label-free detection of ssDNA immobilized on it based on the resistance change of the nanohybrid [94].

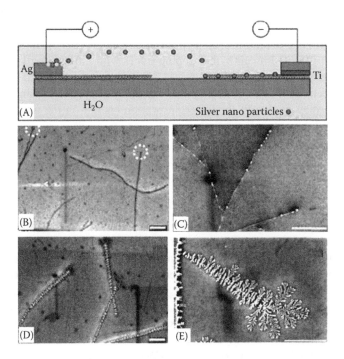

FIGURE 8.7 Attachment of silver nanoparticles on SWNTs. (A) A schematic presentation of silver nanoparticle attachment under an electric field. FESEM micrographs representing the SWNT network decorated with AgNPs at different conditions (B–E). The scale bar represents 1 μm length. (Reproduced S. Sahoo et al., *Journal of the American Chemical Society*, 133, 4005–4009, 2011. Copyright © 2011 American Chemical Society. With permission.)

Metal oxides have also been decorated on CNTs in order to improve the electrocatalytic activity of CNTs. For example, CuO has been modified on CNTs by oxidation of sputtered Cu on CNTs [95]. The CuO-CNT electrode showed much higher electrocatalytic activity and lower overvoltage toward glucose oxidation than the bare CNT electrode; therefore it can be used for sensitive detection of glucose with a linear range up to 1.2 mM [95]. MnO_2 has been electrodeposited on CNTs [96]. The MnO_2-CNT nanocomposite also displayed high electrocatalytic activity toward the oxidation of glucose in alkaline solutions and was highly resistant toward poisoning by chloride ions. This glucose biosensor has a linear dependence on the glucose concentration up to 28 mM with a sensitivity of 33.19 μA mM^{-1} [96]. ZnO has also been modified on the CNT surface by electrodeposition, and the resultant ZnO-CNT-modified electrode can be used for detection of hydroxylamine [97]. This biosensor exhibits a linear response from 0.4 to 1.9 × 10^4 μM with an estimated detection limit of 0.12 μM. Lee and coworkers reported the in situ growth of single-crystalline copper sulfide nanocrystals on MWNTs by the solvothermal method [98]. The morphology of the Cu_2S can be adjusted from spherical particles (~4 nm) to triangular plates (~12 nm) by increasing the concentration of the precursors. This Cu_2S-MWNT hybrid structure responded more sensitively toward the amperometric detection of glucose after the incorporation of GOx. The linear range is from 10 μM to 1.0 mM, with the detection limit of 10 μM [98]. In addition to metal/metal oxide nanoparticle modification, it has also been reported recently that doping CNTs with nitrogen will remarkably improve the electrochemical properties of CNTs. The nitrogen-doped CNTs show enhanced electrocatalytic activity for the oxidation of NO, H_2O_2, and reduction of O_2, which may be due to the lower adsorption energy on the nitrogen-doped CNTs [99,100].

8.3.2 Biosensing Based on the Direct Electron Transfer between Protein and CNT Electrode

Direct electron transfer (DET) between biomacromolecules and an electrode is important for the fabrication of a mediator-free biosensor, a biofuel cell, and the understanding of the intrinsic behavior of enzymes/proteins. It has been reported that CNTs offer more efficient ways of communicating between sensing electrodes and the redox-active sites of biomacromolecules, which are usually embedded deep inside surrounding peptides [23,101]. The high aspect ratio and small diameter of CNTs make them suitable for penetrating to the internal electroactive sites of biomacromolecules to achieve the DET. For example, although the redox center of GOx is buried deep in a protective glycoprotein shell, direct electrochemistry of GOx has been realized on CNTs and nitrogen-doped CNTs, which can be used for mediator-free detection of glucose [23]. Direct electrochemistry of horseradish peroxidase (HRP), microperoxidase 11, cytochrome c, laccase, etc., has also been realized on CNT or metal nanoparticle-modified CNT electrodes [101–107]. These enzyme-CNT electrodes can be used for detection of H_2O_2 and O_2. For the strategies of enzyme/protein immobilization on CNTs, both physical adsorption (hydrophobic interaction) and covalent coupling have been used; however, the effects of these two strategies on the efficiency of the DET of enzymes/proteins still need further study.

8.3.3 CNT-Based Immunosensor, Aptasensor, and DNA Sensor

In addition to small biomolecules, CNTs have also been used for the fabrication of biosensors for macrobiomolecules such as protein, DNA, etc. Yu and coworkers reported the use of CNTs for immunoassay [108]. In their work, CNTs were used both as electrodes that coupled primary antibodies (Ab1) and as "vectors" that hosted secondary antibodies (Ab2) and HRP. Amplified sensing signals resulted from the large surface area of CNTs, which can bind a large number of Ab1 on the electrode and a large number of HRP in the vectors. After the formation of the sandwich structure (Figure 8.8), the CNT-based electrode can be used for the detection of prostate-specific antigen (PSA) by measuring the electrochemical voltage derived from the reaction between the added H_2O_2 and the HRP on the CNTs. This approach could increase the detection sensitivity for PSA some 10–100 times compared with the commercial clinical immunoassays presently available [108]. Another CNT-based immunoassay via formation of the sandwich structure has also been reported [18,109–113]. Usually, CNTs

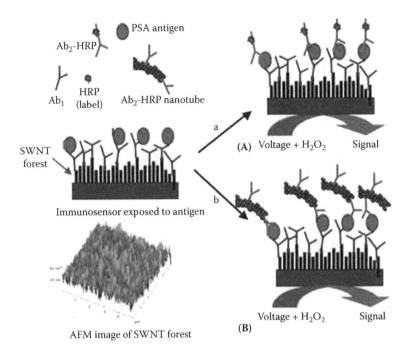

FIGURE 8.8   Illustration of detection principles of SWNT immunosensors. On the bottom left is a tapping mode atomic force microscope image of a SWNT forest that serves as the immunosensor platform. Above this on the left is a cartoon of a SWNT immunosensor that has been equilibrated with an antigen, along with the biomaterials used for fabrication (HRP is the enzyme label). Picture A on the right shows the immunosensor after treating with a conventional HRP-Ab2 providing one label per binding event. Picture B on the right shows the immunosensor after treating with HRPCNT-Ab2 to obtain amplification by providing numerous enzyme labels per binding event. The final detection step involves immersing the immunosensor after secondary antibody attachment into a buffer containing the mediator in an electrochemical cell, applying voltage, and injecting a small amount of hydrogen peroxide. (Reproduced from X. Yu et al., *Journal of the American Chemical Society*, 128, 11199–11205, 2006. Copyright © 2006 American Chemical Society. With permission.)

were used for electrode modification and binding Ab1, and then on the other end of the sandwich, the signals were amplified by nanostructured materials or enzymes. Mahmoud and coworkers have developed biosensors for HIV-1 protease (HIV-1 PR) using CNT-based electrodes [114]. First, a gold electrode was modified with thiolated CNTs and gold nanoparticles. Thiol-modified ferrocene-pepstatin was then bound to the nanoparticles. The pepstatin can bind the protease molecule, which decreases the signal and shifts the oxidation potential for ferrocene by blocking penetration of the supporting electrolyte (Figure 8.9). An estimated detection limit of this electrode is ca. 0.8 pM [114]. Another strategy is to use electrochemical impedance spectroscopy to investigate the same electrode [115]. When protease binds to the ferrocene-pepstatin, the charge transfer resistance of the electrode is changed. These approaches can be used to perform competitive assays for protease inhibitor drugs because if the protease is bound to a drug, it will not bind the electrode. CNTs have also been used as a label

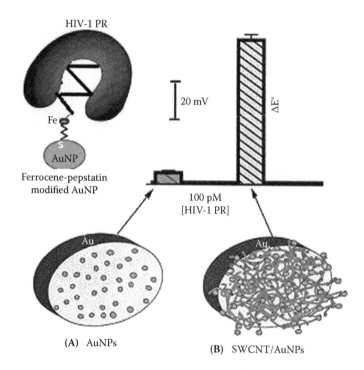

FIGURE 8.9 Schematic representation of the two detection protocols for HIV-1 PR using Fc-pepstatin-modified surfaces: (A) AuNP and (B) SWCNT/AuNP-modified gold electrodes at 100 pM of the enzyme. (Reproduced from K.A. Mahmoud et al., *ACS Nano*, 2, 1051–1057, 2008. Copyright © 2008 American Chemical Society. With permission.)

for signal amplification on the far end of the sandwich structure in the immunoassay. For example, Lai and coworkers fabricated a GOx-functionalized Au nanoparticle/CNT nanocomposite as a label for signal amplification [116]. For the fabrication of the biosensor, first, colloidal Prussian blue, Au nanoparticles, and antibody 1 were coated layer by layer on carbon electrodes. Then, the GOx-functionalized nanocomposites modified with antibody 2 were used for the fabrication of a sandwich-type immunoassay. The signal was obtained by detecting the produced H_2O_2 by GOx-catalyzed reaction. The sensor exhibits detection limits of 1.4 and 2.2 pg ml^{-1} for carcinoembryonic antigen and α-fetoprotein, respectively [116].

Another interesting strategy in an electrochemical CNT-based biosensor is the use of aptamer. Aptamer is an oligonucleotide sequence that has affinity for a variety of specific biomolecular targets, such as drugs, proteins, and other relevant molecules. Aptamer also holds great potential for applications in novel therapies, and is considered a highly suitable receptor for selective detection of a wide range of molecular targets, including bacteria [117,118]. Moreover, aptamer can self-assemble on the CNT surface through π-π stacking between the nucleic acid bases of aptamer and the CNT walls. Quite recently, based on this understanding, Guo and coworkers reported a CNT-based electrochemical aptasensor for thrombin detection [119]. Aptamer for thrombin was initially bound on SWNTs to form a stable aptamer-CNT

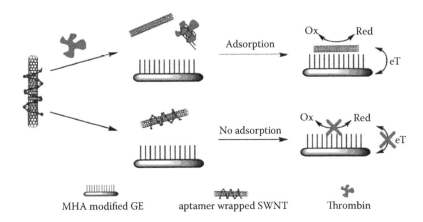

MHA modified GE   aptamer wrapped SWNT   Thrombin

FIGURE 8.10 Electrochemical biosensing strategy for thrombin using aptamer-wrapped SWNT as electrochemical labels. (Reproduced from K. Guo et al., *Electrochemistry Communications*, 13, 707–710, 2011. Copyright © 2011 Elsevier B.V. With permission.)

solution (Figure 8.10). After the addition of thrombin, the aptamer would be removed from CNTs by the strong interaction between thrombin and its aptamer. The bare CNTs were unstable and would be adsorbed on a monolayer of 16-mercaptohexadecanoic acid (MHA)-modified gold electrode. The adsorbed CNTs would mediate the electron transfer between the electrode and electroactive species to give a larger redox current. This strategy exploits the dispersion ability of ssDNA for CNTs, the stronger interaction between thrombin and its aptamer, and the fast electron transfer ability of CNTs. A detection limit of 50 pM thrombin was achieved [119]. Zelada-Guillén and coworkers reported a novel potentiometric biosensor based on aptamer-SWNTs, which allowed selective detection of one single colony-forming unit (CFU) of *Salmonella* Typhi in close to real time [120]. The aptamer was modified with a five-carbon spacer and an amine group at the 3′ end and was covalently immobilized into a layer of previously carboxylated SWNTs by a well-known carbodiimide-mediated wet chemistry approach. The hybrid material aptamer-SWNT acts as both the sensing and the transducing layer of the biosensor. In the absence of the target analyte, the aptamers are self-assembled on CNTs through π-π stacking between the bases and CNT walls. The presence of the target bacteria causes a conformational change of the aptamer, which separates the phosphate groups from the CNT sidewalls, inducing a charge change to the CNTs and the subsequent change of the recorded potential. This study demonstrated the strong potential of CNTs for the fabrication of aptamer-based microbiological diagnostic sensors [120].

Interest in the detection of DNA has grown rapidly because of its importance in drug discovery, diagnosis and treatment of genetic disease, antibioterrorism, etc. The combination of the unique electric, chemical, mechanical, and 3D spatial properties of CNTs with DNA hybridization offers the possibility of fabricating DNA biosensors with simplicity, high sensitivity, and multiplexing. For CNT-based electrochemical DNA biosensors, the large surface area and electrocatalytic properties of CNTs can be used for sensitive detection of intrinsic electroactive residues of target DNA

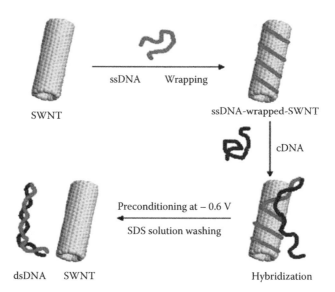

FIGURE 8.11   Schematic diagrams illustrating the interaction between SWNT and DNA. (Reproduced from X.Z. Zhang et al., *Analytical Chemistry*, 81, 6006–6012, 2009. Copyright © 2009 American Chemical Society. With permission.)

(e.g., guanine [121]), enzymatic products from enzyme labels [122], or electroactive labels (e.g., ferrocene [123]). The large surface area and 3D spatial properties of CNTs make them excellent carriers to load a large amount of enzymes or other electroactive species for amplifying the electrochemical signals. Zhang and coworkers reported a reusable DNA sensor based on CNTs [124]. They took advantage of the property of ssDNA to bind SWNTs and the electrocatalytic activity of aligned SWNT toward the oxidation of guanine bases (Figure 8.11). In the absence of complementary DNA (cDNA), the ssDNA-wrapped CNTs would give a sensitive differential pulse voltammetric (DPV) response due to guanine bases' electroxidation. In the presence of cDNA, upon hybridization, the dsDNA can be removed from the CNTs by a preconditioning step (applying a negative potential), resulting in the reduced DPV response. The sensor is label-free and can be regenerated easily by sonication. Moreover, this biosensor can be modified to detect multiple target DNAs with good reproducibility. A linear range of 40–110 nM with a detection limit of 20 nM was obtained [124].

In addition, DNA or protein adsorption onto CNTs will lead to changes in the electronic properties of CNTs, which can be employed for constructing CNT-based field-effect transistors (FETs). These functionalized FETs, coupled with advanced sensor array techniques, are very promising in CNT-based bioassays. Some details of CNT-based FETs can be found in recent review articles [125].

8.4   OPTICAL BIOSENSORS

Compared with electrochemical biosensors, the CNT-based optical biosensors (by exploiting the optical properties of CNTs) were not reported so extensively. However, optical-based systems are very useful for developing entirely nanoscaled biosensors that could operate in confined environments such as a cell's inside. Such systems typically rely on either the use of CNTs on which a classical sandwich-type optical assay is performed [126], the ability of CNTs to quench fluorescence [127], or the near-infrared (NIR) photoluminescence (fluorescence) exhibited by semiconducting CNTs [128,129]. The NIR luminescence of semiconducting SWNTs is particularly interesting for biosensing because NIR radiation is not absorbed by biological tissue, and hence can be used within biological samples or organisms.

8.4.1 BIOSENSING BASED ON FLUORESCENCE QUENCHING AND NIR FLUORESCENCE PROPERTIES OF CNTs

The ability of CNTs to quench fluorescence has been explored by a number of research groups. A couple of notable examples include work by Yang's group [127,130] and Satishkumar's group [131]. Yang et al. used the preference for ssDNA to wrap around SWNTs in comparison with the case of the related duplexes. SWNTs and the sample that may contain the cDNA were added to oligonucleotides labeled with the fluorophore 6-carboxyfluorescein in solution. If no cDNA is present, the fluorescently labeled DNA will wrap around the CNTs and the fluorescence will be quenched. If cDNA is present in the sample, hybridization with the fluorescently labeled probe DNA will give a rigid duplex that does not wrap around CNTs, and hence a fluorescence signal will be observed. However, the strategy used by Satishkumar and coworkers was somewhat different; they employed a dye-ligand conjugate in which the dye was complexed with the SWNTs, thus causing its fluorescence to be quenched [131]. Interaction of the CNT-bound receptor ligand and the analyte caused the displacement of the dye-ligand conjugate from CNTs and the recovery of fluorescence. This approach resulted in nanomolar sensitivity. Recently, Ouyang et al. reported an aptasensor based on the quenching capability of SWNTs [132]. First, aptamer was used to disperse CNTs by binding on the CNT surface. In the absence of lysozyme (the analyte), the SWNTs were wrapped by the ssDNA (aptamer), so that they were well dispersed and remained in the supernatant, providing the quenching substrate for the Eu^{3+} chelates. While in the presence of lysozyme, interaction of the aptamer with lysozyme made it unable to disperse the SWNTs; after centrifugation to separate the SWNTs, the Eu^{3+} complex in solution emitted a strong luminescence [132]. This approach has a limit of detection as low as 0.9 nM, which is about 60-fold lower than those of commonly used fluorescent aptasensors.

Heller and coworkers studied NIR fluorescence of semiconducting SWNTs wrapped by DNA for biosensing [133]. The transition of the DNA secondary structure from an analogous B-to-Z conformation results in a change of the dielectric environment of the SWNTs with a concomitant shift in the wavelength of the SWNT fluorescence. In this study, the shift in optical properties of CNTs by the change in DNA structure was used to detect metal ions that could induce such changes in DNA structure. It is known that divalent metal ions such as Hg^{2+}, Co^{2+}, Ca^{2+}, and Mg^{2+} can cause DNA structure transition from B to Z. Thus the DNA-wrapped CNT-based biosensors were able to detect all these metal ions with the sensitivity

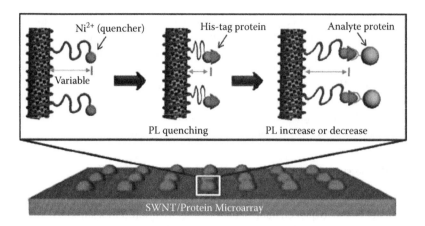

FIGURE 8.12 Signal transduction mechanism for label-free detection of protein-protein interactions: A NIR fluorescence change from the SWNT occurs when the distance between the Ni^{2+} quencher and SWNT is altered upon analyte protein binding. (Reproduced from J.-H. Ahn et al., *Nano Letters*, 2011. DOI 10.1021/nl201033d. Copyright © 2011 American Chemical Society. With permission.)

decreasing in the order $Hg^{2+} > Co^{2+} > Ca^{2+} > Mg^{2+}$ [133]. Recently, based on a NIR fluorescent SWNT/protein microarray, they also developed a label-free approach for selective protein recognition [134]. First, a microarray of Ni^{2+}-chelated chitosan-modified SWNTs was fabricated (Figure 8.12). Then, hexahistidine-tagged capture proteins were directly expressed on the microarray by cell-free synthesis. The Ni^{2+} ion acted as a proximity quencher with the Ni^{2+}/SWNT distance altered upon docking of analyte proteins. This approach can discern single protein binding events with the detection limit down to 10 pM for an observation time of 600 s [134]. Gao and coworkers used the changes in the structure of ssDNA wrapped around SWNTs for the detection of Hg^{2+} ions by circular dichroism. It was observed that the Hg^{2+} ions interacted with the bases of ssDNA causing the interaction between DNA and SWNTs to weaken, with a resultant decrease in the circular dichroism signal induced by the association of the CNTs with the DNA [135].

The detection of DNA hybridization on the surface of solution-suspended SWNTs through a SWNT band gap fluorescence modulation has also been reported by Strano and coworkers. The detection limit was as low as 6 nM for 24-mer DNA [136,137]. Utilizing a similar noncovalent modification strategy, they also demonstrated signal transduction via fluorescence quenching for measuring glucose concentrations at physiologically relevant conditions, from the micromolar to millimolar range [138]. In another work, they studied the fluorescence quenching of collagen-modified SWNTs by three different reagents (H_2O_2, H^+, and $Fe(CN)_6^{3-}$). It was observed that H_2O_2 has the highest quenching equilibrium constant of 1.59 at 20 µM, whereas H^+ is so insensitive that a similar equilibrium constant is obtained with a concentration of 0.1 M [139]. Satishkumar et al. demonstrated that small-molecule quenchers may be removed from SWNT surfaces by avidin and albumin in a specific and nonspecific manner, respectively, with detection limits in the micromolar range [131].

Heller and coworkers also reported optical detection of small molecules by monitoring their interaction with ssDNA-wrapped SWNTs through the fluorescence of SWNTs [37]. This study is very exciting because it is an extension of the concept to multimodal optical sensing. In this work, six genotoxic analytes, including chemotherapeutic alkylating agents and reactive oxygen species (such as H_2O_2, singlet oxygen, and hydroxyl radicals), could be simultaneously detected [37]. The detection of different analytes on the same ssDNA-wrapped SWNTs is based on the differing optical responses of (6, 5) and (7, 5) SWNTs to different analytes. Due to the different effects of analytes on the optical signature of CNTs, simultaneous detection of multiple analytes was achieved. Some sequence specificity was also reported that sequences with more guanine bases are more susceptible to singlet oxygen, while metal ion responses are greater for DNA sequences with stronger metal binding. The final aspect of this study illustrated the ability of the DNA-SWNTs to detect drugs and reactive oxygen species inside living cells without being genotoxic [37]. This optical detection system looks very promising because it has nanoscale dimension and can detect multiple analytes sensitively in a biological environment. However, several problems remain challenging. While only semiconducting SWNTs exhibit band gap photoluminescence, one has to separate and isolate these CNTs from other nonfluorescing CNT isomers. Moreover, the quantum yield of those CNTs depended on their chemical environment, and processing is required to avoid quenching and maximize quantum yield [140]. Signal transduction via band gap modulation and quenching suffers from the limits of spectral resolution as well as photoluminescence intensity. These limitations restrict their use for analytes at relatively high concentrations.

8.4.2 Biosensing Based on the Raman Scattering of SWNTs

Chen and coworkers have used the intense Raman scattering cross section of SWNTs for immunoassay [141]. Compared with the traditionally used fluorophores, the Raman tags have some advantages. For example, the Raman scattering spectra of SWNTs are simple with strong and well-defined Lorentzian peaks of interest. The Raman scattering spectra of SWNTs are easily distinguishable from noise, and no auto-scattering is observed for conventional assay surfaces or reagents [141]. The photobleaching of SWNT Raman tags is not observed even under extraordinarily high laser powers.

Surface-enhanced Raman scattering (SERS) [142] is a technique that may be applied to sensitive detection of Raman active molecules when they are near a metal nanostructure surface (usually gold, silver, or copper nanostructures) with appropriately tuned surface plasmons [143,144]. The combination of the intense resonance enhancement of SWNT Raman tags with SERS presents the opportunity to extend the detection limit of traditional fluorescence assays from ~1 pM to the femtomolar level or lower [145]. In one of the detection strategies, a nanostructured gold-coated assay substrate was first fabricated, and then the substrate surface was bound with analyte and a SWNT Raman tag. This strategy yielded a nearly 100-fold increase in SWNT Raman scattering intensity (Figure 8.13) [141], and this can be developed for sensitive and selective detection of a model analyte with a detection limit of 1 fM (ca. three orders of magnitude lower than common fluorescence methods). Moreover,

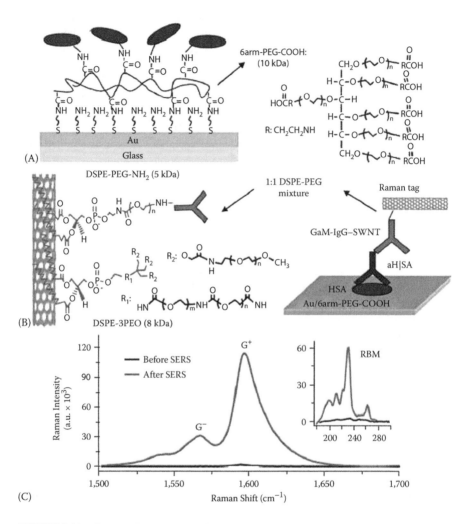

FIGURE 8.13 CNTs as Raman labels for protein microarray detection. (A) Surface chemistry used to immobilize proteins on gold-coated glass slides for Raman detection of analytes by SWNT Raman tags. A self-assembled monolayer of cysteamine on gold was covalently linked to six-arm, branched poly(ethylene glycol)-carboxylate (6arm-PEG-COOH, right) to minimize nonspecific protein binding. Terminal carboxylate groups immobilize proteins. (B) Sandwich assay scheme. Immobilized proteins in a surface spot were used to capture an analyte (antibody) from a serum sample. Detection of the analyte by Raman scattering measurement was carried out after incubation of SWNTs conjugated to goat anti-mouse antibody (GaM-IgG–SWNTs), specific to the captured analyte. SWNTs were functionalized by (DSPE-3PEO) and (DSPE-PEG5000-NH₂) (left). (C) Raman spectra of the SWNT G-mode and radial breathing mode (RBM, inset) regions before and after SERS enhancement. (Reproduced from Z. Chen et al., *Nature Biotechnology*, 26, 1285–1292, 2008. Copyright © 2008 Nature Publishing Group. With permission.)

by taking advantage of highly multiplexable microarray technology and isotopically labeled SWNTs (composed of pure ^{12}C and ^{13}C, respectively), multicolor simultaneous detection of multiple analytes using one single excitation source has also been demonstrated (Figure 8.14) [146]. In addition to immunoassay, the SWNT Raman tags can also be used in other binding events, such as biotin-streptavidin binding, protein A/G-IgG interaction, and DNA hybridization for various biomolecule detections [141].

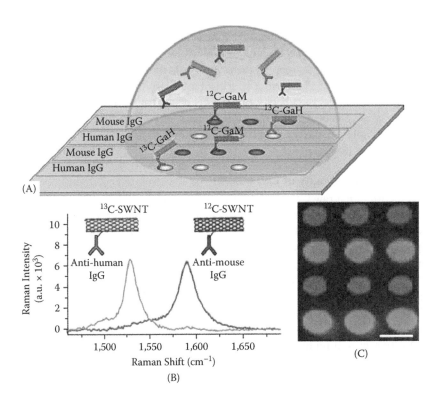

FIGURE 8.14 Multicolor SWNT Raman labels for multiplexed protein detection. (A) Two-layer, direct, microarray format protein detection with distinct Raman labels based upon pure ^{12}C and ^{13}C SWNT tags. ^{12}C and ^{13}C SWNTs were conjugated to GaM and GaH-IgGs, respectively, providing specific binding to complementary IgGs of mouse or human origin, even during mixed incubation with analyte (as shown). (B) G-mode Raman scattering spectra of ^{12}C and ^{13}C SWNT Raman tags are easily resolvable, have nearly identical scattering intensities, and are excited simultaneously with a 785 nm laser. This allows rapid, multiplexed protein detection. (C) Raman scattering map of integrated ^{12}C and ^{13}C SWNT G-mode scattering above baseline, demonstrating easily resolved, multiplexed IgG detection based upon multicolor SWNT Raman labels. Scale bar, 500 mm. (Reproduced from H. Dumortier et al., *Nano Letters*, 6, 1522–1528, 2006. Copyright © 2006 American Chemical Society. With permission.)

8.5 NANOMEDICINE CNTS USED IN DRUG DELIVERY AND TUMOR THERAPY

8.5.1 TOXICITY OF CNTS

Safety is a prerequisite for any material used in medicine. A large number of studies have been carried out in the past several years to explore the potential toxic effect of CNTs. The results indicated that the toxic effect of CNTs depended strongly on the type of CNTs and their functionalization approaches. Both in vitro and in vivo studies conducted by various groups showed no obvious toxicity of properly functionalized CNTs [42,59,146]. On the other hand, raw CNTs were demonstrated to be toxic to mice after inhalation into the lung [147–150]. It has also been shown that unfunctionalized long MWNTs may pose a carcinogenic risk in mice [151]. As a result of the wide variety of reports, the use of CNTs for biomedical applications has aroused much concern. Therefore it is critical to clarify the toxicity issue of CNTs. Currently it seems that the toxicity of CNTs is dependent on the material preparation, especially geometry and surface functionalization. Well-functionalized CNTs with biocompatible surface coatings have been shown to be nontoxic both in vitro to cells and in vivo in mice.

8.5.1.1 In Vitro Toxicity of CNTs

It was reported that raw CNTs inhibit the proliferation of HEK 293 cell [152] and can induce cell cycle arrest and increase apoptosis/necrosis of human skin fibroblasts [153]. For example, Bottini et al. reported that oxidized MWNTs can induce apoptosis of T lymphocytes [154]. It is known that simple oxidation is not enough to make CNTs soluble and stable in saline and cell media, and thus does not represent a biocompatible functionalization. Sayes et al. further reported that the toxicity of CNTs was dependent on the density of functionalization with minimal toxicity for those heavily functionalized with the highest density of phenyl-SO_3X groups [155]. These results are understandable because CNTs without proper functionalization have a highly hydrophobic surface, and thus may aggregate in the cell culture and interact with cells by binding to various biological species through hydrophobic interactions. In addition, other factors, such as surfactant and metal catalyst, remaining during the preparation process may also contribute to the observed toxicity of CNTs in vitro [156,157]. On the other hand, many groups have reported that well-functionalized CNTs that are stable in serum show no observed toxicity in in vitro cellular uptake experiments [22,41,158,159]. Dai and coworkers observed that cells exposed to PEGylated SWNTs showed neither enhanced apoptosis/neurosis nor reduced proliferation of various cell lines [41,158]. Prato and coworkers also reported that covalently functionalized CNTs by 1,3-dipolar cycloaddition are safe for the tested cell lines [146,160]. CNTs coated by DNA, amphiphilic helical peptides, serum proteins, etc., have also been proved to be safe to cells [159,161]. It is concluded that raw CNTs and not well-functionalized CNTs show toxicity to cells, while properly functionalized CNTs appear to be safe even at relatively high dosages.

8.5.1.2 In Vivo Toxicity of CNTs

The toxicity of CNTs has also been investigated in animals. When raw CNTs were intratracheally instilled into animals, they exhibited obvious pulmonary toxicity, such as unusual inflammation and fibrotic reactions due to the aggregation of raw CNTs in the lung airways, or the modification of systemic immunity by modulating dendritic cell function [147–150,162]. However, it has also been reported that 30 days after their intratracheal administration to mice, nanoscale-dispersed SWNTs by biocompatible copolymer exhibit minimal toxicity and are suitable for biomedical applications [163]. Toxicities observed by intratracheal instillation of large amounts of raw CNTs may have little relevance to the toxicology profile of functionalized soluble CNTs for biomedical applications, especially when they are administered through other routes, such as intraperitoneal and intravenous injections, by which lung airways are not exposed to CNTs. In a pilot study, Poland et al. noticed asbestos-like pathogenic behaviors such as mesothelioma associated with exposing the meso-thelial lining of the body cavity of mice to large MWNTs (length 10–50 μm, diameter 80–160 nm) following intraperitoneal injection [151]. Despite the importance of this finding for potential negative effect of CNTs to human health, it should be noted that the MWNTs used in this study were simply sonicated in 0.5% bovine serum albumin (BSA) solutions without careful surface functionalization. It is also observed that the toxicity of CNTs is length dependent. Shorter and smaller CNTs with length of 1–20 μm and diameter of 10–14 nm exhibit no obvious toxic effect, indicating that the tox-icologies of CNTs are related to their sizes [151]. Quite recently, Wang and cowork-ers developed a chronic exposure model in which human lung epithelial BEAS-2B cells were continuously exposed to low doses of SWNTs in culture over a prolonged time period [164]. After such chronic exposure, the cells were evaluated for malig-nant transformation in vitro and tumorigenicity in vivo using a xenograft mouse model. Their result indicates that the chronic exposure causes malignant transforma-tion of human lung epithelial cells, and the transformed cells induce tumorigenesis in mice and exhibit an apoptosis-resistant phenotype characteristic of cancer cells [164]. These results strengthen the safety concern for CNT exposure and support the prudent adoption of prevention strategies and implementation of exposure control.

On the other hand, Schipper's group and Liu's group have demonstrated that well-functionalized biocompatible CNTs should be safe for in vivo biological applica-tions. When mice were intravenously injected with covalently and noncovalently PEGylated SWNTs (~3 mg kg^{-1}), it was found that the blood chemistries and his-tological observations were normal after 4 months [42,165]. Yang et al. reported in a 3-month toxicity study that Tween-80-modified SWNTs exhibited low toxici-ties to mice at a very high dose (~40 mg kg^{-1}) following IV administration. Such toxicity may be due to the oxidative stress induced by SWNTs accumulated in the liver and lungs [166]. The toxicity observed was dose dependent and appeared to be less obvious at lower doses (2 and 16 mg kg^{-1}). They also reported that cova-lently PEGylated SWNTs, which have good aqueous stability and biocompatibility, exhibited an ultralong blood circulation half-life in mice, and no acute toxicity has been observed even at a high dose (24 mg kg^{-1}) [167]. Silva and coworkers observed that RNA-wrapped oxidized double-walled CNTs can be taken up in vitro and then

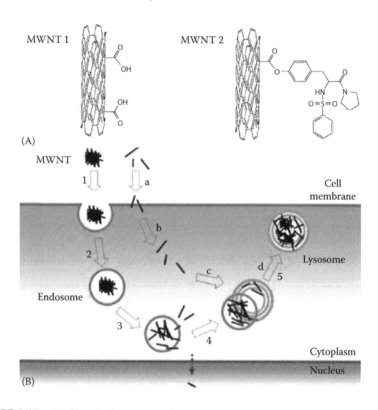

FIGURE 8.15 (A) Chemical structure of two modified MWNTs. (B) A working model for cell uptake of MWNTs. Numbers 1–5 and letters a–d indicate different steps in two possible cellular translocation pathways of MWCNTs. The bundled MWNTs bind to cell membranes (1) and are subsequently internalized into cells (2) inside endosomes. In the endosomes, bundles release single MWNTs that penetrate endosomal membranes and enter cytoplasm (3). Both residual bundled MWNTs in endosomes and free MWNTs in the cytoplasm are recruited into lysosomes for excretion (4, 5). Single MWNTs enter cells through direct membrane penetration (a) to enter cytoplasm (b). They are recruited into lysosomes for excretion (c, d). Short MWNTs are also able to enter the nucleus. (Reproduced from N. Gao et al., *ACS Nano*, 5, 4581–4591, 2011. Copyright © 2011 American Chemical Society. With permission.)

released by cells over a 24 h time period with no toxicity or activation of stress responses. The cellular handling of CNTs is certainly dependent on the functionalization method, RNA wrapping of CNTs, and incubation conditions [168]. Gao and coworkers have reported that proper surface modification of MWNTs can alleviate nuclear factor kappa-light-chain-enhancer of activated B cells (NF-κB) activation and reduce the immunotoxicity of MWNTs [169]. In their work, two kinds of surface-modified MWNTs (MWNT 1 and MWNT 2; see Figure 8.15A) have been studied in both mice and macrophages. It was observed that MWNT 2 caused less immune perturbations than MWNT 1 because MWNT 2 preferred to bind scavenger receptor and alleviate NF-κB activation and immunotoxicity [169]. Heister and coworkers examined various parameters (such as pH and salt concentration of the medium, and length, aspect ratio, surface charge, and functionalization of CNTs) on the dispersion stability and toxicity of CNTs both in vitro and in vivo [170]. They found that factors such as a short aspect ratio, presence of oxidation debris and serum proteins, low salt concentration, and an appropriate pH can improve the dispersion stability of CNTs [170]. Moreover, covalent surface functionalization with amine-terminated poly(ethylene glycol) was demonstrated to stabilize CNT dispersions in various media and to reduce deleterious effects on cultured cells.

Although much work has been performed, to fully address the toxicity concern of CNTs, further work, such as the testing of different animals, different routes of administration, and better modification of CNTs, is still needed.

8.5.2 CNTs for Drug Delivery and Tumor Therapy

8.5.2.1 CNT-Based Drug Delivery

Ever since the finding by Kam's group [20] and Pantarotto's group [171] that functionalized CNTs are able to enter cells by themselves without obvious toxicity, CNTs have been studied for drug delivery. The large surface area and facile surface functionalization enable CNTs to carry various cargoes, including small drug molecules and biomacromolecules such as protein, DNA, RNA, etc. Therefore the behavior of CNTs in living cells, including cell entrance, subcellular locations, and excretion, is crucial for their application. However, the mechanism of cell uptake and the cellular fate of CNTs is not fully understood, and current descriptions are still controversial, especially on cell uptake of CNTs, their intracellular translocation, and subcellular localization. Despite the popular viewpoint that CNTs are taken up by cells through clathrin-dependent endocytosis [20,161,172], energy-independent cell uptake is also reported [21,173]. The different mechanism of cell uptake could be due to the different functionalization and size of the CNTs. Another controversy is on subcellular locations of CNTs. Some described that CNTs go into cells without entering the nucleus [174,175], while others reported that SWNTs entered the cell nucleus [171,176,177] and this entrance might be reversible [177]. In a recent work [178], based on ultra-structural observation of cell uptake of CNTs into human embryonic kidney epithelial cells, Gao and coworkers proposed a model for CNTs' cellular circulation path (Figure 8.15B). Single MWNTs (both positively and negatively charged) enter cells through direct penetration, while MWNT bundles enter cells through endocytosis. MWNT bundles in endosomes may release single nanotubes that then penetrate the endosome membrane and escape into cytoplasm. Short MWNTs can also enter the cell nucleus. All kinds of MWNTs are recruited into the lysosome for excretion.

Small drug molecules are usually covalently conjugated to CNTs for delivery. In a work reported by Bianco and coworkers, fluorescent dyes and drug molecules were simultaneously linked to 1,3-dipolar cycloaddition-functionalized CNTs via amide bonds for the delivery of anticancer drugs into cells [160,179]. The multifunctionalized CNTs preserve the activity of the binded drug molecules and present no toxic effect. Liu and coworkers reported that paclitaxel (PTX, a chemotherapy drug), a commonly used anticancer drug, was conjugated to PEG-modified SWNTs via a cleavable ester bond for drug delivery in vivo [180]. The SWNT-PTX conjugate affords higher efficacy in suppressing tumor growth than clinical Taxol® in a murine 4T1 breast cancer model, which is due to the prolonged blood circulation and 10-fold higher tumor PTX uptake by SWNT delivery likely through enhanced permeability and retention [180]. Wu and coworkers covalently combined antitumor agent 10-hydroxycamptothecin (HCPT) with MWNTs using hydrophilic diaminotriethylene glycol as the linker between CNTs and the drug molecules [181]. This MWNT-HCPT conjugate exhibits superior antitumor activities both in vitro and in vivo compared with clinical HCPT formulation, owing to the enhanced cellular uptake, prolonged blood circulation, and HCPT concentrating action (multivalent presentation of HCPT molecules on a single nanotube) of the conjugates [181]. Feazell et al. reported that PEGylated SWNTs were used to deliver a prodrug of the cytotoxic platinum(II), a platinum(IV) complex, into cancer cells [182]. The platinum(IV) prodrug compounds were activated after being reduced to the active platinum(II) form. Through peptide linkages, the platinum(IV) complexes were loaded on SWNTs. The prodrug-loaded SWNTs were then taken into cancer cells by endocytosis and resided in cell endosomes where the reduced pH would induce reductive release of the platinum(II) for killing the cancer cells. Doxorubicin, a commonly used cancer chemotherapy drug, has also been loaded on PEGylated SWNTs via π-π stacking [41]. Because of the high surface area of SWNTs, the drug loading is very high (4 g of drug per 1 g of CNTs). Moreover, the loading is pH dependent and favorable for drug release in tumor microenvironments with acidic pH [41]. Quite recently, Su and coworkers reported a very interesting work about small-molecule drug delivery using CNTs as carrier [183]. In their work, drug molecules (indole) that are encapsulated in the hollow chambers of CNTs are well protected during the transportation. After labeling the drug-loaded CNTs with EphB4 binding peptides, the labeled CNTs can selectively target EphB4 expressing cell and release indole onto cell surfaces by near-infrared irradiation. The irradiation can increase the molecule diffusion and release the encapsulated drug molecules [183].

For targeted delivery of drugs, CNTs need to be modified with both targeting ligands and drug molecules [41,184]. Folic acid [184], peptides [41,185,186], and antibodies [187,188] have been used to target CNTs to specific types of cells or tumors. Dai and coworkers reported that after the conjugation of folic acid-linked Pt(IV) prodrug molecules to PEGylated SWNTs [184], the modified CNTs exhibited enhanced toxicity to folate receptor positive cells but not to folate receptor negative cells due to the targeting effect of folic acid. In other research, Liu and coworkers reported the conjugation of targeting molecules (Arg-Gly-Asp (RGD) peptide that will be upregulated on various solid tumor cells and tumor vasculatures) to the coating molecules (PL PEG-amine) on CNTs, and the aromatic drugs (doxorubicin) were loaded on CNTs by π-π stacking [41].

In addition to small drug molecules, the delivery of biomacromolecules, including proteins, DNA, and RNA using CNTs, has also been reported. Proteins can be conjugated or noncovalently absorbed on CNTs for delivery [20,189]. After being translocated into cells by CNTs, proteins become bioactive once they are released [189]. DNA can be loaded on CNTs by both electrostatic attraction and covalent binding. For example, positively charged CNTs have been used to bind DNA for gene transfection [190,191]. Amine-terminated CNTs obtained by 1,3-dipolar cycloaddition have also been used for covalent binding of DNA. Liu and coworkers also reported the small interfering RNA delivery by CNTs. RNA modified on CNTs by a cleavable disulfide bond was successfully delivered into cells. They demonstrated that the RNA-CNTs were applicable to those hard-to-transfect human T cells and primary cells, which were resistant to delivery by conventional cationic liposome-based transfection agents [158]. They also observed that the cell uptake of CNTs was dependent on CNT surface functionalization. CNTs with a more hydrophobic surface have a stronger interaction with hydrophobic cell membrane domains, and thus have a higher cellular uptake, which is favorable for the delivery [158]. These results indicate that for functionalized CNTs for drug delivery, one should consider the aqueous solubility, biocompatibility, and also the ability of CNTs to be uptaken by cells.

8.5.2.2 In Vivo Tumor Therapy

For CNT-based tumor therapy, targeting CNTs to tumors is a very important step. Both passive targeting, which relies on the enhanced permeability and retention effect of cancerous tumors, and active targeting, which is based on tumor-targeting ligands, have been studied for different drug deliveries. Zhang and coworkers first reported CNT-based tumor therapy by using $-CONH-(CH_2)_6-NH_3^+Cl^-$-functionalized SWNTs to deliver therapeutic siRNA into cancer cells [192]. In this work, the CNT-siRNA was directly injected into tumors instead of by systemic administration [192]. Liu and coworkers reported a passive targeting delivery of PTX by CNTs. PTX was loaded on PEG-functionalized CNTs by a cleavable ester bond [180]. This CNT-based conjugate exhibited improved treatment efficacy over the clinical Cremophor-based PTX formulation, Taxol in a 4T1 murine breast cancer model in mice. The enhanced treatment efficacy was due to the longer blood circulation half-life and higher tumor uptake of CNTs-PTX than those of simple PEGylated PTX and Taxol [180]. It was also shown that efficient tumor targeting was achieved by conjugating a RGD peptide to PEGylated SWNTs [185]. The PEGylated SWNTs conjugated with both RGD peptide and radiolabels (64Cu-DOTA) were intravenously injected into glioblastoma U87MG tumor-bearing mice and were monitored by micro-positron emission tomography (micro-PET) over time. RGD-conjugated SWNTs exhibited a higher tumor uptake (~13% of injected dose per gram tissue (%ID/g)) than plain SWNTs without RGD (4%–5% ID/g). It was also observed that efficient tumor targeting could only be realized when SWNTs were coated with long PEG (SWNT-PEG5400-RGD) but not with short PEG (SWNT-PEG2000-RGD). The latter had a short blood circulation time, which lowered the possibility of being trapped in tumors or binding the tumor receptors. This indicates that surface functionalization of SWNTs is very significant for tumor targeting in vivo [185]. McDevitt and coworkers reported active tumor-targeting CNTs by covalently attaching multiple copies of tumor-specific monoclonal antibodies, radio-metal-ion chelates, and fluorescent probes [188]. The specific reactivity of the CNT-based nanocomposite was evaluated both in vitro by cell-based immunoreactivity assays and in vivo via biodistribution in a murine xenograft model of lymphoma. The nanocomposites were found to be specifically reactive with the human cancer cells they were designed to target [188]. Chakravarty and coworkers reported a CNT-based cancer therapy by using cancer antibody (targeting element)-functionalized SWNTs for thermal ablation of tumor cells [193].

Before the antibody functionalization, biotinylated polar lipids were used to prepare stable and biocompatible CNT dispersion. This novel approach exploits the heating of SWNTs when absorbing energy from NIR light (tissue is relatively transparent to NIR). The approach is also promising for precisely selective treatment because the antibody-functionalized SWNTs only target the specific cancer cells, and only targeted cells are killed after exposure to NIR irradiation [193]. The research is still preliminary; nevertheless, these results demonstrate that well-functionalized CNTs are very promising for application in drug delivery and tumor therapy.

8.6 CONCLUSIONS

Although not discussed in this review, it is worth mentioning that the fabrication of CNTs plays an important role in their future biosensing and biomedical applications. So far, in most cases, CNTs have been used in a heterogeneous mixture of nanotubes with different lengths, diameters, and chiralities. The heterogeneity of CNTs would affect their electrochemical, optical, and electronic properties. More importantly, the impurity would greatly inhibit their biomedical application. It has been demonstrated that the toxicity, in vitro cellular uptake, as well as in vivo pharmacokinetics of CNTs vary with the CNT size. Thus it is important to obtain and test CNTs with narrow size distributions.

For the application of CNTs in electrochemical biosensors, the origin of the enhancement still needs to be clarified, though CNTs show enhanced electrocatalytic activities in many studies. In theory, CNTs are pure carbon; in reality, they almost always contain some impurities, such as metallic compounds or nanoparticles derived from the catalysts used in CNT growth. After careful removal of these metallic nanoparticle impurities, however, it is suggested that the electrochemical properties of CNTs might be no better than edge planes of highly ordered pyrolytic graphite (HOPG) [194,195]. It has also been reported that the electrochemical activity of CNTs is dependent on their oxidation state and surface modification. The oxidation treatment would open the ends of CNTs and introduce defects in the sidewalls, which would affect the electrochemical activity of CNTs. For electrochemical biosensing applications of CNTs, another challenge is how to incorporate CNTs into bulk electrodes for the best effect. The most commonly used method is randomly distributing CNTs on the electrode surface. The prevalence of this approach is mainly because it is easy to operate, not necessarily because it offers the best performance. However, the use of aligned CNTs by growing aligned CNTs directly from a surface is an especially interesting development [196–200]. This kind of electrode exhibits faster heterogeneous electron transfer than randomly distributed arrays due to the tips of CNTs facilitating more rapid electron transfer than sidewalls [201]. Another strategy is to use a single CNT as a nanoelectrode for in-body biosensing. This is probably the most attractive design of CNT-based electrodes. However, the fabrication of this kind of electrode is still in its early stage [202,203].

Although numerous encouraging results using CNTs in biomedical applications have been published in the past several years, much more work is still needed before CNTs can be utilized in the clinic. The most important issue to be addressed is still the concern of their long-term toxicity. Though it has been shown that well-functionalized CNTs are not toxic, further systematic investigations are still required. Therefore it is still urgent to find better ways for functionalization of CNTs to minimize their toxic effects. For CNT-based drug delivery, although some progress has been made, in vivo targeted delivery remains a challenge. CNTs should first be modified to be completely biocompatible, and then conjugated with targeting ligands, allowing enhanced cellular uptake via receptor-mediated endocytosis without loss of optimal SWNT in vivo characteristics. Further development of suitable bioconjugation chemistry on CNTs may create versatile SWNT-based bioconjugates for actively targeted in vivo drug and gene delivery.

ACKNOWLEDGMENTS

This work was financially supported by National Natural Science Foundation of China (Nos. 31200598 and 21275152), the Hundred-Talent-Project (no. KSCX2-YW-BR-7), and the Knowledge Innovation Project in Biotechnology (no. KSCX2-EW-J-10-6), Chinese Academy of Sciences.

REFERENCES

1. S. Iijima. Helical microtubules of graphitic carbon. *Nature*, 354, 56–58, 1991.
2. H.J. Dai. Carbon nanotubes: Synthesis, integration, and properties. *Accounts of Chemical Research*, 35, 1035–1044, 2002.
3. J.M. Schnorr and T.M. Swager. Emerging applications of carbon nanotubes. *Chemistry of Materials*, 23, 646–657, 2011.
4. M. Ouyang, J.-L. Huang, and C.M. Lieber. Fundamental electronic properties and applications of single-walled carbon nanotubes. *Accounts of Chemical Research*, 35, 1018–1025, 2002.
5. S. Niyogi, M.A. Hamon, H. Hu, B. Zhao, P. Bhowmik, R. Sen, M.E. Itkis, and R.C. Haddon. Chemistry of single-walled carbon nanotubes. *Accounts of Chemical Research*, 35, 1105–1113, 2002.
6. P.M. Ajayan. Nanotubes from carbon. *Chemical Reviews*, 99, 1787–1799, 1999.
7. D. Tasis, N. Tagmatarchis, A. Bianco, and M. Prato. Chemistry of carbon nanotubes. *Chemical Reviews*, 106, 1105–1136, 2006.
8. H. Ago, K. Petritsch, M.S.P. Shaffer, A.H. Windle, and R.H. Friend. Composites of carbon nanotubes and conjugated polymers for photovoltaic devices. *Advanced Materials*, 11, 1281–1285, 1999.
9. A. Javey, J. Guo, Q. Wang, M. Lundstrom, and H.J. Dai. Ballistic carbon nanotube field-effect transistors. *Nature*, 424, 654–657, 2003.
10. Q. Cao and J.A. Rogers. Random networks and aligned arrays of single-walled carbon nanotubes for electronic device applications. *Nano Research*, 1, 259–272, 2008.
11. S.S. Fan, M.G. Chapline, N.R. Franklin, T.W. Tombler, A.M. Cassell, and H.J. Dai. Self-oriented regular arrays of carbon nanotubes and their field emission properties. *Science*, 283, 512–514, 1999.
12. A.C. Dillon, K.M. Jones, T.A. Bekkedahl, C.H. Kiang, D.S. Bethune, and M.J. Heben. Storage of hydrogen in single-walled carbon nanotubes. *Nature*, 386, 377–379, 1997.
13. M. Trojanowicz. Analytical applications of carbon nanotubes: A review. *TrAC Trends in Analytical Chemistry*, 25, 480–489, 2006.
14. S.H. Yeom, B.H. Kang, K.J. Kim, and S.W. Kang. Nanostructures in biosensor—A review. *Frontiers in Bioscience—Landmark*, 16, 997–1023, 2011.

15. Z. Liu, S. Tabakman, K. Welsher, and H.J. Dai. Carbon nanotubes in biology and medicine: In vitro and in vivo detection, imaging and drug delivery. *Nano Research*, 2, 85–120, 2009.

16. W.R. Yang, K.R. Ratinac, S.P. Ringer, P. Thordarson, J.J. Gooding, and F. Braet. Carbon nanomaterials in biosensors: Should you use nanotubes or graphene? *Angewandte Chemie International Edition*, 49, 2114–2138, 2010.

17. Y.H. Lin, W. Yantasee, and J. Wang. Carbon nanotubes (CNTs) for the development of electrochemical biosensors. *Frontiers in Bioscience*, 10, 492–505, 2005.

18. C.B. Jacobs, M.J. Peairs, and B.J. Venton. Review: Carbon nanotube based electrochemical sensors for biomolecules. *Analytica Chimica Acta*, 662, 105–127, 2010.

19. R.J. Chen, S. Bangsaruntip, K.A. Drouvalakis, N.W.S. Kam, M. Shim, Y.M. Li, W. Kim, P.J. Utz, and H.J. Dai. Noncovalent functionalization of carbon nanotubes for highly specific electronic biosensors. *Proceedings of the National Academy of Sciences of the United States of America*, 100, 4984–4989, 2003.

20. N.W.S. Kam, T.C. Jessop, P.A. Wender, and H.J. Dai. Nanotube molecular transporters: Internalization of carbon nanotube-protein conjugates into mammalian cells. *Journal of the American Chemical Society*, 126, 6850–6851, 2004.

21. A. Bianco, K. Kostarelos, C.D. Partidos, and M. Prato. Biomedical applications of functionalised carbon nanotubes. *Chemical Communications*, 571–577, 2005.

22. P. Cherukuri, S.M. Bachilo, S.H. Litovsky, and R.B. Weisman. Near-infrared fluorescence microscopy of single-walled carbon nanotubes in phagocytic cells. *Journal of the American Chemical Society*, 126, 15638–15639, 2004.

23. S.Y. Deng, G.Q. Jian, J.P. Lei, Z. Hu, and H.X. Ju. A glucose biosensor based on direct electrochemistry of glucose oxidase immobilized on nitrogen-doped carbon nanotubes. *Biosensors and Bioelectronics*, 25, 373–377, 2009.

24. Q. Zhao, D.P. Zhan, H.Y. Ma, M.Q. Zhang, Y.F. Zhao, P. Jing, Z.W. Zhu, X.H. Wan, Y.H. Shao, and Q.K. Zhuang. Direct proteins electrochemistry based on ionic liquid mediated carbon nanotube modified glassy carbon electrode. *Frontiers in Bioscience*, 10, 326–334, 2005.

25. G.C. Zhao, Z.Z. Yin, and X.W. Wei. A reagentless biosensor of nitric oxide based on direct electron transfer process of cytochrome C on multi-walled carbon nanotube. *Frontiers in Bioscience*, 10, 2005–2010, 2005.

26. J.J. Zhu, J.Z. Xu, Z. Hu, and H.Y. Chen. Reagentless electrochemical biosensor based on the multi-wall carbon nanotubes and nanogold particles composite film. *Frontiers in Bioscience*, 10, 521–529, 2005.

27. J.J. Gooding, R. Wibowo, J.Q. Liu, W.R. Yang, D. Losic, S. Orbons, F.J. Mearns, J.G. Shapter, and D.B. Hibbert. Protein electrochemistry using aligned carbon nanotube arrays. *Journal of the American Chemical Society*, 125, 9006–9007, 2003.

28. X. Yu, D. Chattopadhyay, I. Galeska, F. Papadimitrakopoulos, and J.F. Rusling. Peroxidase activity of enzymes bound to the ends of single-wall carbon nanotube forest electrodes. *Electrochemistry Communications*, 5, 408–411, 2003.

29. F. Patolsky, Y. Weizmann, and I. Willner. Long-range electrical contacting of redox enzymes by SWCNT connectors. *Angewandte Chemie International Edition*, 43, 2113–2117, 2004.

30. Q. Liu, X.B. Lu, J. Li, X. Yao, and J.H. Li. Direct electrochemistry of glucose oxidase and electrochemical biosensing of glucose on quantum dots/carbon nanotubes electrodes. *Biosensors and Bioelectronics*, 22, 3203–3209, 2007.

31. C.X. Cai and J. Chen. Direct electron transfer of glucose oxidase promoted by carbon nanotubes. *Analytical Biochemistry*, 332, 75–83, 2004.

32. A. Vaze, N. Hussain, C. Tang, D. Leech, and J. Rusling. Biocatalytic anode for glucose oxidation utilizing carbon nanotubes for direct electron transfer with glucose oxidase. *Electrochemistry Communications*, 11, 2004–2007, 2009.

33. M. Tominaga, S. Nomura, and I. Taniguchi. Bioelectrocatalytic current based on direct heterogeneous electron transfer reaction of glucose oxidase adsorbed onto multi-walled carbon nanotubes synthesized on platinum electrode surfaces. *Electrochemistry Communications*, 10, 888–890, 2008.

34. M.K. Wang, Y. Shen, Y. Liu, T. Wang, F. Zhao, B.F. Liu, and S.J. Dong. Direct electro-chemistry of microperoxidase 11 using carbon nanotube modified electrodes. *Journal of Electroanalytical Chemistry*, 578, 121–127, 2005.

35. J. Li, H.T. Ng, A. Cassell, W. Fan, H. Chen, Q. Ye, J. Koehne, J. Han, and M. Meyyappan. Carbon nanotube nanoelectrode array for ultrasensitive DNA detection. *Nano Letters*, 3, 597–602, 2003.

36. J. Koehne, H. Chen, J. Li, A.M. Cassell, Q. Ye, H.T. Ng, J. Han, and M. Meyyappan. Ultrasensitive label-free DNA analysis using an electronic chip based on carbon nano-tube nanoelectrode arrays. *Nanotechnology*, 14, 1239–1245, 2003.

37. D.A. Heller, H. Jin, B.M. Martinez, D. Patel, B.M. Miller, T.K. Yeung, P.V. Jena, C. Hobartner, T. Ha, S.K. Silverman, and M.S. Strano. Multimodal optical sensing and analyte specificity using single-walled carbon nanotubes. *Nature Nanotechnology*, 4, 114–120, 2009.

38. I.D. Rosca, F. Watari, M. Uo, and T. Akaska. Oxidation of multiwalled carbon nanotubes by nitric acid. *Carbon*, 43, 3124–3131, 2005.

39. S.S. Karajanagi, H.C. Yang, P. Asuri, E. Sellitto, J.S. Dordick, and R.S. Kane. Protein-assisted solubilization of single-walled carbon nanotubes. *Langmuir*, 22, 1392–1395, 2006.

40. A.H. Liu, I. Honma, M. Ichihara, and H.S. Zhou. Poly(acrylic acid)-wrapped multi-walled carbon nanotubes composite solubilization in water: Definitive spectroscopic properties. *Nanotechnology*, 17, 2845–2849, 2006.

41. Z. Liu, X.M. Sun, N. Nakayama-Ratchford, and H.J. Dai. Supramolecular chemistry on water-soluble carbon nanotubes for drug loading and delivery. *ACS Nano*, 1, 50–56, 2007.

42. M.L. Schipper, N. Nakayama-Ratchford, C.R. Davis, N.W.S. Kam, P. Chu, Z. Liu, X.M. Sun, H.J. Dai, and S.S. Gambhir. A pilot toxicology study of single-walled carbon nano-tubes in a small sample of mice. *Nature Nanotechnology*, 3, 216–221, 2008.

43. B. Zhao, H. Hu, A.P. Yu, D. Perea, and R.C. Haddon. Synthesis and characterization of water soluble single-walled carbon nanotube graft copolymers. *Journal of the American Chemical Society*, 127, 8197–8203, 2005.

44. N. Karousis, N. Tagmatarchis, and D. Tasis. Current progress on the chemical modifica-tion of carbon nanotubes.*Chemical Reviews*, 110, 5366–5397, 2010.

45. L.L. Zeng, L.B. Alemany, C.L. Edwards, and A.R. Barron. Demonstration of cova-lent sidewall functionalization of single wall carbon nanotubes by NMR spectroscopy: Side chain length dependence on the observation of the sidewall sp(3) carbons. *Nano Research*, 1, 72–88, 2008.

46. K.M. Lee, L.C. Li, and L.M. Dai. Asymmetric end-functionalization of multi-walled carbon nanotubes. *Journal of the American Chemical Society*, 127, 4122–4123, 2005.

47. M.J. Moghaddam, S. Taylor, M. Gao, S.M. Huang, L.M. Dai, and M.J. McCall. Highly efficient binding of DNA on the sidewalls and tips of carbon nanotubes using photo-chemistry. *Nano Letters*, 4, 89–93, 2004.

48. K.S. Coleman, S.R. Bailey, S. Fogden, and M.L.H. Green. Functionalization of single-walled carbon nanotubes via the Bingel reaction. *Journal of the American Chemical Society*, 125, 8722–8723, 2003.

49. T. Umeyama, N. Tezuka, M. Fujita, Y. Matano, N. Takeda, K. Murakoshi, K. Yoshida, S. Isoda, and H. Imahori. Retention of intrinsic electronic properties of soluble single-walled carbon nanotubes after a significant degree of sidewall functionalization by the Bingel reaction. *Journal of Physical Chemistry C*, 111, 9734–9741, 2007.

50. V. Georgakilas, K. Kordatos, M. Prato, D.M. Guldi, M. Holzinger, and A. Hirsch. Organic functionalization of carbon nanotubes. *Journal of the American Chemical Society*, 124, 760–761, 2002.

51. N. Tagmatarchis and M. Prato. Functionalization of carbon nanotubes via 1,3-dipolar cycloadditions. *Journal of Materials Chemistry*, 14, 437–439, 2004.
52. A. Star, J.F. Stoddart, D. Steuerman, M. Diehl, A. Boukai, E.W. Wong, X. Yang, S.W. Chung, H. Choi, and J.R. Heath. Preparation and properties of polymer-wrapped single-walled carbon nanotubes. *Angewandte Chemie International Edition*, 40, 1721–1725, 2001.
53. J. Chen, H.Y. Liu, W.A. Weimer, M.D. Halls, D.H. Waldeck, and G.C. Walker. Noncovalent engineering of carbon nanotube surfaces by rigid, functional conjugated polymers. *Journal of the American Chemical Society*, 124, 9034–9035, 2002.
54. C.A. Mitchell, J.L. Bahr, S. Arepalli, J.M. Tour, and R. Krishnamoorti. Dispersion of functionalized carbon nanotubes in polystyrene. *Macromolecules*, 35, 8825–8830, 2002.
55. Y.J. Kang and T.A. Taton. Micelle-encapsulated carbon nanotubes: A route to nanotube composites. *Journal of the American Chemical Society*, 125, 5650–5651, 2003.
56. C. Richard, F. Balavoine, P. Schultz, T.W. Ebbesen, and C. Mioskowski. Supramolecular self-assembly of lipid derivatives on carbon nanotubes. *Science*, 300, 775–778, 2003.
57. H. Wang, W. Zhou, D.L. Ho, K.I. Winey, J.E. Fischer, C.J. Glinka, and E.K. Hobbie. Dispersing single-walled carbon nanotubes with surfactants: A small angle neutron scattering study. *Nano Letters*, 4, 1789–1793, 2004.
58. R.J. Chen, Y.G. Zhang, D.W. Wang, and H.J. Dai. Noncovalent sidewall functionalization of single-walled carbon nanotubes for protein immobilization. *Journal of the American Chemical Society*, 123, 3838–3839, 2001.
59. P. Wu, X. Chen, N. Hu, U.C. Tam, O. Blixt, A. Zettl, and C.R. Bertozzi. Biocompatible carbon nanotubes generated by functionalization with glycodendrimers. *Angewandte Chemie International Edition*, 47, 5022–5025, 2008.
60. C. Vijayakumar, B. Balan, M.-J. Kim, and M. Takeuchi. Noncovalent functionalization of SWNTs with azobenzene-containing polymers: Solubility, stability, and enhancement of photoresponsive properties. *Journal of Physical Chemistry C*, 115, 4533–4539, 2011.
61. D.M. Guldi, H. Taieb, G.M.A. Rahman, N. Tagmatarchis, and M. Prato. Novel photoactive single-walled carbon nanotube-porphyrin polymer wraps: Efficient and long-lived intracomplex charge separation. *Advanced Materials*, 17, 871–875, 2005.
62. N.W.S. Kam, Z. Liu, and H.J. Dai. Functionalization of carbon nanotubes via cleavable disulfide bonds for efficient intracellular delivery of siRNA and potent gene silencing. *Journal of the American Chemical Society*, 127, 12492–12493, 2005.
63. N.W.S. Kam, M. O'Connell, J.A. Wisdom, and H.J. Dai. Carbon nanotubes as multifunctional biological transporters and near-infrared agents for selective cancer cell destruction. *Proceedings of the National Academy of Sciences of the United States of America*, 102, 11600–11605, 2005.
64. A.H. Liu, T. Watanabe, I. Honma, J. Wang, and H.S. Zhou. Effect of solution pH and ionic strength on the stability of poly(acrylic acid)-encapsulated multiwalled carbon nanotubes aqueous dispersion and its application for NADH sensor. *Biosensors and Bioelectronics*, 22, 694–699, 2006.
65. A. Liu, I. Honma, and H. Zhou. Simultaneous voltammetric detection of dopamine and uric acid at their physiological level in the presence of ascorbic acid using poly(acrylic acid)-multiwalled carbon-nanotube composite-covered glassy-carbon electrode. *Biosensors and Bioelectronics*, 23, 74–80, 2007.
66. T. Fukushima, K. Asaka, A. Kosaka, and T. Aida. Fully plastic actuator through layer-by-layer casting with ionic-liquid-based bucky gel. *Angewandte Chemie International Edition*, 44, 2410–2413, 2005.
67. J.Y. Wang, H.B. Chu, and Y. Li. Why single-walled carbon nanotubes can be dispersed in imidazolium-based ionic liquids. *ACS Nano*, 2, 2540–2546, 2008.
68. M. Zheng, A. Jagota, E.D. Semke, B.A. Diner, R.S. Mclean, S.R. Lustig, R.E. Richardson, and N.G. Tassi. DNA-assisted dispersion and separation of carbon nanotubes. *Nature Materials*, 2, 338–342, 2003.

69. X.M. Tu and M. Zheng. A DNA-based approach to the carbon nanotube sorting problem. *Nano Research*, 1, 185–194, 2008.
70. H.K. Moon, C. Il Chang, D.K. Lee, and H.C. Choi. Effect of nucleases on the cellular internalization of fluorescent labeled DNA-functionalized single-walled carbon nanotubes. *Nano Research*, 1, 351–360, 2008.
71. E. Katz and I. Willner. Biomolecule-functionalized carbon nanotubes: Applications in nanobioelectronics. *Chemphyschem*, 5, 1085–1104, 2004.
72. J.J. Gooding. Nanostructuring electrodes with carbon nanotubes: A review on electrochemistry and applications for sensing. *Electrochimica Acta*, 50, 3049–3060, 2005.
73. R.L. McCreery. Advanced carbon electrode materials for molecular electrochemistry. *Chemical Reviews*, 108, 2646–2687, 2008.
74. J.-J. Zhang, M.-M. Gu, T.-T. Zheng, and J.-J. Zhu. Synthesis of gelatin-stabilized gold nanoparticles and assembly of carboxylic single-walled carbon nanotubes/Au composites for cytosensing and drug uptake. *Analytical Chemistry*, 81, 6641–6648, 2009.
75. J. Wang, M. Musameh, and Y.H. Lin. Solubilization of carbon nanotubes by Nafion toward the preparation of amperometric biosensors. *Journal of the American Chemical Society*, 125, 2408–2409, 2003.
76. M. Wooten and W. Gorski. Facilitation of NADH electro-oxidation at treated carbon nanotubes. *Analytical Chemistry*, 82, 1299–1304, 2010.
77. M. Musameh, J. Wang, A. Merkoci, and Y.H. Lin. Low-potential stable NADH detection at carbon-nanotube-modified glassy carbon electrodes. *Electrochemistry Communications*, 4, 743–746, 2002.
78. J. Wang and M. Musameh. Carbon nanotube/teflon composite electrochemical sensors and biosensors. *Analytical Chemistry*, 75, 2075–2079, 2003.
79. J. Wang and Y.H. Lin. Functionalized carbon nanotubes and nanofibers for biosensing applications. *TrAC Trends in Analytical Chemistry*, 27, 619–626, 2008.
80. R.P. Deo, J. Wang, I. Block, A. Mulchandani, K.A. Joshi, M. Trojanowicz, F. Scholz, W. Chen, and Y.H. Lin. Determination of organophosphate pesticides at a carbon nanotube/organophosphorus hydrolase electrochemical biosensor. *Analytica Chimica Acta*, 530, 185–189, 2005.
81. V.B. Kandimalla and H.X. Ju. Binding of acetylcholinesterase to a multiwall carbon nanotube-cross-linked chitosan composite for flow-injection amperometric detection of an organophosphorous insecticide. *Chemistry—A European Journal*, 12, 1074–1080, 2006.
82. B.H. Wu, Y.J. Kuang, X.H. Zhang, and J.H. Chen. Noble metal nanoparticles/carbon nanotubes nanohybrids: Synthesis and applications. *Nano Today*, 6, 75–90, 2011.
83. S. Mubeen, T. Zhang, N. Chartuprayoon, Y. Rheem, A. Mulchandani, N.V. Myung, and M.A. Deshusses. Sensitive detection of H_2S using gold nanoparticle decorated single-walled carbon nanotubes. *Analytical Chemistry*, 82, 250–257, 2009.
84. L. Meng, J. Jin, G. Yang, T. Lu, H. Zhang, and C. Cai. Nonenzymatic electrochemical detection of glucose based on palladium–single-walled carbon nanotube hybrid nanostructures. *Analytical Chemistry*, 81, 7271–7280, 2009.
85. B.H. Wu, D. Hu, Y.J. Kuang, Y.M. Yu, X.H. Zhang, and J.H. Chen. High dispersion of platinum-ruthenium nanoparticles on the 3,4,9,10-perylene tetracarboxylic acid-functionalized carbon nanotubes for methanol electro-oxidation. *Chemical Communications*, 47, 5253–5255, 2011.
86. B.H. Wu, D. Hu, Y.M. Yu, Y.J. Kuang, X.H. Zhang, and J.H. Chen. Stabilization of platinum nanoparticles dispersed on carbon nanotubes by ionic liquid polymer. *Chemical Communications*, 46, 7954–7956, 2010.
87. S.J. Guo, S.J. Dong, and E.K. Wang. Constructing carbon nanotube/Pt nanoparticle hybrids using an imidazolium-salt-based ionic liquid as a linker. *Advanced Materials*, 22, 1269–1272, 2010.

88. Z.J. Wang, Q.X. Zhang, D. Kuehner, X.Y. Xu, A. Ivaska, and L. Niu. The synthesis of ionic-liquid-functionalized multiwalled carbon nanotubes decorated with highly dispersed Au nanoparticles and their use in oxygen reduction by electrocatalysis. *Carbon*, 46, 1687–1692, 2008.

89. L.T. Qu and L.M. Dai. Substrate-enhanced electroless deposition of metal nanoparticles on carbon nanotubes. *Journal of the American Chemical Society*, 127, 10806–10807, 2005.

90. L.T. Qu, L.M. Dai, and E. Osawa. Shape/size-control led syntheses of metal nanoparticles for site-selective modification of carbon nanotubes. *Journal of the American Chemical Society*, 128, 5523–5532, 2006.

91. B. Ritz, H. Heller, A. Myalitsin, A. Kornowski, F.J. Martin-Martinez, S. Melchor, J.A. Dobado, B.H. Juárez, H. Weller, and C. Klinke. Reversible attachment of platinum alloy nanoparticles to nonfunctionalized carbon nanotubes. *ACS Nano*, 4, 2438–2444, 2010.

92. G. Wei, F. Xu, Z. Li, and K.D. Jandt. Protein-promoted synthesis of Pt nanoparticles on carbon nanotubes for electrocatalytic nanohybrids with enhanced glucose sensing. *Journal of Physical Chemistry C*, 115, 11453–11460, 2011.

93. J.C. Claussen, A.D. Franklin, A. Ul Haque, D.M. Porterfield, and T.S. Fisher. Electrochemical biosensor of nanocube-augmented carbon nanotube networks. *ACS Nano*, 3, 37–44, 2009.

94. S. Sahoo, S. Husale, S. Karna, S.K. Nayak, and P.M. Ajayan. Controlled assembly of Ag nanoparticles and carbon nanotube hybrid structures for biosensing. *Journal of the American Chemical Society*, 133, 4005–4009, 2011.

95. L.C. Jiang and W.D. Zhang. A highly sensitive nonenzymatic glucose sensor based on CuO nanoparticles-modified carbon nanotube electrode. *Biosensors and Bioelectronics*, 25, 1402–1407, 2010.

96. J. Chen, W.D. Zhang, and J.S. Ye. Nonenzymatic electrochemical glucose sensor based on MnO_2/MWNTs nanocomposite. *Electrochemistry Communications*, 10, 1268–1271, 2008.

97. C.H. Zhang, G.F. Wang, M. Liu, Y.H. Feng, Z.D. Zhang, and B. Fang. A hydroxylamine electrochemical sensor based on electrodeposition of porous ZnO nanofilms onto carbon nanotubes films modified electrode. *Electrochimica Acta*, 55, 2835–2840, 2010.

98. H. Lee, S.W. Yoon, E.J. Kim, and J. Park. In-situ growth of copper sulfide nanocrystals on multiwalled carbon nanotubes and their application as novel solar cell and amperometric glucose sensor materials. *Nano Letters*, 7, 778–784, 2007.

99. X. Xu, L. Yang, S. Jiang, Z. Hu, and S. Liu. High reaction activity of nitrogen-doped carbon nanotubes toward the electrooxidation of nitric oxide. *Chemical Communications*, 47, 7137–7139, 2011.

100. X. Xu, S. Jiang, Z. Hu, and S. Liu. Nitrogen-doped carbon nanotubes: High electrocatalytic activity toward the oxidation of hydrogen peroxide and its application for biosensing. *ACS Nano*, 4, 4292–4298, 2010.

101. A. Zebda, C. Gondran, A. Le Goff, M. Holzinger, P. Cinquin, and S. Cosnier. Mediatorless high-power glucose biofuel cells based on compressed carbon nanotube-enzyme electrodes. *Nature Communications*, 2011. DOI 10.1038/ncomms1365.

102. M.K. Wang, F. Zhao, Y. Liu, and S.J. Dong. Direct electrochemistry of microperoxidase at Pt microelectrodes modified with carbon nanotubes. *Biosensors and Bioelectronics*, 21, 159–166, 2005.

103. F.H. Li, Z.H. Wang, C.S. Shan, J.F. Song, D.X. Han, and L. Niu. Preparation of gold nanoparticles/functionalized multiwalled carbon nanotube nanocomposites and its glucose biosensing application. *Biosensors and Bioelectronics*, 24, 1765–1770, 2009.

104. H.L. Pang, J. Liu, D. Hu, X.H. Zhang, and J.H. Chen. Immobilization of laccase onto 1-aminopyrene functionalized carbon nanotubes and their electrocatalytic activity for oxygen reduction. *Electrochimica Acta*, 55, 6611–6616, 2010.

105. Y. Liu, J.P. Lei, and H.X. Ju. Amperometric sensor for hydrogen peroxide based on electric wire composed of horseradish peroxidase and toluidine blue-multiwalled carbon nanotubes nanocomposite. *Talanta*, 74, 965–970, 2008.

106. Y.D. Zhao, W.D. Zhang, H. Chen, Q.M. Luo, and S.F.Y. Li. Direct electrochemistry of horseradish peroxidase at carbon nanotube powder microelectrode. *Sensors and Actuators B—Chemical*, 87, 168–172, 2002.

107. S. Liu, B. Lin, X. Yang, and Q. Zhang. Carbon-nanotube-enhanced direct electron-transfer reactivity of hemoglobin immobilized on polyurethane elastomer film. *Journal of Physical Chemistry B*, 111, 1182–1188, 2007.

108. X. Yu, B. Munge, V. Patel, G. Jensen, A. Bhirde, J.D. Gong, S.N. Kim, J. Gillespie, J.S. Gutkind, F. Papadimitrakopoulos, and J.F. Rusling. Carbon nanotube amplification strategies for highly sensitive immunodetection of cancer biomarkers. *Journal of the American Chemical Society*, 128, 11199–11205, 2006.

109. S.K. Vashist, D. Zheng, K. Al-Rubeaan, J.H.T. Luong, and F.S. Sheu. Advances in carbon nanotube based electrochemical sensors for bioanalytical applications. *Biotechnology Advances*, 29, 169–188, 2011.

110. A. Merkoci, M. Pumera, X. Llopis, B. Perez, M. del Valle, and S. Alegret. New materials for electrochemical sensing VI: Carbon nanotubes. *TrAC Trends in Analytical Chemistry*, 24, 826–838, 2005.

111. J.P. Kim, B.Y. Lee, J. Lee, S. Hong, and S.J. Sim. Enhancement of sensitivity and specificity by surface modification of carbon nanotubes in diagnosis of prostate cancer based on carbon nanotube field effect transistors. *Biosensors and Bioelectronics*, 24, 3372–3378, 2009.

112. S. Viswanathan, C. Rani, A.V. Anand, and J.A.A. Ho. Disposable electrochemical immunosensor for carcinoembryonic antigen using ferrocene liposomes and MWCNT screen-printed electrode. *Biosensors and Bioelectronics*, 24, 1984–1989, 2009.

113. B.S. Munge, C.E. Krause, R. Malhotra, V. Patel, J.S. Gutkind, and J.F. Rusling. Electrochemical immunosensors for interleukin-6. Comparison of carbon nanotube forest and gold nanoparticle platforms. *Electrochemistry Communications*, 11, 1009–1012, 2009.

114. K.A. Mahmoud, S. Hrapovic, and J.H.T. Luong. Picomolar detection of protease using peptide/single walled carbon nanotube/gold nanoparticle-modified electrode. *ACS Nano*, 2, 1051–1057, 2008.

115. K.A. Mahmoud and J.H.T. Luong. Impedance method for detecting HIV-1 protease and screening for its inhibitors using ferrocene-peptide conjugate/Au nanoparticle/single-walled carbon nanotube modified electrode. *Analytical Chemistry*, 80, 7056–7062, 2008.

116. G. Lai, F. Yan, and H. Ju. Dual signal amplification of glucose oxidase-functionalized nanocomposites as a trace label for ultrasensitive simultaneous multiplexed electrochemical detection of tumor markers. *Analytical Chemistry*, 81, 9730–9736, 2009.

117. S.M. Shamah, J.M. Healy, and S.T. Cload. Complex target SELEX. *Accounts of Chemical Research*, 41, 130–138, 2008.

118. H.J. Qiu, Y.L. Sun, X.R. Huang, and Y.B. Qu. A sensitive nanoporous gold-based electrochemical aptasensor for thrombin detection. *Colloids and Surface B*, 79, 304–308, 2010.

119. K. Guo, Y. Wang, H. Chen, J. Ji, S. Zhang, J. Kong, and B. Liu. An aptamer-SWNT biosensor for sensitive detection of protein via mediated signal transduction. *Electrochemistry Communications*, 13, 707–710, 2011.

120. G.A. Zelada-Guillén, J. Riu, A. Düzgün, and F.X. Rius. Immediate detection of living bacteria at ultralow concentrations using a carbon nanotube based potentiometric aptasensor. *Angewandte Chemie International Edition*, 48, 7334–7337, 2009.

121. A. Erdem, P. Papakonstantinou, and H. Murphy. Direct DNA hybridization at disposable graphite electrodes modified with carbon nanotubes. *Analytical Chemistry*, 78, 6656–6659, 2006.

122. J. Wang, G.D. Liu, and M.R. Jan. Ultrasensitive electrical biosensing of proteins and DNA: Carbon-nanotube derived amplification of the recognition and transduction events. *Journal of the American Chemical Society*, 126, 3010–3011, 2004.

123. B. Seiwert and U. Karst. Ferrocene-based derivatization in analytical chemistry. *Analytical and Bioanalytical Chemistry*, 390, 181–200, 2008.

124. X.Z. Zhang, K. Jiao, S.F. Liu, and Y.W. Hu. Readily reusable electrochemical DNA hybridization biosensor based on the interaction of DNA with single-walled carbon nanotubes. *Analytical Chemistry*, 81, 6006–6012, 2009.

125. A. Star, B.L. Allen, and P.D. Kichambare. Carbon nanotube field-effect-transistor-based biosensors. *Advanced Materials*, 19, 1439–1451, 2007.

126. D.X. Cui, B.F. Pan, H. Zhang, F. Gao, R. Wu, J.P. Wang, R. He, and T. Asahi. Self-assembly of quantum dots and carbon nanotubes for ultrasensitive DNA and antigen detection. *Analytical Chemistry*, 80, 7996–8001, 2008.

127. R.H. Yang, J.Y. Jin, Y. Chen, N. Shao, H.Z. Kang, Z. Xiao, Z.W. Tang, Y.R. Wu, Z. Zhu, and W.H. Tan. Carbon nanotube-quenched fluorescent oligonucleotides: Probes that fluoresce upon hybridization. *Journal of the American Chemical Society*, 130, 8351–8358, 2008.

128. M.J. O'Connell, S.M. Bachilo, C.B. Huffman, V.C. Moore, M.S. Strano, E.H. Haroz, K.L. Rialon, P.J. Boul, W.H. Noon, C. Kittrell, J.P. Ma, R.H. Hauge, R.B. Weisman, and R.E. Smalley. Band gap fluorescence from individual single-walled carbon nanotubes. *Science*, 297, 593–596, 2002.

129. P. Avouris, M. Freitag, and V. Perebeinos. Carbon-nanotube photonics and optoelectronics. *Nature Photonics*, 2, 341–350, 2008.

130. R.H. Yang, Z.W. Tang, J.L. Yan, H.Z. Kang, Y.M. Kim, Z. Zhu, and W.H. Tan. Noncovalent assembly of carbon nanotubes and single-stranded DNA: An effective sensing platform for probing biomolecular interactions. *Analytical Chemistry*, 80, 7408–7413, 2008.

131. B.C. Satishkumar, L.O. Brown, Y. Gao, C.C. Wang, H.L. Wang, and S.K. Doorn. Reversible fluorescence quenching in carbon nanotubes for biomolecular sensing. *Nature Nanotechnology*, 2, 560–564, 2007.

132. X. Ouyang, R. Yu, J. Jin, J. Li, R. Yang, W. Tan, and J. Yuan. New strategy for label-free and time-resolved luminescent assay of protein: Conjugate Eu^{3+} complex and aptamer-wrapped carbon nanotubes. *Analytical Chemistry*, 83, 782–789, 2011.

133. D.A. Heller, E.S. Jeng, T.K. Yeung, B.M. Martinez, A.E. Moll, J.B. Gastala, and M.S. Strano. Optical detection of DNA conformational polymorphism on single-walled carbon nanotubes. *Science*, 311, 508–511, 2006.

134. J.-H. Ahn, J.-H. Kim, N.F. Reuel, P.W. Barone, A.A. Boghossian, J. Zhang, H. Yoon, A.C. Chang, A.J. Hilmer, and M.S. Strano. Label-free, single protein detection on a near-infrared fluorescent single-walled carbon nanotube/protein microarray fabricated by cell-free synthesis. *Nano Letters*, 2011. DOI 10.1021/nl201033d.

135. X.Y. Gao, G.M. Xing, Y.L. Yang, X.L. Shi, R. Liu, W.G. Chu, L. Jing, F. Zhao, C. Ye, H. Yuan, X.H. Fang, C. Wang, and Y.L. Zhao. Detection of trace Hg^{2+} via induced circular dichroism of DNA wrapped around single-walled carbon nanotubes. *Journal of the American Chemical Society*, 130, 9190–9191, 2008.

136. M.S. Strano, E.S. Jeng, P.W. Barone, and J.D. Nelson. Hybridization kinetics and thermodynamics of DNA adsorbed to individually dispersed single-walled carbon nanotubes. *Small*, 3, 1602–1609, 2007.

137. E.S. Jeng, A.E. Moll, A.C. Roy, J.B. Gastala, and M.S. Strano. Detection of DNA hybridization using the near-infrared band-gap fluorescence of single-walled carbon nanotubes. *Nano Letters*, 6, 371–375, 2006.

138. P.W. Barone, S. Baik, D.A. Heller, and M.S. Strano. Near-infrared optical sensors based on single-walled carbon nanotubes. *Nature Materials*, 4, 86–U16, 2005.

139. H. Jin, D.A. Heller, J.H. Kim, and M.S. Strano. Stochastic analysis of stepwise fluorescence quenching reactions on single-walled carbon nanotubes: Single molecule sensors. *Nano Letters*, 8, 4299–4304, 2008.

140. T. Hertel, J. Crochet, and M. Clemens. Quantum yield heterogeneities of aqueous single-wall carbon nanotube suspensions. *Journal of the American Chemical Society*, 129, 8058–8059, 2007.
141. Z. Chen, S.M. Tabakman, A.P. Goodwin, M.G. Kattah, D. Daranciang, X.R. Wang, G.Y. Zhang, X.L. Li, Z. Liu, P.J. Utz, K.L. Jiang, S.S. Fan, and H.J. Dai. Protein microarrays with carbon nanotubes as multicolor Raman labels. *Nature Biotechnology*, 26, 1285–1292, 2008.
142. S.M. Nie and S.R. Emery. Probing single molecules and single nanoparticles by surface-enhanced Raman scattering. *Science*, 275, 1102–1106, 1997.
143. V. Espina, E.C. Woodhouse, J. Wulfkuhle, H.D. Asmussen, E.F. Petricoin, and L.A. Liotta. Protein microarray detection strategies: Focus on direct detection technologies. *Journal of Immunological Methods*, 290, 121–133, 2004.
144. H. Qiu, Z. Zhang, X. Huang, and Y. Qu. Dealloying Ag-Al alloy to prepare nanoporous silver as a substrate for surface-enhanced Raman scattering: Effects of structural evolution and surface modification. *ChemPhysChem*, 12, 2118–2123, 2011.
145. A. Prakash, P. Mallick, J. Whiteaker, H.D. Zhang, A. Paulovich, M. Flory, H. Lee, R. Aebersold, and B. Schwikowski. Signal maps for mass spectrometry-based comparative proteomics. *Molecular and Cellular Proteomics*, 5, 423–432, 2006.
146. H. Dumortier, S. Lacotte, G. Pastorin, R. Marega, W. Wu, D. Bonifazi, J.P. Briand, M. Prato, S. Muller, and A. Bianco. Functionalized carbon nanotubes are non-cytotoxic and preserve the functionality of primary immune cells. *Nano Letters*, 6, 1522–1528, 2006.
147. C.W. Lam, J.T. James, R. McCluskey, and R.L. Hunter. Pulmonary toxicity of single-wall carbon nanotubes in mice 7 and 90 days after intratracheal instillation. *Toxicological Sciences*, 77, 126–134, 2004.
148. D.B. Warheit, B.R. Laurence, K.L. Reed, D.H. Roach, G.A.M. Reynolds, and T.R. Webb. Comparative pulmonary toxicity assessment of single-wall carbon nanotubes in rats. *Toxicological Sciences*, 77, 117–125, 2004.
149. A.A. Shvedova, E.R. Kisin, R. Mercer, A.R. Murray, V.J. Johnson, A.I. Potapovich, Y.Y. Tyurina, O. Gorelik, S. Arepalli, D. Schwegler-Berry, A.F. Hubbs, J. Antonini, D.E. Evans, B.K. Ku, D. Ramsey, A. Maynard, V.E. Kagan, V. Castranova, and P. Baron. Unusual inflammatory and fibrogenic pulmonary responses to single-walled carbon nanotubes in mice. *American Journal of Physiology—Lung Cellular and Molecular Physiology*, 289, L698–L708, 2005.
150. J. Muller, F. Huaux, N. Moreau, P. Misson, J.F. Heilier, M. Delos, M. Arras, A. Fonseca, J.B. Nagy, and D. Lison. Respiratory toxicity of multi-wall carbon nanotubes. *Toxicology and Applied Pharmacology*, 207, 221–231, 2005.
151. C.A. Poland, R. Duffin, I. Kinloch, A. Maynard, W.A.H. Wallace, A. Seaton, V. Stone, S. Brown, W. MacNee, and K. Donaldson. Carbon nanotubes introduced into the abdominal cavity of mice show asbestos-like pathogenicity in a pilot study. *Nature Nanotechnology*, 3, 423–428, 2008.
152. D.X. Cui, F.R. Tian, C.S. Ozkan, M. Wang, and H.J. Gao. Effect of single wall carbon nanotubes on human HEK293 cells. *Toxicology Letters*, 155, 73–85, 2005.
153. L.H. Ding, J. Stilwell, T.T. Zhang, O. Elboudwarej, H.J. Jiang, J.P. Selegue, P.A. Cooke, J.W. Gray, and F.Q.F. Chen. Molecular characterization of the cytotoxic mechanism of multiwall carbon nanotubes and nano-onions on human skin fibroblast. *Nano Letters*, 5, 2448–2464, 2005.
154. M. Bottini, S. Bruckner, K. Nika, N. Bottini, S. Bellucci, A. Magrini, A. Bergamaschi, and T. Mustelin. Multi-walled carbon nanotubes induce T lymphocyte apoptosis. *Toxicology Letters*, 160, 121–126, 2006.

155. C.M. Sayes, F. Liang, J.L. Hudson, J. Mendez, W.H. Guo, J.M. Beach, V.C. Moore, C.D. Doyle, J.L. West, W.E. Billups, K.D. Ausman, and V.L. Colvin. Functionalization density dependence of single-walled carbon nanotubes cytotoxicity in vitro. *Toxicology Letters*, 161, 135–142, 2006.

156. L. Dong, K.L. Joseph, C.M. Witkowski, and M.M. Craig. Cytotoxicity of single-walled carbon nanotubes suspended in various surfactants. *Nanotechnology*, 19, 255702, 2008.

157. D.L. Plata, P.M. Gschwend, and C.M. Reddy. Industrially synthesized single-walled carbon nanotubes: Compositional data for users, environmental risk assessments, and source apportionment. *Nanotechnology*, 19, 185706, 2008.

158. Z. Liu, M. Winters, M. Holodniy, and H.J. Dai. siRNA delivery into human T cells and primary cells with carbon-nanotube transporters. *Angewandte Chemie International Edition*, 46, 2023–2027, 2007.

159. H.N. Yehia, R.K. Draper, C. Mikoryak, E. Walker, P. Bajaj, I.H. Musselman, M.C. Daigrepont, G.R. Dieckmann, and P. Pantano. Single-walled carbon nanotube interactions with HeLa cells. *Journal of Nanobiotechnology*, 5, 8, 2007.

160. W. Wu, S. Wieckowski, G. Pastorin, M. Benincasa, C. Klumpp, J.P. Briand, R. Gennaro, M. Prato, and A. Bianco. Targeted delivery of amphotericin B to cells by using functionalized carbon nanotubes. *Angewandte Chemie International Edition*, 44, 6358–6362, 2005.

161. S.F. Chin, R.H. Baughman, A.B. Dalton, G.R. Dieckmann, R.K. Draper, C. Mikoryak, I.H. Musselman, V.Z. Poenitzsch, H. Xie, and P. Pantano. Amphiphilic helical peptide enhances the uptake of single-walled carbon nanotubes by living cells. *Experimental Biology and Medicine*, 232, 1236–1244, 2007.

162. A.V. Tkach, G.V. Shurin, M.R. Shurin, E.R. Kisin, A.R. Murray, S.-H. Young, A. Star, B. Fadeel, V.E. Kagan, and A.A. Shvedova. Direct effects of carbon nanotubes on dendritic cells induce immune suppression upon pulmonary exposure. *ACS Nano*, 2011. DOI 10.1021/nn2014479.

163. G.K.M. Mutlu, G.R.S. Budinger, A.A. Green, D. Urich, S. Soberanes, S.E. Chiarella, G.F. Alheid, D.R. McCrimmon, I. Szleifer, and M.C. Hersam. Biocompatible nanoscale dispersion of single-walled carbon nanotubes minimizes in vivo pulmonary toxicity. *Nano Letters*, 10, 1664–1670, 2010.

164. L. Wang, S. Luanpitpong, V. Castranova, W. Tse, Y. Lu, V. Pongrakhananon, and Y. Rojanasakul. Carbon nanotubes induce malignant transformation and tumorigenesis of human lung epithelial cells. *Nano Letters*, 2011. DOI 10.1021/nl2011214.

165. Z. Liu, C. Davis, W.B. Cai, L. He, X.Y. Chen, and H.J. Dai. Circulation and long-term fate of functionalized, biocompatible single-walled carbon nanotubes in mice probed by Raman spectroscopy. *Proceedings of the National Academy of Sciences of the United States of America*, 105, 1410–1415, 2008.

166. S.T. Yang, X. Wang, G. Jia, Y.Q. Gu, T.C. Wang, H.Y. Nie, C.C. Ge, H.F. Wang, and Y.F. Liu. Long-term accumulation and low toxicity of single-walled carbon nanotubes in intravenously exposed mice. *Toxicology Letters*, 181, 182–189, 2008.

167. S.T. Yang, K.A.S. Fernando, J.H. Liu, J. Wang, H.F. Sun, Y.F. Liu, M. Chen, Y.P. Huang, X. Wang, H.F. Wang, and Y.P. Sun. Covalently PEGylated carbon nanotubes with stealth character in vivo. *Small*, 4, 940–944, 2008.

168. V. Neves, E. Heister, S. Costa, C. Tîlmaciu, E. Borowiak-Palen, C.E. Giusca, E. Flahaut, B. Soula, H.M. Coley, J. McFadden, and S.R.P. Silva. Uptake and release of double-walled carbon nanotubes by mammalian cells. *Advanced Functional Materials*, 20, 3272–3279, 2010.

169. N. Gao, Q. Zhang, Q. Mu, Y. Bai, L. Li, H. Zhou, E.R. Butch, T.B. Powell, S.E. Snyder, G. Jiang, and B. Yan. Steering carbon nanotubes to scavenger receptor recognition by nanotube surface chemistry modification partially alleviates NFκB activation and reduces its immunotoxicity. *ACS Nano*, 5, 4581–4591, 2011.

170. E. Heister, C. Lamprecht, V. Neves, C. Tilmaciu, L. Datas, E. Flahaut, B. Soula, P. Hinterdorfer, H.M. Coley, S.R.P. Silva, and J. McFadden. Higher dispersion efficacy of functionalized carbon nanotubes in chemical and biological environments. *ACS Nano*, 4, 2615–2626, 2010.

171. D. Pantarotto, J.P. Briand, M. Prato, and A. Bianco. Translocation of bioactive peptides across cell membranes by carbon nanotubes. *Chemical Communications*, 16–17, 2004.

172. H. Jin, D.A. Heller, and M.S. Strano. Single-particle tracking of endocytosis and exocytosis of single-walled carbon nanotubes in NIH-3T3 cells. *Nano Letters*, 8, 1577–1585, 2008.

173. K. Kostarelos, L. Lacerda, G. Pastorin, W. Wu, S. Wieckowski, J. Luangsivilay, S. Godefroy, D. Pantarotto, J.P. Briand, S. Muller, M. Prato, and A. Bianco. Cellular uptake of functionalized carbon nanotubes is independent of functional group and cell type. *Nature Nanotechnology*, 2, 108–113, 2007.

174. L. Lacerda, G. Pastorin, D. Gathercole, J. Buddle, M. Prato, A. Bianco, and K. Kostarelos. Intracellular trafficking of carbon nanotubes by confocal laser scanning microscopy. *Advanced Materials*, 19, 1480–1484, 2007.

175. D.W. Zhang, C.Q. Yi, J.C. Zhang, Y. Chen, X.S. Yao, and M.S. Yang. The effects of carbon nanotubes on the proliferation and differentiation of primary osteoblasts. *Nanotechnology*, 18, 9, 2007.

176. E. Mooney, P. Dockery, U. Greiser, M. Murphy, and V. Barron. Carbon nanotubes and mesenchymal stem cells: Biocompatibility, proliferation and differentiation. *Nano Letters*, 8, 2137–2143, 2008.

177. J.P. Cheng, K.A.S. Fernando, L.M. Veca, Y.P. Sun, A.I. Lamond, Y.W. Lam, and S.H. Cheng. Reversible accumulation of PEGylated single-walled carbon nanotubes in the mammalian nucleus. *ACS Nano*, 2, 2085–2094, 2008.

178. Q. Mu, D.L. Broughton, and B. Yan. Endosomal leakage and nuclear translocation of multiwalled carbon nanotubes: Developing a model for cell uptake. *Nano Letters*, 9, 4370–4375, 2009.

179. G. Pastorin, W. Wu, S. Wieckowski, J.P. Briand, K. Kostarelos, M. Prato, and A. Bianco. Double functionalisation of carbon nanotubes for multimodal drug delivery. *Chemical Communications*, 1182–1184, 2006.

180. Z. Liu, K. Chen, C. Davis, S. Sherlock, Q.Z. Cao, X.Y. Chen, and H.J. Dai. Drug delivery with carbon nanotubes for in vivo cancer treatment. *Cancer Research*, 68, 6652–6660, 2008.

181. W. Wu, R. Li, X. Bian, Z. Zhu, D. Ding, X. Li, Z. Jia, X. Jiang, and Y. Hu. Covalently combining carbon nanotubes with anticancer agent: Preparation and antitumor activity. *ACS Nano*, 3, 2740–2750, 2009.

182. R.P. Feazell, N. Nakayama-Ratchford, H. Dai, and S.J. Lippard. Soluble single-walled carbon nanotubes as longboat delivery systems for platinum(IV) anticancer drug design. *Journal of the American Chemical Society*, 129, 8438–8439, 2007.

183. Z. Su, S. Zhu, A.D. Donkor, C. Tzoganakis, and J.F. Honek. Controllable delivery of small-molecule compounds to targeted cells utilizing carbon nanotubes. *Journal of the American Chemical Society*, 133, 6874–6877, 2011.

184. H.J. Dai, S. Dhar, Z. Liu, J. Thomale, and S.J. Lippard. Targeted single-wall carbon nanotube-mediated Pt(IV) prodrug delivery using folate as a homing device. *Journal of the American Chemical Society*, 130, 11467–11476, 2008.

185. Z. Liu, W.B. Cai, L.N. He, N. Nakayama, K. Chen, X.M. Sun, X.Y. Chen, and H.J. Dai. In vivo biodistribution and highly efficient tumour targeting of carbon nanotubes in mice. *Nature Nanotechnology*, 2, 47–52, 2007.

186. C.H. Villa, T. Dao, I. Ahearn, N. Fehrenbacher, E. Casey, D.A. Rey, T. Korontsvit, V. Zakhaleva, C.A. Batt, M.R. Philips, and D.A. Scheinberg. Single-walled carbon nanotubes deliver peptide antigen into dendritic cells and enhance IgG responses to tumor-associated antigens. *ACS Nano*, 2011. DOI 10.1021/nn200182x.

187. Z.A. Liu, X.L. Li, S.M. Tabakman, K.L. Jiang, S.S. Fan, and H.J. Dai. Multiplexed multicolor Raman imaging of live cells with isotopically modified single walled carbon nanotubes. *Journal of the American Chemical Society*, 130, 13540–13541, 2008.

188. M.R. McDevitt, D. Chattopadhyay, B.J. Kappel, J.S. Jaggi, S.R. Schiffman, C. Antczak, J.T. Njardarson, R. Brentjens, and D.A. Scheinberg. Tumor targeting with antibody-functionalized, radiolabeled carbon nanotubes. *Journal of Nuclear Medicine*, 48, 1180–1189, 2007.

189. N.W.S. Kam and H.J. Dai. Carbon nanotubes as intracellular protein transporters: Generality and biological functionality. *Journal of the American Chemical Society*, 127, 6021–6026, 2005.

190. D. Pantarotto, R. Singh, D. McCarthy, M. Erhardt, J.P. Briand, M. Prato, K. Kostarelos, and A. Bianco. Functionalized carbon nanotubes for plasmid DNA gene delivery. *Angewandte Chemie International Edition*, 43, 5242–5246, 2004.

191. L.Z. Gao, L. Nie, T.H. Wang, Y.J. Qin, Z.X. Guo, D.L. Yang, and X.Y. Yan. Carbon nanotube delivery of the GFP gene into mammalian cells. *Chembiochem*, 7, 239–242, 2006.

192. Z.H. Zhang, X.Y. Yang, Y. Zhang, B. Zeng, Z.J. Wang, T.H. Zhu, R.B.S. Roden, Y.S. Chen, and R.C. Yang. Delivery of telomerase reverse transcriptase small interfering RNA in complex with positively charged single-walled carbon nanotubes suppresses tumor growth. *Clinical Cancer Research*, 12, 4933–4939, 2006.

193. P. Chakravarty, R. Marches, N.S. Zimmerman, A.D.E. Swafford, P. Bajaj, I.H. Musselman, P. Pantano, R.K. Draper, and E.S. Vitetta. Thermal ablation of tumor cells with anti body-functionalized single-walled carbon nanotubes. *Proceedings of the National Academy of Sciences of the United States of America*, 105, 8697–8702, 2008.

194. R.R. Moore, C.E. Banks, and R.G. Compton. Basal plane pyrolytic graphite modified electrodes: Comparison of carbon nanotubes and graphite powder as electrocatalysts. *Analytical Chemistry*, 76, 2677–2682, 2004.

195. M. Pumera. The electrochemistry of carbon nanotubes: Fundamentals and applications. *Chemistry—A European Journal*, 15, 4970–4978, 2009.

196. L.M. Dai and A.W.H. Mau. Controlled synthesis and modification of carbon nanotubes and C-60: Carbon nanostructures for advanced polymer composite materials. *Advanced Materials*, 13, 899–913, 2001.

197. L.T. Qu, F. Du, and L.M. Dai. Preferential syntheses of semiconducting vertically aligned single-walled carbon nanotubes for direct use in FETs. *Nano Letters*, 8, 2682–2687, 2008.

198. B.Q. Wei, R. Vajtai, Y. Jung, J. Ward, R. Zhang, G. Ramanath, and P.M. Ajayan. Organized assembly of carbon nanotubes—Cunning refinements help to customize the architecture of nanotube structures. *Nature*, 416, 495–496, 2002.

199. L.M. Dai, A. Patil, X.Y. Gong, Z.X. Guo, L.Q. Liu, Y. Liu, and D.B. Zhu. Aligned nanotubes. *ChemPhysChem*, 4, 1150–1169, 2003.

200. R. Malhotra, V. Patel, J.P. Vaqué, J.S. Gutkind, and J.F. Rusling. Ultrasensitive electrochemical immunosensor for oral cancer biomarker IL-6 using carbon nanotube forest electrodes and multilabel amplification. *Analytical Chemistry*, 82, 3118–3123, 2010.

201. J.J. Gooding, A. Chou, J.Q. Liu, D. Losic, J.G. Shapter, and D.B. Hibbert. The effects of the lengths and orientations of single-walled carbon nanotubes on the electrochemistry of nanotube-modified electrodes. *Electrochemistry Communications*, 9, 1677–1683, 2007.

202. J.K. Campbell, L. Sun, and R.M. Crooks. Electrochemistry using single carbon nanotubes. *Journal of the American Chemical Society*, 121, 3779–3780, 1999.

203. I. Heller, J. Kong, K.A. Williams, C. Dekker, and S.G. Lemay. Electrochemistry at single-walled carbon nanotubes: The role of band structure and quantum capacitance. *Journal of the American Chemical Society*, 128, 7353–7359, 2006.

9 Synthesis of Higher Diamondoids by Pulsed Laser Ablation Plasmas in Supercritical Fluids

Sven Stauss, Sho Nakahara, Toru Kato,
Takehiko Sasaki, and Kazuo Terashima

CONTENTS

9.1 INTRODUCTION

Metallic, semiconducting, and carbon nanomaterials and nanoparticles are increasingly finding applications in a variety of fields, ranging from biotechnology and medicine to opto- and nanoelectronics. One of the main difficulties in the synthesis of nanoparticles and their application is to limit the size of particles to the nanometer range and to realize very narrow size distributions. However, this is often difficult to achieve by conventional materials processing techniques, making it necessary to develop new approaches to overcome these obstacles.

Supercritical fluids (SCFs) are increasingly finding applications as high-pressure media in analytic chemistry [1] or for chemical synthesis [2]. SCFs allow their properties, e.g., thermal capacity, thermal conductivity, density fluctuation, dielectric constant, diffusivity, etc., to be tuned continuously by adjusting either the temperature, pressure, or both. On the other hand, plasmas represent highly reactive media containing electrons, excited neutral radicals, and positively and negatively charged atomic and molecular ions, allowing us to realize reactions that are far from thermodynamic equilibrium [3,4]. Plasmas generated in supercritical fluids combine the advantageous transport properties of SCFs with the high reactivity of plasmas. This allows the synthesis of novel nanomaterials or facilitates the tailoring of existing nanomaterials that could be used in a variety of applications, such as medicine, pharmaceutics, biotechnology, nanooptics, and nanoelectronics.

Here we review the current state of research on the synthesis of a special class of carbon nanomaterials that recently has gained (renewed) interest in the scientific community, diamondoids [5,6], by pulsed laser ablation (PLA) plasmas generated in SCFs. We briefly describe the structure and physical properties of diamondoids and their possible applications before introducing the techniques used for realizing chemical reactions by PLA in high-pressure and SCF media. Finally, we present the latest results of the synthesis of diamondoids by PLA in supercritical Xe (scXe) and supercritical CO_2 (scCO$_2$). It is anticipated that this overview will further increase the interest in this fascinating class of carbon nanomaterials and also further fuel the application of pulsed laser plasmas generated in SCFs for the synthesis of nanomaterials.

9.1.1 STRUCTURE AND PHYSICAL PROPERTIES OF DIAMONDOIDS

Macroscopically, carbon can exist under different forms or allotropes: amorphous carbon, graphite, hexagonal diamond (Lonsdaleite), and cubic diamond. In the last 25 years, other members of carbon allotropes at the nanoscale have been discovered [7]: carbyne and polyyne [8], fullerenes [9], carbon nanotubes [10], graphene [11], nanodiamonds, and diamondoids [5].

The ability of carbon to form different chemical bonds that result in this large variety of nanomaterials is due to the electronic structure of carbon.

The 2s and 2p atomic orbitals of carbon can be mixed in different combinations that can lead to the formation of either sp^1-, sp^2-, or sp^3-hybridized orbitals. Figure 9.1 illustrates the carbon nanostructures formed as a result of the three types of hybridization. For sp^1 hybridizations, diacetylene (or butadiene) can be considered the basic block for the assembly of linear chains, cumulene, or polyyne. The linking of fragments of ovalene (sp^2) conceptually leads to the formation of fullerenes, whereas corannulene can be considered to be a basic unit for the creation of carbon nanotubes (CNTs) and graphenes.

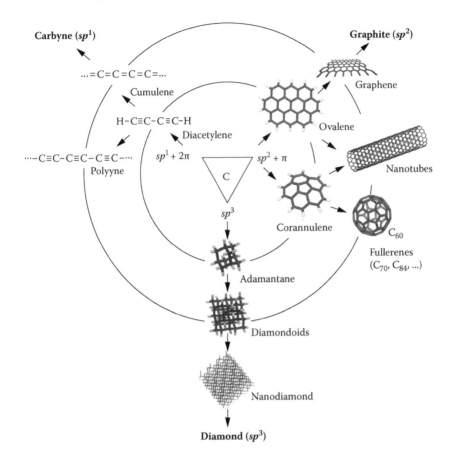

FIGURE 9.1 Overview of carbon nanostructures formed as a consequence of the different hybridizations of C, sp^1, sp^2, and sp^3. sp^1 hybridization leads to the formation of linear chains, cumulene, or polyyne. For sp^2-hybridized carbon nanostructures, corannulene can be considered a building block for fullerene, whereas ovalene can be regarded as a precursor for carbon nanotubes or graphene. In contrast to fullerenes, carbon nanotubes, and graphene, the C-C bonds in adamantane are sp^3 hybridized and the fusion of a number of these building blocks conceptually leads to the formation of diamondoids and, as the number of cages increases further, nanodiamonds. The molecular structures in this and following figures have been rendered by visual molecular dynamics (VMD). (From W. Humphrey et al., *J. Mol. Graph.*, 14(1), 33, 1996. With permission.)

Finally, sp^3 hybridization leads to the formation of a rigid cage structure consisting of carbon atoms.

Therefore the main difference between the other carbon nanomaterials, fullerenes, CNTs, and graphene that are comprised of sp^2-hybridized bonds between carbon atoms and diamondoids is that for the latter, their structure consists of $C(sp^3)$-$C(sp^3)$ bonds. As illustrated in Figure 9.2, the carbon cages comprising the diamondoids can be superimposed on a diamond lattice. The terminations of the carbon cage structure are well defined and usually consist of hydrogen atoms. The first member

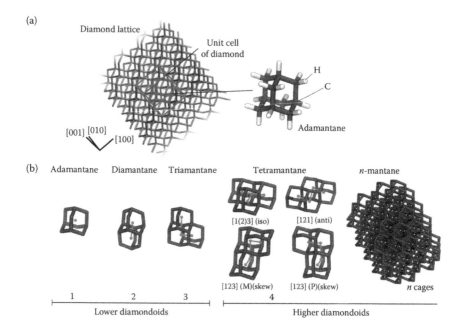

FIGURE 9.2 Relation of the diamondoid structure and diamondoid lattice and molecular structures of lower and higher diamondoids. (a) The first member of diamondoids, adamantane, can be regarded as the basic unit of diamond, and diamondoids can be superimposed on a diamond lattice. (b) Structures of lower diamondoids adamantane, diamantane, triamantane, and their relation to a 279-carbon tetrahedral diamond lattice. For higher clarity, the H-terminations are not shown except for one adamantane. For the diamondoids from $n = 1$ to $n = 4$ the dual graphs for the systematic classification of diamondoids introduced by Balaban and Schleyer [13] are also indicated.

of the diamondoid series is admantane ($C_{10}H_{16}$, number of cages $n = 1$), and larger diamondoids can conceptually be formed by fusing increasingly larger numbers of these building blocks. The next members in the diamondoid series are diamantane ($n = 2$) and triamantane ($n = 3$). From tetramantane ($n = 4$), the possibilities for fusing additional diamondoid cages grow and the number of isomers increases. This leads to different diamondoid series with the molecular formulae $C_{4n+6}H_{4n+12}$, $C_{4n+5}H_{4n+10}$, $C_{4n+2}H_{4n+6}$, $C_{4n+4}H_{4n+8}$, $C_{4n+1}H_{4n+4}$, $C_{4n-2}H_{4n}$, $C_{4n-5}H_{4n-4}$, and $C_{4n-3}H_{4n-2}$ [5].

In order to facilitate the classification of diamondoids, a nomenclature based on graph theory was developed [13]. For this, the centers of the diamondoid cages are connected by lines, the shortest possible path of such a dual graph giving the name of the diamondoid, together with the Greek expression for the number of fused adamantane cages and the suffix -*mantane*. Figure 9.2(b) shows the structure of the lower diamondoids adamantane, diamantine, and triamantane, and the first higher-order diamondoid, tetramantane, with their corresponding dual graphs and notation. As illustrated for $n = 4$, the cages can be fused in different combinations, leading to the formation of the [1(2)3](anti), [121](iso), [123](M)(skew), and [123](P)(skew) structural isomers of tetramantane.

While the fusion of increasingly higher numbers of adamantane cages approaches the size of nanodiamonds and ultimately diamond, it is important to make a clear distinction between nanodiamonds and diamondoids. Usually the term *nanodiamond* is used in a broader context for a variety of structures that contain diamond nanocrystals, present, for instance, in interstellar dust and meteorites, diamond particles nucleated in the gas phase or on substrates, and nanocrystalline films [7]. Consequently, nanodiamonds may have a wide range of size scales, size distributions, and also consist of mixtures containing other elements (i.e., doped nanodiamonds). In contrast, diamondoids "are chemically well defined, of high purity, and structurally well characterized" [6]. Moreover, while the interior of nanodiamonds is composed of sp^3-bonded networks, their surface can be terminated by a variety of functional groups or consist of reconstructed surfaces formed by sp^2-hybridized graphitic networks [14]. In contrast with nanodiamonds, diamondoids have a well-defined structure and the sp^3-bonded carbon cages are terminated by H-atoms or other functional groups. For this reason, diamondoids are sometimes also called diamond molecules.

Therefore despite advances in the purification and functionalization, the major disadvantages of nanodiamonds in comparison to diamondoids are their large size and usually broad size distribution [6] that might limit their application in nanodevices.

In the following sections, we will give a short overview of the possible applications of diamondoids and their derivatives and describe the methods that have been used for obtaining them, either by isolation from natural gas and oil reservoirs or by synthesis using pulsed laser plasmas generated in supercritical xenon and CO_2.

9.1.2 APPLICATIONS OF DIAMONDOIDS AND DIAMONDOID DERIVATIVES

In recent years, diamondoids have elicited increasing interest as possible future nanomaterials for use in a variety of applications.

Work by several groups suggests that the special properties of diamondoids, i.e., high thermal stability, well-defined structure, and nontoxicity, make them interesting for applications in various fields, pharmaceutics, medicine, biotechnology, and nano- and optoelectronics [15].

Because of their nontoxicity and high thermal stability, diamondoids could be used in medical applications, e.g., as biomarkers or as scaffolds for new pharmaceutical drugs. For instance, two derivatives of adamantane, memantine and amantadine, shown in Figure 9.3(a), have been used to treat Alzheimer's [17] and Parkinson's [18] diseases, respectively, while another adamantane derivative, rimantadine, has shown antiviral activity [19].

Concerning the use of diamondoids in optoelectronics, it has been reported that large-area self-assembled monolayers (SAMs) of a functionalized diamondoid, [121] tetramantane-6-thiol, could be used for highly monochromatic electron photoemission [20]. The structure of such SAMs is illustrated in Figure 9.4(a). Recent work has shown that the diamondoid monolayer structure and thiol substitution can be controlled, permitting us to change the electronic structure of the SAMs [21].

SAMs of [121]tetramantane-6-thiol on Au or Ag films possess quantum yields larger than unity, and the reason for monochromatic photoelectron emission of

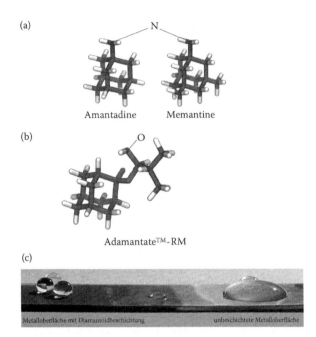

(a)

Amantadine   Memantine

(b)

Adamantate™-RM

(c)

Metalloberfläche mit Diamantoidbeschichtung   unbeschichtete Metalloberfläche

FIGURE 9.3 Examples of current applications of diamondoids. (a) Amantadine and memantine used for treating Alzheimer's and Parkinson's diseases, respectively. (b) Derivatives of adamantane, such as Adamantate™, are used as additives in photoresists to increase the thermal resistance. (c) Inclusion of diamondoids reduces the wettability of metallic coatings (left) when compared to a non-treated metal surface (right). (The photograph (c) is reprinted with permission from H. Schwertfeger and P.R. Schreiner, *Chem. Unserer Zeit,* 44(4), 248–253, 2010, Copyright 2010, John Wiley & Sons.)

diamondoid SAMs is that like diamonds, diamondoids possess a negative electron affinity (NEA) [22].

In-depth studies on the mechanisms of monochromatic photoelectron emission revealed that the NEA of diamondoids results from unoccupied states in the lowest unoccupied molecular orbitals (LUMOs), which, when populated by an electron, directly lead to spontaneous electron emission [23].

The detailed mechanism is considered to consist of several steps (see Figure 9.4(b)). Electrons of the metal substrate are excited by photons to unoccupied energy levels above the Fermi level, E_F, which lies in the middle of the energy gap of the diamondoid. The vacuum level (E_{vac}) is below the conduction band minimum of the diamondoid, which is characteristic of NEA materials.

In the second step, the excited electrons attain thermal equilibrium, producing more electrons with lower energy. Then electrons with energy above the conduction band minimum transfer to the diamondoids. These electrons further lower their energies by exciting phonons in the diamondoid molecules, finally accumulating at the bottom of the conduction band. In the last step, electrons accumulated at the bottom of the conduction band emit into vacuum spontaneously. The main electron emission takes place at the low-kinetic energy threshold. At higher kinetic energy levels, the photoelectron yield is lower.

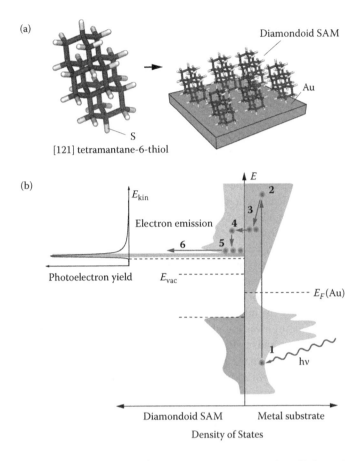

FIGURE 9.4 Use of diamondoid SAMs as possible sources for efficient photoelectron emitters. (a) Structure of [121]tetramantane-6-thiol and SAM deposited on Au surface. (b) Schematic of the currently understood processes leading to spontaneous photoelectron emission: (1) Photons excite electrons in the metal substrate, (2) the electrons pass to unoccupied states above the Fermi level (E_F), (3) energy transfer of the excited electrons produces additional electrons with lower energies, (4) electrons with energies above the conduction band minimum of the diamondoid transfer to the diamondoid molecule, (5) these electrons further lower their energies by exciting phonons in the diamondoid molecule, accumulating at the bottom of the conduction band, and (6) because of NEA the electrons accumulated at the bottom of the conduction band emit into vacuum spontaneously.

In contrast to Si and Ge nanoparticles, where the LUMO is core confined, the LUMO of diamondoids is a delocalized surface state [24]. The charge transfer mechanism, illustrated in Figure 9.4(b), that allows the electron to be promoted from the metal into the diamondoid LUMO is very efficient, electrons being emitted almost instantaneously with an upper bound of a few femtoseconds. This suggests that if suitable pulsed light sources are available, diamondoid SAMs could be used as ultrashort (~fs), high-brilliance pulsed electron sources for use in electron microscopy, electron beam lithography, and for field-emission flat-panel displays and photocathodes.

Investigations of diamondoids up to [121]tetramantane also showed that the dielectric constant κ is about half that of bulk diamond (κ ~ 5.6), ranging from 2.46 for adamantane to 2.68 for tetramantane [25]. This suggests that diamondoids could be new candidates as low-κ materials for microelectronics applications.

By density functional theory (DFT) simulations, the effect of different doping strategies on the band gap of diamondoids has been studied [26]. The approaches that were investigated were the enlargement and changing of the shapes/morphologies of the particles, the effect of CH bond substitutions with various functional groups (external doping), and finally, the effect of incorporating heteroatom functionalities of one or more CH or CH_2 fragments (interstitial or internal doping). The authors found that increasing the size of diamondoids from $C_{10}H_{16}$ (adamantane) to about 2 nm ($C_{286}H_{144}$) leads to a decrease of the band gap from 9.4 to 6.7 eV, which is still larger than the band gap of bulk diamond (5.5 eV). It was also found that it is the size, and not the morphology of the diamondoids, that influences the band gap most. The authors also discovered that band gap tuning through external (by CH bond substitution) or internal (by replacing CH or CH_2 groups) doping is nonadditive for the same dopant. This means that doubling the number of same functional groups attached to a diamondoid does not necessarily lead to a linear decrease in its band gap. Functionalization by attaching electron-donating and electron-withdrawing groups (push-pull doping) was found to be most effective and permitted us to reduce the band gap of diamondoids to that of bulk diamond.

The effect of clustering on the band gap has also been investigated. DFT calculations showed that in contrast to bulk diamond, crystalline adamantane has a direct band gap. As the transition of electrons across the band gap can accompany the absorption and emission of photons, the efficiency of this process can be expected to be far better in crystalline diamondoid materials than in indirect band gap materials [27].

In summary, the well-defined structure of diamondoids, which distinguishes them from nanodiamonds, makes them predestined as molecular building blocks in a wide range of applications. However, while certain physical properties of diamondoids have already been demonstrated experimentally, the main problem is that they are not readily accessible. Especially higher diamondoids with cage numbers $n \geq 4$ are still not available in large enough quantities to be used at an industrial scale.

In the following section, the methods used until now for obtaining higher diamondoids by isolation from natural gas and oil reservoirs are presented.

9.1.3 NATURAL OCCURRENCE OF DIAMONDOIDS AND THEIR ISOLATION

Diamondoids present in natural sources such as gas and oil reservoirs are believed to have formed from kerogen macromolecules in sedimentary rocks. The presence of mineral catalysts such as aluminosilicates present in such sediments seems to promote the formation of diamondoids [28]. The term *diamondoid* was coined by the German chemist Decker, who tried to synthesize diamonds. The structure of the first diamondoid, adamantane, was predicted in 1933 by Kleinfeller and Frercks. Adamantane was isolated from oil reservoirs near Hodonin, Czechoslovakia, the same year, while the next diamondoid member, diamantane, was isolated in 1966 [29].

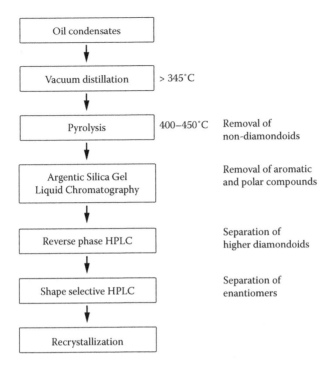

FIGURE 9.5 Schematic of the separation process for obtaining single crystals of higher diamondoids [5]. Oil condensates are further concentrated by vacuum distillation at temperatures above 345°C. Volatile components and non-diamondoids are cracked by pyrolysis conducted in the temperature range of 400–450°C. In a next step, polar and aromatic compounds are removed by silica-gel-based column chromatography. Individual types of higher diamondoids are isolated by reverse phase HPLC, enantiomeric diamondoids being separated by shape-selective HPLC. The separated fractions are finally recrystallized.

Diamondoids were often considered a nuisance by oil mining companies, because of the clogging of oil pipes. In the 1990s, Chevron Oil looked for methods to solve these problems and for how to isolate diamondoids from petroleum. Dahl et al. [5] then gave the first proof for the existence of higher diamondoids up to undecamantane ($n = 11$). They quickly realized that these compounds could be very interesting for other applications and patented methods for the extraction of diamondoids.

The isolation of higher diamondoids and their purification has to be done in several steps, as indicated in Figure 9.5. First, oil condensates are concentrated by vacuum distillation at temperatures above 345°C. Pyrolysis at temperatures between 400 and 450°C permits removal of volatile components and non-diamondoids. Remaining polar and aromatic compounds are then removed by silica-gel-based column chromatography. To separate individual diamondoids and enantiomers, reverse phase high-performance liquid chromatography (HPLC) followed by shape-selective HPLC is used. The separated fractions can then be recrystallized.

In addition to the isolation and purification processes developed [5] and briefly described above, Iwasa et al. [30] presented a vapor transport technique where a two-zone furnace is used. With this approach, it is possible to both obtain diamondoid single crystals with volumes up to ~ 1 cm³ and reduce the number of impurities considerably. In the next section, we briefly give an overview on conventional synthesis approaches of diamondoids and their limitations.

9.1.4 CONVENTIONAL SYNTHESIS OF DIAMONDOIDS AND ITS LIMITATIONS

A brief history on the synthesis of lower diamondoids up to tetramantane is given in the paper by Schwertfeger et al. [6].

Following the fabrication of lower diamondoids, the organic synthesis of higher diamondoids was the topic of intensive efforts up to the early 1980s [29]. However, it could be demonstrated that while lower diamondoids can be synthesized by carbocation equilibration reactions, this approach does not work for higher diamondoids. The main problems encountered were the lack of suitable precursors, the rapidly growing number of possible isomers, intermediates being trapped in local energy minima, and the formation of unwanted side products. As a result, research on carbocation-mediated syntheses of diamondoids beyond tetramantane was stopped at the beginning of the 1980s [31], and up to now, only one of the higher diamondoids, [121]tetramantane, has been synthesized, but the reaction was complex and the reaction yields very low [32].

The interest in diamondoids was only renewed when it was reported that diamondoids with a number of cages (n) up to 11 were isolated from crude oil [5]. However, the quantities in which diamondoids are available decreases inversely to their size.

In order to fabricate diamondoids, especially those with a large number of cages, and to make them available in larger quantities, it is necessary to develop alternative synthesis methods that allow us to synthesize and functionalize diamondoids. Two approaches that have shown promise are electric discharge and pulsed laser plasmas generated in SCFs. In the following section, we discuss the generation of pulsed laser plasmas in supercritical fluids and their application for the fabrication of nanomaterials, especially diamondoids.

9.2 GENERATION OF PULSED LASER PLASMAS IN SUPERCRITICAL FLUIDS

9.2.1 BASIC CHARACTERISTICS AND CURRENT APPLICATIONS OF SUPERCRITICAL FLUIDS

As discussed in the previous section, one of the major obstacles in conventional organic synthesis of diamondoids is that reactions have to be realized close to thermodynamic equilibrium.

In contrast, owing to the presence of reactive species, i.e., radicals, ions, and electrons present in plasmas, reactions far from thermodynamic equilibrium can be realized. Here we will not discuss the properties of plasmas in detail, as there is a vast amount of literature available [3,4,33].

First, we will give a short introduction on supercritical fluids (SCFs), their main physical properties, and some of their current applications.

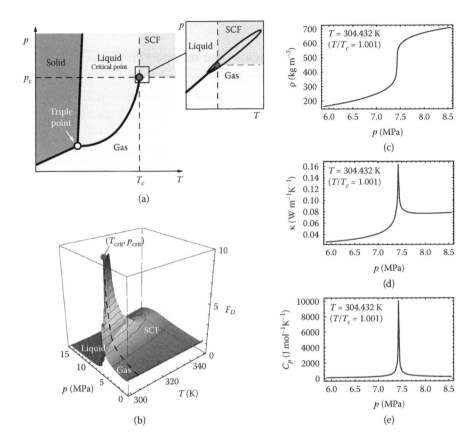

FIGURE 9.6 Phase diagram and variation of physical properties of supercritical fluids around the critical point. (a) Phase diagram of a pure substance, the domains corresponding to solid, liquid, gaseous, and supercritical phases. Above the critical point, the interface between liquid and gaseous phases disappears. The inset shows the compressible region around the critical point. (b) Density fluctuation of CO_2 near the critical point. (c) Variation of the density ρ, (d) thermal conductivity κ, and (e) thermal capacity at constant pressure C_p at a temperature of 304.432 K (T/T_{crit} = 1.001) for CO_2. While in the supercritical domain, the density ρ changes continuously with pressure, and the density fluctuation F_D, the thermal conductivity κ, and thermal capacity C_p reach their maximum near the critical point. (Data retrieved from NIST Chemistry Webbook, *Thermophysical Properties of Fluid Systems*, Technical Report, NIST, 2011.)

SCFs are defined as media in a state of temperature (T) and pressure (p) above their critical values, $T \geq T_{crit}$ and $p \geq p_{crit}$, as shown in the phase diagram of Figure 9.6(a). While for fluids below the critical temperature and critical pressure, the liquid and gaseous phases are clearly separated, at temperatures and pressures above the critical point, the two phases cannot be discriminated anymore.

Therefore from a macroscopic point of view, SCFs represent an intermediate state between liquids and gases. As such, SCFs possess high density, high diffusivity, high solubility, and low surface tension. In addition, there is a continuity in their

TABLE 9.1

Comparison of Average Physical and Transport Properties of Liquid, SCF, and Gaseous States

Property	Liquid	SCF	Gas
Density (kg m^{-3})	600–1600	200–900	0.6–2.0
Viscosity ($\times 10^5$ kgm^{-1} s^{-1})	20–300	1–9	1–3
Diffusion coefficient ($\times 10^8$ m^2 s^{-1})	0.02–0.2	1–40	1000–4000
Thermal conductivity ($\times 10^3$ Wm^{-1} K^{-1})	80	20–250	4–30
Surface tension (dyne cm^{-1})	20–450	0	0

Note: Most of the thermophysical properties of SCFs are intermediate between those of gases and liquids, but there are also properties that are higher in SCFs when compared to liquids or gases. For example, the thermal conductivity κ reaches a maximum near the critical point.

thermophysical properties, and their density, heat capacity, and dielectric constant can be varied continuously by changing the pressure, temperature, or both. For example, while water at ambient pressure and temperature is a polar solvent and consequently does not dissolve apolar substances easily, the dielectric constant of supercritical H_2O (scH_2O) can be varied so that nonpolar organic substances can be dissolved in scH_2O.

However, not all properties of SCFs are intermediate between those of liquids and gases. In the vicinity of the critical point, the thermal conductivity (κ), specific heat (C_p), and compressibility (β) attain a maximum (Figure 9.6(d) and (e)), and they are significantly higher compared to the respective values of the gaseous or liquid states (cf. [34–37] for data on thermal conductivity and specific heat).

Table 9.1 summarizes typical properties of liquid, supercritical, and gaseous phases, while Table 9.2 lists the values of critical temperature T_{crit}, pressure p_{crit}, and density ρ_{crit} for elements that are typically used as SC media.

In Figure 9.6(c–e), the variation of the density ρ, the thermal conductivity κ, and the heat capacity C_p of CO_2 as a function of pressure between 5.90 and 8.55 MPa ($p/p_{crit} = 0.8 – 1.16$) at a temperature of 304.432 K ($T/T_{crit} = 1.001$) are shown.

These unique physical characteristics derive from the microscopic fluid structure—molecular clustering—of SCFs. While macroscopically, a SCF appears to be homogeneous, on a microscale the structure of the fluid is very heterogeneous. Especially in close vicinity of the critical point, the competition between weak attractive forces and thermal motion leads to rapid and repeated aggregation and redispersion of clusters.

In SCFs, such interactions between atoms or molecules are mainly governed by weak van der Waals forces. Molecular dynamics studies on solvent clustering in SCF solutions have shown that the average lifetime of a cluster, i.e., the exchange of particles in a cluster shell, is on the order of picoseconds (ps; 10^{-12} s) [40].

One consequence of molecular clustering is an increase in the density fluctuation (F_D). The density fluctuation is a measure of local density enhancements and is defined by Equation (9.1) ([41], p. 96):

TABLE 9.2
Critical Constants (Temperature T_{crit}, Pressure p_{crit}, and Density ρ_{crit}) of Most Common Media Used in Supercritical Applications

Chemical Formula	T_{crit} (K)	p_{crit} (MPa)	ρ_{crit} (g cm⁻³)
He	5.19	0.227	0.070
Ne	44.4	2.76	0.481
Ar	150.87	4.898	0.533
Kr	209.41	5.50	0.921
Xe	289.77	5.841	1.113
H_2	32.97	1.293	0.031
H_2O	647.14	22.06	0.322
CO_2	304.13	7.375	0.468
N_2	126.21	3.39	0.311
O_2	154.59	5.043	0.438

Source: Reprinted with permission from S. Nakahara et al., *J. Appl. Phys.*, 109(12), 123304, Copyright 2011, American Institute of Physics.
Note: Because the critical point of CO_2 is close to ambient conditions, it is currently the most commonly used SCF. Due to its very high reactivity, scH_2O is increasingly used for the oxidization of toxic waste.

$$F_D = \frac{\langle\langle N \rangle\rangle}{\langle N \rangle} = \frac{\langle\left(N - \langle N \rangle\right)^2\rangle}{\langle N \rangle} \tag{9.1}$$

where N is the total number of particles in the system and $\langle N \rangle$ is the average number of particles. For estimating F_D, one can use its relation to the isothermal compressibility given by

$$Z_T = -\frac{1}{V}\left(\frac{\partial V}{\partial p}\right)_{T,N} \tag{9.2}$$

and the isothermal compressibility of a perfect gas

$$Z_T^0 = \frac{1}{nk_BT} \tag{9.3}$$

where $n = \langle N \rangle/V$ and one can show that [41]

$$F_D = \frac{Z_T}{Z_T^0} \tag{9.4}$$

The plot in Figure 9.6(b) shows the variation of F_D for CO_2 as a function of pressure and temperature near the critical point. Near the critical point, the value of F_D reaches a maximum. Besides this maximum, the region of increased F_D also extends into both liquid and gaseous regions around the critical point. A consequence of the increase of the density fluctuation is that around the critical point, the compressibility of a SCF is greatly enhanced.

The unique properties of SCFs make them attractive for a wide range of applications. For instance, the solvating power of SCFs permits replacement of environmentally much more harmful liquid organic solvents by supercritical CO_2 (scCO_2) or water (scH_2O). In addition, in general SCFs are noncarcinogenic, nontoxic, nonmutagenic, nonflammable, and thermodynamically stable [2].

Owing to their temporal and spatial density fluctuations, high solubility, and density, SCFs are used as solvents in a variety of chemical processes, extraction of components from petroleum using supercritical pentane, or extraction of drugs from plant materials [42].

The use of SCFs may be advantageous for reactions involved in fuels processing, biomass conversion, biocatalysis, homogeneous and heterogeneous catalysis, environmental control, polymerization, and materials and chemical synthesis [43].

In the vicinity of the critical point, where the thermal conductivity [34–36] and specific heat [37] attain their maximum values, it is expected that new types of reactions involving clusters can be used, and that by changing the conditions of the SCF medium, the selectivity of reactions can be adjusted.

Furthermore, it has been shown that the use of media in a supercritical state near the critical point can lead to enhanced reaction rates and selectivity of chemical reactions. This has been attributed to the enhanced molecular clustering and density fluctuation near the critical point [44–46].

Because of their high dissolving power, SCFs have also shown excellent potential for chemical analysis applications. Supercritical fluid chromatography (SFC), using scCO_2 as a mobile phase, is especially suited for the separation of compounds that are sensitive to organic solvents commonly used in HPLC [47,48]. SFC often shows superior performance compared to conventional HPLC because, besides avoiding the use of organic mobile phases such as acetonitrile, the diffusion coefficients are an order of magnitude higher than in liquids (cf. Table 9.1). Consequently, the transfer of solutes through the separation column encounters less resistance, with the result that separations may be realized more rapidly or with higher resolution in comparison to HPLC.

In addition to chemical synthesis and as solvents in chemical analysis, SCFs have also shown promising potential for use in materials processing.

Among the main commercial applications of SCF media are textile dyeing and food processing, e.g., for the decaffeination of coffee [49,50]. Furthermore, in recent years, the use of SCFs, in particular scCO_2, for reducing water and organic solvents in the microelectronics industry has been investigated [51].

Because of the high dissolving power, scCO_2 can be used for photoresist stripping or for the cleaning of substrates. For instance, Namatsu et al. [52,53] demonstrated the drying of microstructured silicon and photoresist patterns without pattern collapse. The pattern collapse in high aspect ratio Si patterns is avoided owing to the low surface tension of the supercritical medium, which reduces the stresses due to capillary forces on opposite trenches.

Another application of SCFs in materials processing is chemical fluid deposition (CFD). CFD involves the reduction of organometallic compounds that are soluble in SCFs, such as $scCO_2$, using hydrogen, which is soluble in $scCO_2$, as a reducing agent. Owing to the low surface tension, CFD allows deposition of conformal thin films of various materials. Compared to thermal CVD where typically temperatures of about 600°C are required, CFD processes can be realized at more modest temperatures. In addition, CFD allows a wider choice of precursors, the reason being that in contrast to CVD, where all species have to be in the gas phase, also precursors with lower vapor pressures can be used. Thus far CFD has been used to deposit Pt films by hydrogenolysis of dimethyl-(cyclooctadiene) platinum(II) ($CODPtMe_2$) on Si, on polymer substrates, and in nanoporous Al_2O_3 in $scCO_2$ ($p = 15.5$ MPa) at modest temperatures of 80°C [54]. Blackburn et al. [55] used CFD for the conformal deposition of Cu and Ni films in high aspect ratio trenches etched in silicon. It has been shown that besides noble and near-noble metals (Cu, Ni, Pt, Pd, Ru), CFD could also be used to deposit alloys and oxides such as RuO_x, TiO_2, Al_2O_3, and ZnO [56].

In recent years, SCFs have been used increasingly for the design of advanced nanostructured materials [57]. In particular, SCFs have been used for the production of a wide range of nanomaterials or the functionalization of nanomaterials: nanofibers, nanorods, nanowires, nanotubes, conformational films, core-shell structures, supported nanoparticles, polymers impregnated with nanoparticles, and organic coatings of particles [58]. Such SCF-based technologies have also been proposed for polymer processing for pharmaceutical and medical applications [59].

All these applications that take advantage of the specific properties of SCFs have turned out to be generally more flexible, simpler, and with a reduced environmental impact because the SCF can be completely eliminated at the end of the process.

However, the main drawback of chemical synthesis in SCFs is that they have to be conducted near thermodynamic equilibrium, which often makes it necessary to work at high temperatures. In the next section, it will be shown that plasmas generated in SCFs offer the possibility of realizing reactions far from thermodynamic equilibrium. This permits the synthesis of nanomaterials that cannot be obtained easily by conventional synthesis techniques in SCFs.

9.2.2 GENERATION OF PULSED LASER PLASMAS IN SUPERCRITICAL FLUIDS

While SCFs show higher reaction rates than synthesis in the liquid or gas phase, it has been suggested that the combination of plasmas with SCFs could lead to even higher efficiency for realizing chemical reactions. The reason is the formation of highly reactive species, ions, electrons, and radicals in the plasma that would allow realization of reactions far from thermodynamic equilibrium.

There are two main approaches for generating plasmas in SCFs: The first is by electric discharges, and the second by pulsed laser ablation.

It has been shown that plasmas generated in SCFs show interesting properties. For instance, it has been discovered that by using very small electrodes with gap distances below 10 μm, the breakdown voltage reaches a minimum near the critical point [60]. This behavior is contrary to what could be expected from classic discharge theory. This decrease of the breakdown voltage has been the same irrespective of whether it was an atomic gas such as Xe, a nonpolar molecule such as CO_2, or a polar molecule such as H_2O [61].

Following these experiments, microplasmas in SCFs, mainly $scCO_2$ and scXe, have been used for the fabrication of nanomaterials. For example, sp^2-hybridized carbon nanomaterials, e.g., CNTs, nanohorns, and nano-onions, were synthesized by generating either short pulse direct-current (pulse duration: 400 μs) plasmas or low-temperature dielectric barrier discharges (DBDs) in neat $scCO_2$ [62,63,64].

In contrast to SCF processes that were used for the synthesis of CNTs [65,66] and fullerenes [65], conducted at either high temperature or high pressure and using catalysts, plasmas generated in $scCO_2$ allowed fabrication of CNTs at milder conditions (i.e., temperatures close to room temperature) and without the use of any catalyst.

However, while plasmas in SCFs enable reactions without any catalysts and far from thermodynamic equilibrium, the generation of electric discharges in very high pressure or even SCFs is not very straightforward. The breakdown voltage depends a lot on the electrode geometry (electrode gap, dielectric layer thickness) and the conditions (pressure, temperature, and composition) of the SCFs. In contrast, pulsed lasers allow the formation of dense plasmas under various conditions, from high vacuum, liquid, and SCFs.

By using pulsed lasers, the generation of a plasma in high-pressure or supercritical conditions is easier and more independent of the medium.

Plasmas can be generated either by irradiation of a solid target or by inducing breakdown of the supercritical medium by focusing the laser beam in a small spot.

Pulsed laser ablation (PLA) by plasmas that are generated by irradiating a solid target has been used since the appearance of Ruby lasers in the early 1960s. The PLA process was studied in detail for the first time by Patil and coworkers [67]. They found that the important chemical reactions take place at the interface between the plasma plume and surrounding medium [68].

There are two main mechanisms of plasma formation by PLA: The first is by nanosecond lasers, while the second is for femto- or picosecond lasers. It is generally a highly nonequilibrium process, and the heating, forming of the plasma plume, and material ejection occur after the laser pulse [69–72].

In Figure 9.7, the mechanisms leading to plasma formation by irradiation of a target by different types of pulsed lasers are shown. While the detailed mechanisms are still not understood yet, depending on the type of laser, continuous wave (CW), nanosecond, or pico- or femtosecond, the mechanisms leading to the removal of material or plasma formation are different. In the case of CW lasers, material is removed primarily by melting, which creates a large heat-affected zone (HAZ), and material ejection is mainly dominated by thermal processes [73]. In nanosecond lasers, there are three main stages that lead to the formation of a plasma. In the first, laser photons couple with both electrons and phonons of the target material. The photon-electron coupling then results in an immediate rise of the electron temperature, leading to vaporization of the target. Compared to CW lasers, the HAZ created by nanosecond pulsed lasers is smaller.

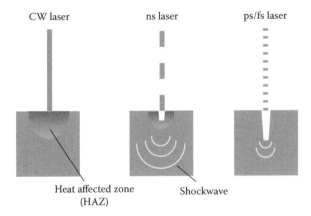

CW laser   ns laser   ps/fs laser

Heat affected zone   Shockwave
(HAZ)

FIGURE 9.7 Mechanisms of plasma formation and material removal by pulsed laser ablation. Depending on the type of laser, the interaction between the laser light and the target material varies. CW lasers remove material primarily by melting, which creates a large heat-affected zone (HAZ). Nanosecond laser pulses create a smaller HAZ, and material is removed by melt expulsion driven by the vapor pressure and recoil pressure. With ultrafast pulses of the order of pico- or femtoseconds, the laser pulse duration is much shorter than the timescale for energy transfer between free electrons and the material lattice. This leads to very high pressures and temperatures limited to a very shallow zone of only a few micrometers, where the material is vaporized immediately.

With ultrafast pico- and femtosecond pulses, the laser pulse duration is much shorter than the timescale for energy transfer between free electrons, and the material lattice and electrons are excited to only a few or few tens of electron volts. Consequently, the lattice temperature of the target remains unchanged, and the main amount of the laser pulse energy is primarily absorbed in a thin layer of only a few microns close to the surface, where extremely high pressures and temperatures can be attained. The absorbed energy heats the material very quickly past the melting point, directly to the vapor phase with its high kinetic energy, and the material is removed by direct vaporization. Consequently, in the case of pico- and femtosecond pulsed lasers, mainly the photon absorption depth governs the heated volume, the influence of thermal diffusion depth being smaller.

With nano-, pico-, and femtosecond pulsed lasers becoming more and more available, PLA has been gaining increasing attention due to its potential for a wide range of materials processing: deposition of thin solid films, nanocrystal growth, surface cleaning, and the fabrication of microelectronic devices.

Besides the generation of pulsed laser plasmas in vacuum, pulsed laser ablation plasmas have recently found increased use in liquids, for both the synthesis and functionalization of nanomaterials [74].

Pulsed laser ablation in water has also been used for the synthesis of nanodiamonds [75]. The nanodiamond particle consists of a diamond core with diameters ranging from 5 to 15 nm, surrounded by a graphitic shell with a thickness ranging from 3 to 4 nm. It was found that the synthesis of nanodiamonds occurs in a very narrow temporal window, at the early stages of plasma formation. The shock

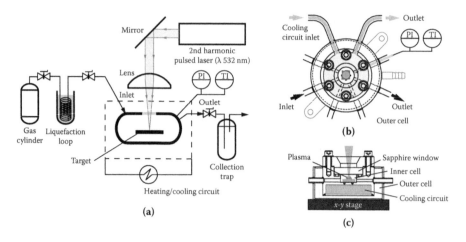

FIGURE 9.8 Experimental setup for realizing diamondoid synthesis by pulsed laser ablation in $scCO_2$ and scXe. (a) Schematic of the experimental setup consisting of a stainless steel high-pressure cell, cooling/heating circuit, liquefaction loop (typically cooled by liquid nitrogen) for condensing the gas, and pressure and temperature indicators (PI and TI). After the end of the experiment, the chamber is purged and the synthesized materials collected in a trap connected to the outlet. (b) Top view of the high-pressure cell and cooling/heating circuit that allows adjustment of the temperature in the inner cell. (c) Side view of the high-pressure cell. The pulsed laser light is focused on a target through a sapphire window, and the high intensity of the laser leads to the rapid heating of a zone close to the laser spot and formation of the plasma. The schematics of the high-pressure cell were provided by Taiatsu Techno Co.

wave duration was discovered to be roughly twice the pulse duration (~10 ns), during which the pressure inside the shock wave reached values between 4.5 and 22.5 GPa. After the shock wave subdued, the pressure values were too low for generating diamondoids, and only sp^2 carbon was formed.

For generating plasmas in SCFs, the use of high-pressure cells that can withstand the high-pressure of SCF media is necessary. The most common high-pressure cells are fabricated out of stainless steel, but for applications needing optical access, cells consisting of quartz are used. When using highly reactive SCFs such as supercritical water, special alloys, e.g., hastelloy, are employed.

A schematic of a high-pressure cell that was used for realizing pulsed laser plasmas in $scCO_2$ and scXe for the synthesis of diamondoids is presented in Figure 9.8.

The high-pressure stainless steel cell consists of two parts, an inner cell that contains the pressurized gas and an outer cell that is connected to a water-circulating circuit and that allows adjustment of the temperature in the inner cell. The temperature and pressure in the inner cell are monitored by a thermocouple and pressure sensor, respectively. A window made out of sapphire is placed on top of the cell for optical access and for focusing the laser beam on the surface of the target that has been placed inside the high-pressure cell.

For condensing Xe and to reach SCF conditions, the gas was flown through a liquefaction loop that was cooled by liquid nitrogen. In the case of CO_2, the gas was compressed using a high-pressure pump. After the experiment, the synthesized materials have to be collected in a trap connected to the outlet of the high-pressure cell.

9.2.3   APPLICATION OF PULSED LASER PLASMAS IN SUPERCRITICAL FLUIDS TO NANOMATERIALS SYNTHESIS

In contrast to PLA in liquids, PLA in SCFs has not been used as frequently up to now. Probably the main difficulty lies with the necessity to use chambers that can withhold the high pressures of SCFs.

While xenon and CO_2 are the two gases whose critical temperatures are closest to room temperature, Xe is about 1000 times more expensive than other SCF media, such as Ar and CO_2 [2]. This makes Xe less attractive as a supercritical medium for both science and industry. On the other hand, $scCO_2$ is one of the most widely used SCFs, having already found applications in various fields. Because of this, most of the PLA in SCFs so far is based on using $scCO_2$.

As one example of PLA in SCF, the fabrication of gold nanoparticles by pulsed laser ablation in $scCO_2$ has been investigated [76]. In addition to noble metal nanoparticles, the same authors also investigated the formation of silicon nanocrystals (Si-nc) by pulsed laser ablation of Si targets [77]. The authors show that the size and crystal structure of the Si-nc (core vs. shell) could be changed by adjusting the density of the SCF medium.

Sasaki and Aoki [78] investigated the effect of the pressure of the supercritical medium on the formation of nanoparticles. They found that the quenching rates near the critical point reach a maximum. Therefore the conditions of the SCF can be adjusted in such a way as to optimize the size of the nanomaterials.

In short, while compared to liquids, PLA in SCFs offers finer tuning of the medium, and therefore can be used to adjust the characteristics of the nanomaterials formed, it is still not widely used. In the next section we describe the application of PLA in scXe and $scCO_2$ for the synthesis of diamondoids.

9.3   SYNTHESIS OF DIAMONDOIDS BY PULSED LASER PLASMAS

Despite the intricacy of synthesizing higher diamondoids consisting of a minimum of four cages, two groups independently reported the synthesis of higher diamondoids using lower diamondoids as precursors using electric discharge plasmas generated in supercritical Xe (scXe) [79–81] and by hot-filament CVD [82].

In these previous studies, dielectric barrier discharges (DBDs) [79–81] were used to produce a highly dense plasma reaction field in scXe. Higher diamondoids consisting of up to 10 cages were synthesized by the generation of plasmas in supercritical Xe (scXe; $T_{crit} = 289.7$ K, $p_{crit} = 5.84$ MPa) with dissolved adamantane ($n = 1$) as a precursor. The key factors for the artificial synthesis of diamondoids were considered to be the high density and highly nonequilibrium reaction field of the plasmas generated in the SCF media, especially in the vicinity of the critical point, and the use of dissolved adamantane as a precursor.

TABLE 9.3
Experimental Conditions for the Pulsed Laser Ablation in scCO$_2$ with and without Cyclohexane

Parameter	scCO$_2$	scCO$_2$ with Cyclohexane
Adamantane concentration (mol/L)	$(1.2–12) \times 10^2$	$(3.1–6.2) \times 10^2$
Cyclohexane concentration (mol/L)	—	7.7×10^1
Temperature T (K)	304.5–305.2	304.5–304.8
Reduced temperature (T/T_{crit})	1.001–1.004	1.001–1.002
Pressure p (MPa)	7.42–7.59	7.54–7.56
Reduced pressure (p/p_{crit})	1.005–1.028	1.022–1.024

Source: Reprinted with permission from S. Nakahara et al., *J. Appl. Phys.*, 109(12), 123304, Copyright 2011, American Institute of Physics.

Note: The addition of cyclohexane facilitates the dissolution of the precursor adamantane in the scCO$_2$.

9.3.1 EXPERIMENTAL PROCEDURE

An HOPG target (mosaic spread: 3.5°–5°) was placed in a batch-type high-pressure cell (inner volume: 3 cm^3), as described in Figure 9.8. Adamantane (purity > 99.0%; Tokyo Chemical Industry) was then dissolved in scCO$_2$ with and without cyclohexane as a cosolvent; see the experimental conditions listed in Table 9.3.

The high-density CO$_2$ was realized by condensation of the CO$_2$ gas inside a liquefaction loop cooled by liquid nitrogen before introducing it into the inner cell. A second-harmonic neodymium-doped yttrium aluminum garnet (Nd:YAG) laser (λ = 532 nm) with a pulse width of 7 ns at a repetition rate of 10 Hz was used for the ablation of the HOPG targets in both scXe [84] and scCO$_2$. The maximum energy was 7 mJ per pulse, giving a maximum fluence of approximately 18 J/cm^2 on the surface of the target. On average, the ablation experiments were conducted for periods ranging from 60 to 240 min. The synthesized products were retrieved by evacuating the gas through a collection trap containing 10 ml of cyclohexane.

In addition, the cyclohexane solutions used for cleansing the cell were also collected. The HOPG target was dipped in the collection trap and sonicated. The concentration of the products was then increased by partial evaporation of the collected organic solvents.

Following the procedure established for the isolation of higher diamondoids [5], to remove non-diamondoids and to improve the analysis of the diamondoids by gas chromatography–mass spectrometry (GC-MS), volatile compounds, and non-diamondoids, the samples were treated by pyrolysis in Ar [83].

After retrieval and concentration of the synthesized products, micro-Raman spectroscopy and GC-MS measurements were conducted for as-collected samples, and both micro-Raman spectroscopy and GC-MS measurements were also conducted for the sample after pyrolysis.

Samples of the synthesized products for assessment by micro-Raman spectroscopy were prepared by spotting small volumes (~ 0.1 ml) of the suspensions on aluminum substrates. Micro-Raman spectroscopy was performed using a second-harmonic Nd:YAG laser (λ = 532 nm) with an excitation power of 12 mW and a spatial resolution of around 1 μm. Spectra were measured using an acquisition time of 1 min. For the GC-MS measurements, the GC was set up in splitless sample injection mode, which, compared to split injection, allows analysis of very small quantities of sample. The MS measurements were conducted by electron-impact mass spectrometry at 70 eV, in both selected ion monitoring (SIM) mode and scan mode on the products dissolved in 1 L of cyclohexane. SIM ions included combinations of m/z varying from 188 to 590 that could correspond to diamondoids with cage numbers varying between n = 2 and 12. Mass scans were acquired over an m/z range from 40 to 1000 [83].

9.3.2 MICRO-RAMAN SPECTROSCOPY

The characteristics of the Raman spectra of diamondoids can be summarized by dividing them into three regions: (1) strong peaks in the low-energy region below 500 cm^{-1}, due to vibrational modes of the CCC structure, such as CCC bend and deformation modes; (2) well-resolved peaks in the range 1000–1500 cm^{-1}, due to CH twist and wagging modes; and (3) strong and broad peaks in the high-energy region [85]. For the high-energy region, it is reported that the stretching mode of sp^3-bonded CH$_x$ shows peaks in the region of 2800–2950 cm^{-1}, while sp^2-bonded CH$_x$ shows peaks in the region of 2980–3060 cm^{-1} [86]. Comparing the measured spectrum with these characteristics might help to identify the synthesized products, even though these characteristics are common in not only diamondoids but also other hydrocarbons, such as paraffin [87]. In Figure 9.9, the Raman spectra of the products obtained by PLA in scCO$_2$ with and without cyclohexane as a cosolvent (both before and after pyrolysis) are presented together with the Raman spectrum of the material obtained by PLA in scXe [84].

For illustrating the effect of increasing cage numbers on the Raman spectra, in Figure 9.9(f) and 9(g) the spectra of commercial adamantane and [12312] hexamantane [85] are shown. The Raman spectra of the products obtained by PLA in scCO$_2$, regardless of the presence of cyclohexane, present almost the same features that are known to be characteristics of Raman spectra of diamondoids, i.e., well-resolved peaks in the range 1000–1500 cm^{-1} and strong peaks in the high-energy region, 2800–2950 cm^{-1}. On the other hand, the peaks in the measured Raman spectra are supposed to originate from a mixture of different hydrocarbons. Moreover, it is known that the Raman scattering signal varies largely for each material; e.g., sp^2-hybridized carbon shows a much stronger Raman signal than that of sp^3-hybridized carbon [86,88]. Therefore it is possible that the peaks from synthesized diamondoids and those from other hydrocarbons, such as paraffin and aromatics, get balanced out, which could explain the lack of strong peaks in the low-energy region below 500 cm^{-1}. The features in the wavenumber range between 100 and 1500 cm^{-1}, and the peaks in the range between 2800 and 2950 cm^{-1}, which are characteristic of sp^3 CH$_x$ stretching modes [86], indicate the presence of diamond-structured hydrocarbons in the synthesized products. The existence of diamond-structured hydrocarbons in the synthesized products was also indicated for PLA in scCO$_2$, both with and without cyclohexane. In all the Raman spectra of the synthesized materials, no indication

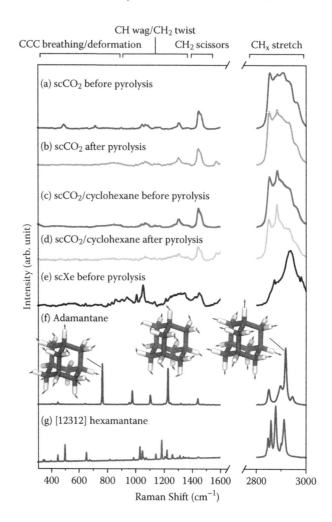

FIGURE 9.9 Micro-Raman spectra of synthesized products obtained by PLA and compared to spectra obtained on diamondoids isolated from crude oil. (a) Products synthesized in scCO$_2$ before pyrolysis. (b) Spectrum obtained on products after pyrolysis. (c) In scCO$_2$ with cyclohexane before pyrolysis. (d) After pyrolysis. (e) In scXe before pyrolysis [83]. For illustrating the effect of increasing cage number, the Raman spectra of (f) commercial adamantane and (g) [12312] hexamantane (taken from [85]) are included as examples. (f) The three strongest Raman active vibration modes for adamantane, corresponding to C-C breathing, CH wag/CH$_2$ twist and CH$_x$ stretching modes, respectively, are also indicated. (Adapted with permission from S. Nakahara et al., *J. Appl. Phys.*, 109(12), 123304, Copyright 2011, American Institute of Physics.)

of the precursor adamantane could be found. This can be explained by the fact that adamantane sublimes even at room temperature, and that for increasing the concentration of the samples, the solutions collected after PLA were evaporated. While the Raman spectra collected from the products obtained by PLA in scCO$_2$ with and without cyclohexane showed similar characteristics, smaller differences that are found in the Raman spectra for the samples obtained in $scCO_2$ with cyclohexane as a cosolvent can be attributed to the possible synthesis of species formed by the dissociation of cyclohexane. In contrast, the differences between the products obtained by PLA in $scCO_2$ and scXe [84] were larger. For example, the strongest intensities in the low-energy region (300–1500 cm^{-1}) appear around 1440 cm^{-1} for PLA in $scCO_2$ with and without cyclohexane, while the most intense Raman peaks appear around 1050 cm^{-1} for PLA in scXe. Moreover, the shapes of the strong peaks in the high-energy region (2800–3000 cm^{-1}) vary considerably. These dissimilarities indicate that the synthesized materials are different for PLA in $scCO_2$ and scXe. The effects of pyrolysis for the samples obtained by PLA in $scCO_2$ and in $scCO_2$/cyclohexane are shown in Figure 9.9(a–d). Compared to the spectra before pyrolysis, the biggest difference can be observed in the high-wavenumber region between 2980 and 3000 cm^{-1}. The intensity of the peaks that are attributed to sp^2 CH_x stretching modes [86] is lower after pyrolysis. In addition, the peaks in the high-wavenumber region become better resolved. This shows that pyrolysis permits reduction of the number of most non-diamondoid hydrocarbons. However, as-collected products from plasmas generated in SCFs are typically a mixture of hydrocarbons, including many kinds of non-diamondoid products, and consequently, micro-Raman spectra of as-collected samples contain peaks of mixed materials. Accordingly, the analysis of the micro-Raman spectra only allows a confirmation of the presence of diamondoids. Therefore except for pure, recrystallized samples obtained by isolation and purification, it is not possible to identify individual diamondoids.

9.3.3   Gas Chromatography–Mass Spectrometry

The technique of choice for the identification of individual synthesized diamondoids is gas chromatography–mass spectrometry (GC-MS). The method allows us to both separate individual diamondoids as a function of the temperature ramp of the GC and identify individual diamondoids or diamondoid derivatives. The main difficulty is to achieve conditions where the diamondoids are put into the gas phase. Consequently, relatively high temperature ramps of up to 360°C are necessary for being able to detect higher diamondoids. Typical features of unsubstituted higher diamondoids found in GC-MS analysis are a strong molecular ion peak M^+ and a few fragment ions, including one at m/z $M^+/2$, which corresponds to the doubly charged molecular ion.

9.3.3.1   Synthesis of Diamantane

In order to prove the synthesis of diamondoids from the precursor adamantane by PLA in scXe and $scCO_2$, the retention times for m/z 136 and m/z 188, which correspond to the molecular weights (MWs) of adamantane and diamantane ($n = 2$), respectively, were compared for products obtained by PLA in $scCO_2$ with and without cyclohexane, and a reference solution containing dissolved commercially available adamantane and diamantane. The comparison revealed the same time lag, 3.5 min, for all experimental conditions. As shown in Figure 9.10, the mass spectra of the GC-MS measurements at the retention times of SIM for m/z 188 for both products from PLA in $scCO_2$ with and without cyclohexane present almost the same fragments as those of the standard diamantane.

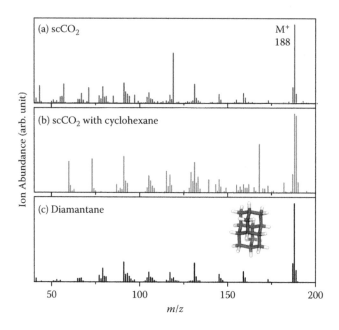

FIGURE 9.10 GC-MS mass spectra acquired at a retention time of t = 7.3 min. (a) Synthesized product obtained in neat scCO$_2$. (b) Product synthesized in scCO$_2$ and cyclohexane. (c) Mass spectrum of diamantane of the standard sample. The inset shows the molecular structure of diamantane. The main features of the mass spectra are the molecular ion peak at m/z 188 and fragment peaks that are characteristic of diamantine. (Adapted with permission from S. Nakahara et al., *J. Appl. Phys.*, 109(12), 123304, Copyright 2011, American Institute of Physics.)

To estimate the production rate of diamantane, a calibration curve has to be used. For this, cyclohexane solutions with varying diamantane concentrations are prepared and the SIM peak areas corresponding to diamantane are measured for each concentration. By comparing the SIM peak areas of the synthesized products, the production rate of diamantane can then be estimated. In the case of the PLA experiments conducted in scCO$_2$ both with and without cyclohexane, the estimated production rate was 0.2 μg/h [83] (see Table 9.4).

9.3.3.2 Synthesis of Diamondoids with $n \geq 3$

In a first approximation, the larger the diamondoid, the longer the retention time. SIM curves and mass spectra indicated the possible synthesis of diamondoids consisting of three or more cages by PLA in scCO$_2$. Besides triamantane (n = 3), peaks were observed that could be attributed to diamondoids consisting of up to 12 cages. The cage numbers of eluted diamondoids obtained by PLA in scCO$_2$ with and without cyclohexane as a function of the experimentally determined GC-MS retention times are listed in Tables 9.5 and 9.6, and Figure 9.11, respectively.

The relative retention times were calculated by dividing the individual retention times by the retention time of diamantane in the range 7.148–7.157 min. The GC-MS signals could be assigned to the following diamondoids: triamantane at

TABLE 9.4

SIM Peak Area and Estimated Diamantane Concentration of the Product Obtained by PLA in scCO$_2$

Supercritical Medium for PLA	Solution for GC-MS Analyses	SIM Peak Area (Arb. Unit)	Diamantane Concentration (mg/L)
scCO$_2$	Sample in 1 ml cyclohexane	2100	0.23
scCO$_2$ with cyclohexane	Sample in 1 ml cyclohexane	1900	0.21

Source: Reprinted with permission from S. Nakahara et al., *J. Appl. Phys.,* 109(12), 123304, Copyright 2011, American Institute of Physics.

TABLE 9.5

List of Possibly Synthesized Diamondoids by PLA in scCO$_2$, with Their Molecular Formulae, MWs, GC-MS Relative Retention Times, and Dual Graph Codes

Cage Number (*n*)	Molecular Formula	M$^+$ (*m/z*) Base Peak	GC-MS Relative Retention Time	Structure
2	C$_{14}$H$_{20}$	188	1.000	Diamantane
3	C$_{18}$H$_{24}$	240	1.189	Triamantane
4	C$_{22}$H$_{28}$	292	1.509	E.g., [121]
6	C$_{29}$H$_{34}$	382	1.986	E.g., [12(1)31]
	C$_{30}$H$_{36}$	396	2.038	E.g., [12121]
7	C$_{30}$H$_{34}$	394	1.843	E.g., [123121]
10	C$_{35}$H$_{36}$	456	2.111	[1231241(2)3]
12	C$_{45}$H$_{46}$	586	2.495	E.g., [12131431234]

Source: Reprinted with permission from S. Nakahara et al., *J. Appl. Phys.,* 109(12), 123304, Copyright 2011, American Institute of Physics.

m/z 240; tetramantane at *m/z* 292; hexamantanes at *m/z* 342, 382, and 396; heptamantane at *m/z* 394; octamantane at *m/z* 446; decamantanes at *m/z* 456 (C$_{45}$H$_{46}$) and 590; and dodecamantane at *m/z* 586. From Figure 9.11 one can see that the GC-MS retention times increase almost linearly with the number of cages, which is in agreement with previous work [89]. Furthermore, for the relative retention times of the diamondoids obtained by PLA in scCO$_2$ both with and without cyclohexane, the authors found good agreement: triamantane (1.188–1.189), tetramantane (1.509–1.510), hexamantanes (1.986–1.987 for a MW of 382 and 2.038–2.039 for a MW of 396), and dodecamantane (2.493–2.495). In addition, some of the observed diamondoids, corresponding to MWs of 382, 586, and 590, were not found in previous works in which the diamondoids were isolated from petroleum [5]. These molecules were detected at high GC temperature, higher than the maximum GC temperature used by another group (593 K). For example, molecular ion

TABLE 9.6

List of Possibly Synthesized Diamondoids by PLA in scCO$_2$ with Cyclohexane as a Cosolvent, with Their Molecular Formulae, MWs, GC-MS Relative Retention Times, and Dual Graph Codes

Cage Number (n)	Molecular Formula	M$^+$ (m/z) Base Peak	GC-MS Relative Retention Time	Structure
2	C$_{14}$H$_{20}$	188	1.000	Diamantane
3	C$_{18}$H$_{24}$	240	1.188	Triamantane
4	C$_{22}$H$_{28}$	292	1.510	E.g., [121]
6	C$_{26}$H$_{30}$	342	1.627	[12312]
	C$_{29}$H$_{34}$	382	1.987	E.g., [12(1)31]
	C$_{30}$H$_{36}$	396	2.039	E.g., [12121]
8	C$_{34}$H$_{38}$	446	2.156	E.g., [1213(1)21]
10	C$_{45}$H$_{50}$	590	2.423	E.g., [12(3)1(2)3(1)23]
12	C$_{45}$H$_{46}$	586	2.493	E.g., [12131431234]

Source: Reprinted with permission from S. Nakahara et al., *J. Appl. Phys.*, 109(12), 123304, Copyright 2011, American Institute of Physics.

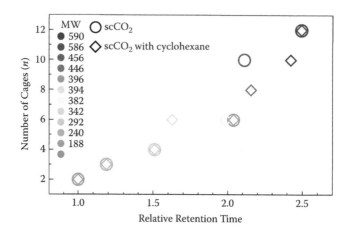

FIGURE 9.11 Variation of the number of diamondoid cages obtained by scCO$_2$ with and without cyclohexane as a function of the GC-MS relative retention time. The retention time of diamantane (t = 7.148–7.157 min) is taken as a reference retention time, and the grayscale map indicates the molecular weights (MWs) of the detected diamondoids. Diamondoids with higher molecular weights need increasingly longer elution times for being detected by MS, the increase is almost linear.

peaks corresponding to a MW of 586 appeared when the GC oven temperature was raised to 620 K. In GC-MS measurements, each diamondoid is expected to elute at a distinct retention time. In contrast, for higher diamondoids, there are MW groups that have many structural isomers. For example, MW 292 (tetramantane) has 4

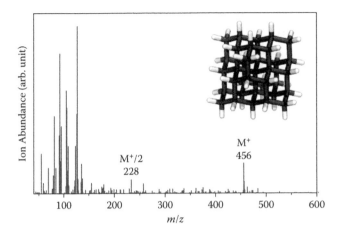

FIGURE 9.12 GC-MS mass spectrum of possible superadamantane (MW 456) obtained by PLA in scCO$_2$. The inset shows the molecular structure of superadamantane. (Adapted with permission from S. Nakahara et al., *J. Appl. Phys.*, 109(12), 123304, Copyright 2011, American Institute of Physics.)

possible isomers and MW 396 (hexamantane) has 28. Therefore in the GC-MS measurements, diamond molecules corresponding to the same MW group but possessing different structures are expected to elute at different retention times.

Depending on the setting of the heating cycle and the resolution of the GC-MS, it is expected that different isomers can appear at the same scan times, even if they elute at different times. Slower heating rates or faster scan rates can help to improve the resolution of GC-MS analyses. Figure 9.12 shows an example of mass spectra corresponding to a MW of 456. This diamondoid is the most compact with $n = 10$, [1231241(2)3] decamantane, also known as superadamantane.

Strong peaks at low m/z in Figure 9.12 are supposed to be due to either other hydrocarbon molecules such as paraffin and aromatics that co-elute at the same retention time, or other materials that originated from the GC column and the septum that appear in the entire GC-MS analysis.

Moreover, non-diamondoids and other unresolved complex mixtures co-eluting with the diamondoids affect the resolution of GC-MS analyses [90].

Other studies on the isolation of diamondoids from petroleum reported that the extremely high thermal stability of diamondoids can be used to separate them from non-diamondoids [5]. All higher diamondoids are thermally stable up to a temperature of 723 K, which is high enough to decompose most other non-diamondoid hydrocarbons [91].

9.3.4 EFFECTS OF PYROLYSIS ON THE SYNTHESIZED PRODUCTS

The mass spectra acquired after the pyrolysis experiments and corresponding to possible higher diamondoids are listed in Tables 9.7 and 9.8.

The observed peaks were different before and after pyrolysis. However, the molecular ion peak with a MW of 456 was observed both before and after pyrolysis.

TABLE 9.7

List of Observed Higher Diamondoids for PLA in scCO$_2$ after Pyrolysis

Cage Number (n)	Molecular Formula	M$^+$ (m/z) Base Peak	GC-MS Relative Retention Time
5	$C_{25}H_{30}$	330	1.702.25
5	$C_{26}H_{32}$	344	2.302.31
7	$C_{30}H_{34}$	394	2.02
7	$C_{34}H_{40}$	448	2.45
9	$C_{34}H_{36}$	444	1.66
9	$C_{37}H_{40}$	484	2.20
10	$C_{35}H_{36}$	456	2.122.13
10	$C_{44}H_{48}$	576	2.57

Source: Reprinted with permission from S. Nakahara et al., *J. Appl. Phys.*, 109(12), 123304, Copyright 2011, American Institute of Physics.

TABLE 9.8

List of Observed Higher Diamondoids for PLA in scCO$_2$ with Cyclohexane after Pyrolysis

Cage Number (n)	Molecular Formula	M$^+$ (m/z) Base Peak	GC-MS Relative Retention Time
4	$C_{22}H_{28}$	292	1.34
5	$C_{26}H_{32}$	344	1.83
6	$C_{30}H_{36}$	396	1.67
7	$C_{32}H_{36}$	420	2.01
7	$C_{34}H_{40}$	448	2.45

Source: Reprinted with permission from S. Nakahara et al., *J. Appl. Phys.*, 109(12), 123304, Copyright 2011, American Institute of Physics.

Moreover, the relative retention times of GC-MS were very close to each other. This result suggests that the synthesized products with a MW of 456 survived after pyrolysis. It is expected that non-diamondoid hydrocarbons of high MW will decompose after thermal treatment at 723 K. Therefore the present results suggest that the peak corresponding to a MW of 456 originated from superadamantane. Many MW groups of diamondoids, such as those with $n = 5$ and 7 cages, were not found before pyrolysis, although they were found after the pyrolysis experiments (see Figure 9.13).

In scCO$_2$, species originating from the decomposition of CO$_2$ could possibly participate in the formation of new materials. Assuming those materials were non-diamondoid hydrocarbons and were decomposed after pyrolysis, this might explain

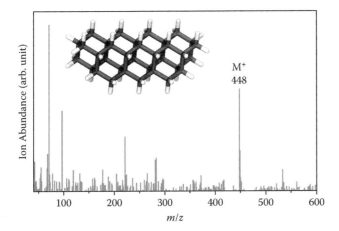

FIGURE 9.13 GC-MS mass spectra of possible diamondoids after pyrolysis: MW 448 (heptamantane, $C_{34}H_{40}$) obtained by PLA in $scCO_2$ with cyclohexane. The inset shows the molecular structure of [121212] heptamantane, which is one of the possible isomers possessing MW 448. (Adapted with permission from S. Nakahara et al., *J. Appl. Phys.*, 109(12), 123304, Copyright 2011, American Institute of Physics.)

why the observed spectra contained peaks corresponding to diamondoids with five and seven cages only after pyrolysis.

9.3.5 COMPARISON BETWEEN PLA IN scCO₂ AND SCXE

There are MW groups that were only synthesized in scXe and not in $scCO_2$ [84]. Moreover, the production rate of diamantane of PLA in $scCO_2$ was 0.2 μg/h, which is lower than that of PLA in scXe. This could be attributed to the lower solubility of adamantane in $scCO_2$ than that in scXe. The production rate of diamantane is approximately the same in neat $scCO_2$ and in the $scCO_2$/cyclohexane mixture. Therefore while the solubility of adamantane in $scCO_2$ can be enhanced using cyclohexane as a cosolvent, this is probably not the only factor for increasing the production rate. In the case of scXe, the molecules of the medium do not participate in the reaction, whereas in the case of $scCO_2$, CO_2 itself can also dissociate and form reaction products. Consequently, it is suspected that the lower reaction yield of diamantane in $scCO_2$ is due to both the lower solubility of adamantane and competing reactions resulting in the formation of non-diamondoids. In contrast, the mass spectrum showing a MW of 456 was only obtained by PLA in pure $scCO_2$. C-C and C-H bonds must be dissociated to allow the formation of successive diamondoids from adamantane. In the previous work where diamondoids were synthesized by DBD in scXe [80], it was suggested that the absence of oxidants resulted in a preferential dissociation of C-C bonds and led to the synthesis of diamondoids with a high H/C ratio. On the contrary, the products obtained from PLA in $scCO_2$ included relatively low H/C ratio diamondoids compared with those in scXe. It is supposed that oxidant species originating from $scCO_2$ might lead to selective dissociation of C-H bonds, enabling the synthesis of low H/C ratio diamondoids, such as superadamantane.

9.4 CONCLUSIONS AND PERSPECTIVES

Here we have reviewed diamondoids, their structure, and physical and chemical properties and given a short overview of their current main and possible future applications. While smaller diamondoids up to tetramantane can be synthesized, synthesis beyond tetramantane has turned out to be not possible by conventional organic chemical methods.

Plasmas generated in high-pressure or even supercritical media present a very reactive environment that allows the fabrication of nanomaterials far from thermodynamical equilibrium.

Compared to electric discharges, the generation of pulsed laser plasmas is relatively easy. PLA was performed in both scXe and scCO$_2$, with and without cyclohexane as a cosolvent, for the synthesis of diamondoids. Raman spectra of the synthesized products point to the presence of sp^3-hybridized materials, including diamondoids. The GC-MS measurements indicate the synthesis of diamantane and possibly other, higher-order diamondoids, including those with a number of cages larger than reported so far ($n = 12$) and superadamantane ($n = 10$). Because oxidant species originating from scCO$_2$ might lead to the selective dissociation of C-H bonds, synthesis of low H/C ratio higher-order diamondoids, such as superadamantane, could be realized. Moreover, there were more newly found higher-order diamondoids with $n = 5$–10 for PLA in scCO$_2$. It is expected that the fraction of non-diamondoid hydrocarbons might be effectively removed by pyrolysis. Therefore PLA in scCO$_2$ is considered to be a promising approach for the facile synthesis of higher-order diamondoids.

As a final conclusion, what are the perspectives for the use of PLA in SCFs for nanomaterials processing, in particular, diamondoids?

In comparison to electric discharges, PLA in SCFs allows generation of plasmas more easily, and it is less dependent on the conditions of the supercritical medium. In combination with SCFs, such plasmas permit realization of reactions that might not be possible to achieve with conventional chemical synthesis methods. The main drawbacks of the PLA approach are the difficulties related to scaling up, but this could be alleviated with further advances in laser research and the possible availability of smaller, cheaper pulsed laser systems.

Another difficulty is that this type of approach needs two main steps, synthesis and separation, of the products. However, one could imagine a continuous flow process, where the separation, isolation, and purification of the products is done directly after the synthesis.

In summary, we are convinced that the combination of PLA in SCFs offers many possibilities for the synthesis of not only diamondoids, but also other classes of nanomaterials.

ACKNOWLEDGMENTS

This work was supported financially in part by a Grant-in-Aid for Scientific Research on Innovative Areas (Frontier Science of Interactions between Plasmas and Nano-interfaces, Grant 21110002) from the Ministry of Education, Culture, Sports, Science, and Technology of Japan. The authors thank the Materials Design and

Characterization Laboratory, Institute for Solid State Physics, University of Tokyo for providing access to the pulsed laser facility, and Prof. M. Suzuki for assistance with the GC-MS measurements and helpful discussions.

REFERENCES

1. T.L. Chester, J.D. Pinkston, and D.E. Raynie. Supercritical fluid chromatography and extraction. *Anal. Chem.*, 68(12), R487–R514, 1996.
2. P.G. Jessop and W. Leitner, Eds. *Chemical Synthesis Using Supercritical Fluids*. Wiley-VCH Verlag GmbH, Berlin, 1999.
3. M.A. Lieberman and A.J. Lichtenberg. *Principles of Plasma Discharges and Materials Processing*. 2nd ed. Wiley-Interscience, New York, 2005.
4. A. Fridman. *Plasma Chemistry*. 1st ed. Cambridge University Press, Cambridge, 2008.
5. J.E. Dahl, S.G. Liu, and R.M.K. Carlson. Isolation and structure of higher diamondoids, nanometer-sized diamond molecules. *Science*, 299(5603), 96–99, 2003.
6. H. Schwertfeger, A.A. Fokin, and P.R. Schreiner. Diamonds are a chemist's best friend: Diamondoid chemistry beyond adamantine. *Angew. Chem. Int. Ed.*, 47(6), 1022–1036, 2008.
7. O.A. Shenderova, V.V. Zhirnov, and D.W. Brenner. Carbon nanostructures. *Crit. Rev. Solid State Mater. Sci.*, 27(3–4), 227–356, 2002.
8. R.R. Tykwinski, W. Chalifoux, S. Eisler, A. Lucotti, M. Tommasini, D. Fazzi, M. Del Zoppo, and G. Zerbi. Toward carbyne: Synthesis and stability of really long polyynes. *Pure Appl. Chem.*, 82(4), 891–904, 2010.
9. H.W. Kroto, J.R. Heath, S.C. Obrien, R.F. Curl, and R.E. Smalley. C-60—Buckminsterfullerene. *Nature*, 318(6042), 162–163, 1985.
10. S. Iijima. Helical microtubules of graphitic carbon. *Nature*, 354(6348), 56–58, 1991.
11. K.S. Novoselov, A.K. Geim, S.V. Morozov, D. Jiang, Y. Zhang, S.V. Dubonos, I.V. Grigorieva, and A.A. Firsov. Electric field effect in atomically thin carbon films. *Science*, 306(5696), 666–669, 2004.
12. W. Humphrey, A. Dalke, and K. Schulten. VMD: Visual molecular dynamics. *J. Mol. Graph.*, 14(1), 33, 1996.
13. A.T. Balaban and P.V. Schleyer. Systematic classification and nomenclature of diamond hydrocarbons.1. Graph-theoretical enumeration of polymantanes. *Tetrahedron*, 34(24), 3599–3609, 1978.
14. J.Y. Raty and G. Galli. Ultradispersity of diamond at the nanoscale. *Nat. Mater.*, 2(12), 792–795, 2003.
15. R.A. Freitas and R.C. Merkle. A minimal toolset for positional diamond mechanosynthesis. *J. Comput. Theor. Nanosci.*, 5(5), 760–861, 2008.
16. H. Schwertfeger and P.R. Schreiner. Future of diamondoids. *Chem. Unserer Zeit*, 44(4), 248–253, 2010.
17. B. Reisberg, R. Doody, A. Stoffler, F. Schmitt, S. Ferris, and H.J. Mobius. Memantine in moderate-to-severe Alzheimer's disease. *N. Engl. J. Med.*, 348(14), 1333–1341, 2003.
18. W.J. Geldenhuys, S.F. Malan, J.R. Bloomquist, A.P. Marchand, and C.J. Van der Schyf. Pharmacology and structure-activity relationships of bioactive polycyclic cage compounds: A focus on pentacycloundecane derivatives. *Med. Res. Rev.*, 25(1), 21–48, 2005.
19. V.G.H. Evidente, C.H. Adler, J.N. Caviness, and K. Gwinn-Hardy. A pilot study on the motor effects of rimantadine in Parkinson's disease. *Clin. Neuropharmacol.*, 22(1), 30–32, 1999.
20. W.L. Yang, J.D. Fabbri, T.M. Willey, J.R.I. Lee, J.E. Dahl, R.M.K. Carlson, P.R. Schreiner, A.A. Fokin, B.A. Tkachenko, N.A. Fokina, W. Meevasana, N. Mannella, K. Tanaka, X.J. Zhou, T. van Buuren, M.A. Kelly, Z. Hussain, N.A. Melosh, and Z.X. Shen. Monochromatic electron photoemission from diamondoid monolayers. *Science*, 316(5830), 1460–1462, 2007.

21. T.M. Willey, J.D. Fabbri, J.R.I. Lee, P.R. Schreiner, A.A. Fokin, B.A. Tkachenko, N.A. Fokina, J.E.P. Dahl, R.M.K. Carlson, A.L. Vance, W.L. Yang, L.J. Terminello, T. van Buuren, and N.A. Melosh. Near-edge x-ray absorption fine structure spectroscopy of diamondoid thiol monolayers on gold. *J. Am. Chem. Soc.*, 130(32), 10536–10544, 2008.

22. W.A. Clay, Z. Liu, W.L. Yang, J.D. Fabbri, J.E. Dahl, R.M.K. Carlson, Y. Sun, P.R. Schreiner, A.A. Fokin, B.A. Tkachenko, N.A. Fokina, P.A. Pianetta, N. Melosh, and Z.X. Shen. Origin of the monochromatic photoemission peak in diamondoid monolayers. *Nano Lett.*, 9(1), 57–61, 2009.

23. S. Roth, D. Leuenberger, J. Osterwalder, J.E. Dahl, R.M.K. Carlson, B.A. Tkachenko, A.A. Fokin, P.R. Schreiner, and M. Hengsberger. Negative-electron-affinity diamondoid monolayers as high-brilliance source for ultrashort electron pulses. *Chem. Phys. Lett.*, 495(1–3), 102–108, 2010.

24. N.D. Drummond, A.J. Williamson, R.J. Needs, and G. Galli. Electron emission from diamondoids: A diffusion quantum Monte Carlo study. *Phys. Rev. Lett.*, 95(9), 096801, 2005.

25. W.A. Clay, T. Sasagawa, M. Kelly, J.E. Dahl, R.M.K. Carlson, N. Melosh, and Z.X. Shen. Diamondoids as low-κ dielectric materials. *Appl. Phys. Lett.*, 93(17), 172901-1–172901-3, 2008.

26. A.A. Fokin and P.R. Schreiner. Band gap tuning in nanodiamonds: First principle computational studies. *Mol. Phys.*, 107(8–12), 823–830, 2009.

27. T. Sasagawa and Z.X. Shen. A route to tunable direct band-gap diamond devices: Electronic structures of nanodiamond crystals. *J. Appl. Phys.*, 104(7), 073704, 2008.

28. Z.B. Wei, J.M. Moldowan, J. Dahl, T.P. Goldstein, and D.M. Jarvie. The catalytic effects of minerals on the formation of diamondoids from kerogen macromolecules. *Org. Geochem.*, 37(11), 1421–1436, 2006.

29. R.I.K. Carlson, J.E.P. Dahl, S.G. Liu, M.M. Olmstead, P.R. Buerki, and R. Gat, Diamond molecules found in petroleum—New members of the H-terminated diamond series. In *Synthesis and Properties of Ultrananocrystalline Diamond*. Springer, Dordrecht, Berlin, 2005, pp. 63–78.

30. A. Iwasa, W.A. Clay, J.E. Dahl, R.M.K. Carlson, Z.X. Shen, and T. Sasagawa. Environmentally friendly refining of diamond-molecules via the growth of large single crystals. *Cryst. Growth Des.*, 10(2), 870–873, 2010.

31. E. Osawa, A. Furusaki, N. Hashiba, T. Matsumoto, V. Singh, Y. Tahara, E. Wiskott, M. Farcasiu, T. Iizuka, N. Tanaka, T. Kan, and P.V. Schleyer. Application of force-field calculations to organic-chemistry. 2. Thermodynamic rearrangements of larger polycyclic-hydrocarbons derived from the 38.5 and 41.5-degrees-c melting dimers of cyclooctatetraene—Crystal and molecular-structures of 5-bromoheptacyclo [8.6.0.02,8.03,13.0 4,11.05,9.012,16] hexadecane (5-bromo-(c2)-bisethanobisnordiamantane), 6,12-dibromoheptacyclo [7.7.0.02,6.03,15.04,12.05,10.011,16] hexadecane, and nonacyclo [11.7 .1.12,18.03,16.04,13.05,10.06,14.07,11.015,20] docosane (bastardane). *J. Org. Chem.*, 45(15), 2985–2995, 1980.

32. W. Burns, M.A. McKervey, T.R.B. Mitchell, and J.J. Rooney. New approach to construction of diamondoid hydrocarbons—Synthesis of anti-tetramantane. *J. Am. Chem. Soc.*, 100(3), 906–911, 1978.

33. Y.P. Raizer. *Gas Discharge Physics*. Springer-Verlag, Berlin, 1991.

34. C.E. Pittman, L.H. Cohen, and H. Meyer. Transport properties of helium near the liquid vapor critical point. 1. Thermal-conductivity of He-3. *J. Low Temp. Phys.*, 46(1–2), 115–135, 1982.

35. V. Vesovic, W.A. Wakeham, G.A. Olchowy, J.V. Sengers, J.T.R. Watson, and J. Millat. The transport-properties of carbon-dioxide. *J. Phys. Chem. Ref. Data*, 19(3), 763–808, 1990.

36. Z.H. Chen, K. Tozaki, and K. Nishikawa. Development of thermal conductivity measurement for fluids which is convenient and effective for samples near the critical point. *Jpn. J. Appl. Phys.*, 38(1AB), L92–L95, 1999.

37. Stanley M. Walas. *Phase Equilibria in Chemical Engineering*. Butterworth-Heinemann, London, 1985.
38. D.R. Lide, Ed. *Handbook of Chemistry and Physics*. Taylor & Francis, Boca Raton, FL, 2006.
39. NIST Chemistry Webbook. *Thermophysical Properties of Fluid Systems*. Technical Report. NIST, 2011.
40. C.C. Liew, H. Inomata, and S. Saito. Molecular-dynamics study on solvent clustering in supercritical-fluid solutions based on particle radial kinetic-energy. *Fluid Phase Equilib.*, 104, 317–329, 1995.
41. H.E. Stanley. *Introduction to Phase Transitions and Critical Phenomena*. Oxford University Press, Oxford, 1971.
42. C.A. Eckert, B.L. Knutson, and P.G. Debenedetti. Supercritical fluids as solvents for chemical and materials processing. *Nature*, 383(6598), 313–318, 1996.
43. P.E. Savage, S. Gopalan, T.I. Mizan, C.J. Martino, and E.E. Brock. Reactions at super-critical conditions—Applications and fundamentals. *AICHE J.*, 41(7), 1723–1778, 1995.
44. T. Toriumi, T. Kawakami, J. Sakai, D. Ogawa, and N. Higashi. (In Japanese). *J. Ind. Chem.*, 49, 1, 1946.
45. Y. Ikushima, S. Ito, T. Asano, T. Yokoyama, N. Saito, K. Hatakeda, and T. Goto. A Diels-Alder reaction in supercritical carbon-dioxide medium. *J. Chem. Eng. Jpn.*, 23(1), 96–98, 1990.
46. R.L. Thompson, R. Glaser, D. Bush, C.L. Liotta, and C.A. Eckert. Rate variations of a hetero-Diels-Alder reaction in supercritical fluid CO_2. *Ind. Eng. Chem. Res.*, 38(11), 4220–4225, 1999.
47. D.R. Gere. Supercritical fluid chromatography. *Science*, 222(4621), 253–259, 1983.
48. M.L. Lee and K.E. Markides. Chromatography with supercritical fluids. *Science*, 235(4794), 1342–1347, 1987.
49. K. Zosel. Separation with supercritical gases—Practical applications. *Angew. Chem. Int. Ed. Engl.*, 17(10), 702–709, 1978.
50. G.A. Montero, C.B. Smith, W.A. Hendrix, and D.L. Butcher. Supercritical fluid technology in textile processing: An overview. *Ind. Eng. Chem. Res.*, 39(12), 4806–4812, 2000.
51. G.L. Weibel and C.K. Ober. An overview of supercritical CO_2 applications in microelectronics processing. *Microelectron. Eng.*, 65(1–2), 145–152, 2003.
52. H. Namatsu, K. Yamazaki, and K. Kurihara. Supercritical drying for nanostructure fabrication without pattern collapse. *Microelectron. Eng.*, 46(1–4), 129–132, 1999.
53. H. Namatsu, K. Yamazaki, and K. Kurihara. Supercritical resist dryer. *J. Vac. Sci. Technol. B*, 18(2), 780–784, 2000.
54. J.J. Watkins, J.M. Blackburn, and T.J. McCarthy. Chemical fluid deposition: Reactive deposition of platinum metal from carbon dioxide solution. *Chem. Mater.*, 11(2), 213–215, 1999.
55. J.M. Blackburn, D.P. Long, A. Cabanas, and J.J. Watkins. Deposition of conformal copper and nickel films from supercritical carbon dioxide. *Science*, 294(5540), 141–145, 2001.
56. E. Kondoh, K. Sasaki, and Y. Nabetani. Deposition of zinc oxide thin films in supercritical carbon dioxide solutions. *Appl. Phys. Express*, 1(6), 061201, 2008.
57. F. Cansell and C. Aymonier. Design of functional nanostructured materials using super-critical fluids. *J. Supercrit. Fluids*, 47(3), 508–516, 2009.
58. E. Reverchon and R. Adami. Nanomaterials and supercritical fluids. *J. Supercrit. Fluids*, 37(1), 1–22, 2006.
59. E. Reverchon, R. Adami, S. Cardea, and G. Della Porta. Supercritical fluids processing of polymers for pharmaceutical and medical applications. *J. Supercrit. Fluids*, 47(3), 484–492, 2009.
60. T. Ito and K. Terashima. Generation of micrometer-scale discharge in a supercritical fluid environment. *Appl. Phys. Lett.*, 80(16), 2854–2856, 2002.

61. M. Sawada, T. Tomai, T. Ito, H. Fujiwara, and K. Terashima. Micrometer-scale discharge in high-pressure H_2O and Xe environments including supercritical fluid. *J. Appl. Phys.*, 100(12), 123304, 2006.
62. T. Ito, K. Katahira, Y. Shimizu, T. Sasaki, N. Koshizaki, and K. Terashima. Carbon and copper nanostructured materials syntheses by plasma discharge in a supercritical fluid environment. *J. Mater. Chem.*, 14(10), 1513–1515, 2004.
63. T. Tomai, T. Ito, and K. Terashima. Generation of dielectric barrier discharge in high-pressure N_2 and CO_2 environments up to supercritical conditions. *Thin Solid Films*, 506, 409–413, 2006.
64. T. Tomai, K. Katahira, H. Kubo, Y. Shimizu, T. Sasaki, N. Koshizaki, and K. Terashima. Carbon materials syntheses using dielectric barrier discharge microplasma in supercritical carbon dioxide environments. *J. Supercrit. Fluids*, 41(3), 404–411, 2007.
65. M. Motiei, Y.R. Hacohen, J. Calderon-Moreno, and A. Gedanken. Preparing carbon nanotubes and nested fullerenes from supercritical CO_2 by a chemical reaction. *J. Am. Chem. Soc.*, 123(35), 8624–8625, 2001.
66. Z.S. Lou, Q.W. Chen, W. Wang, and Y.F. Zhang. Synthesis of carbon nanotubes by reduction of carbon dioxide with metallic lithium. *Carbon*, 41(15), 3063–3067, 2003.
67. P.P. Patil, D.M. Phase, S.A. Kulkarni, S.V. Ghaisas, S.K. Kulkarni, S.M. Kanetkar, S.B. Ogale, and V.G. Bhide. Pulsed-laser induced reactive quenching at a liquid-solid interface—Aqueous oxidation of iron. *Phys. Rev. Lett.*, 58(3), 238–241, 1987.
68. S.B. Ogale, P.P. Patil, D.M. Phase, Y.V. Bhandarkar, S.K. Kulkarni, S. Kulkarni, S.V. Ghaisas, and S.M. Kanetkar. Synthesis of metastable phases via pulsed-laser-induced reactive quenching at liquid-solid interfaces. *Phys. Rev. B*, 36(16), 8237–8250, 1987.
69. T. Sakka, K. Takatani, Y.H. Ogata, and M. Mabuchi. Laser ablation at the solid-liquid interface: Transient absorption of continuous spectral emission by ablated aluminium atoms. *J. Phys. D Appl. Phys.*, 35(1), 65–73, 2002.
70. T. Sakka, K. Saito, and Y.H. Ogata. Emission spectra of the species ablated from a solid target submerged in liquid: Vibrational temperature of c-2 molecules in water-confined geometry. *Appl. Surf. Sci.*, 197, 246–250, 2002.
71. K. Saito, K. Takatani, T. Sakka, and Y.H. Ogata. Observation of the light emitting region produced by pulsed laser irradiation to a solid-liquid interface. *Appl. Surf. Sci.*, 197, 56–60, 2002.
72. K. Saito, T. Sakka, and Y.H. Ogata. Rotational spectra and temperature evaluation of c-2 molecules produced by pulsed laser irradiation to a graphite-water interface. *J. Appl. Phys.*, 94(9), 5530–5536, 2003.
73. M.N.R. Ashfold, F. Claeyssens, G.M. Fuge, and S.J. Henley. Pulsed laser ablation and deposition of thin films. *Chem. Soc. Rev.*, 33(1), 23–31, 2004.
74. V. Amendola and M. Meneghetti. Laser ablation synthesis in solution and size manipulation of noble metal nanoparticles. *Phys. Chem. Chem. Phys.*, 11(20), 3805–3821, 2009.
75. D. Amans, A.C. Chenus, G. Ledoux, C. Dujardin, C. Reynaud, O. Sublemontier, K. Masenelli-Varlot, and O. Guillois. Nanodiamond synthesis by pulsed laser ablation in liquids. *Diam. Relat. Mater.*, 18(2–3), 177–180, 2009.
76. K. Saitow, T. Yamamura, and T. Minami. Gold nanospheres and nanonecklaces generated by laser ablation in supercritical fluid. *J. Phys. Chem. C*, 112(47), 18340–18349, 2008.
77. K. Saitow and T. Yamamura. Effective cooling generates efficient emission: Blue, green, and red light-emitting Si nanocrystals. *J. Phys. Chem. C*, 113(19), 8465–8470, 2009.
78. K. Sasaki and S. Aoki. Temporal variation of two-dimensional temperature in a laser-ablation plume produced from a graphite target. *Appl. Phys. Express*, 1(8), 086001, 2008.
79. H. Kikuchi, S. Stauss, S. Nakahara, K. Matsubara, T. Tomai, T. Sasaki, and K. Terashima. Development of sheet-like dielectric barrier discharge microplasma generated in supercritical fluids and its application to the synthesis of carbon nanomaterials. *J. Supercrit. Fluids*, 55(1), 325–332, 2010.

80. S. Stauss, H. Miyazoe, T. Shizuno, K. Saito, T. Sasaki, and K. Terashima. Synthesis of the higher-order diamondoid hexamantane using low-temperature plasmas generated in supercritical xenon. *Jpn. J. Appl. Phys.*, 49(7), 070213, 2010.

81. T. Shizuno, H. Miyazoe, K. Saito, S. Stauss, M. Suzuki, T. Sasaki, and K. Terashima. Synthesis of diamondoids by supercritical fluid discharge plasma. *Appl. Phys. Express*, 50(3), 030207, 2011.

82. J.E.P. Dahl, J.M. Moldowan, Z. Wei, P.A. Lipton, P. Denisevich, R. Gat, S. Liu, P.R. Schreiner, and R.M.K. Carlson. Synthesis of higher diamondoids and implications for their formation in petroleum. *Angew. Chem. Int. Ed.*, 122(51), 10077, 2010.

83. S. Nakahara, S. Stauss, T. Kato, T. Sasaki, and K. Terashima. Synthesis of higher diamondoids by pulsed laser ablation plasmas in supercritical Co_2. *J. Appl. Phys.*, 109(12), 123304, 2011.

84. S. Nakahara, S. Stauss, H. Miyazoe, T. Shizuno, M. Suzuki, H. Katakoka, T. Sasaki, and K. Terashima. Laser ablation synthesis of diamond molecules in supercritical fluids. *Appl. Phys. Express*, 3(9), 096201, 2010.

85. J. Filik, J.N. Harvey, N.L. Allan, P.W. May, J.E.P. Dahl, S.G. Liu, and R.M.K. Carlson. Raman spectroscopy of diamondoids. *Spectrochim. Acta Pt. A Mol. Biol.*, 64(3), 681–692, 2006.

86. D. Ballutaud, F. Jomard, T. Kociniewski, E. Rzepka, H. Girard, and S. Saada. Sp(3)/sp(2) character of the carbon and hydrogen configuration in micro- and nanocrystalline diamond. *Diam. Relat. Mater.*, 17(4–5), 451–456, 2008.

87. N. Zhang, Z.J. Tian, Y.Y. Leng, H.T. Wang, F.Q. Song, and J.H. Meng. *Sci. China Ser. D Earth Sci.*, 50(8), 1171, 2007.

88. N. Wada, P.J. Gaczi, and S.A. Solin. Diamond-like 3-fold coordinated amorphous-carbon. *J. Non-Cryst. Solids*, 35(6), 543–548, 1980.

89. A.T. Balaban, D.J. Klein, J.E. Dahl, and R.M.K. Carlson. Molecular descriptors for natural diamondoid hydrocarbons and quantitative structure-property relationships for their chromatographic data. *Open Org. Chem. J.*, 1, 13–31, 2007.

90. L. Mansuy, R.P. Philp, and J. Allen. Source identification of oil spills based on the isotopic composition of individual components in weathered oil samples. *Environ. Sci. Technol.*, 31(12), 3417–3425, 1997.

91. B.S. Greensfelder, H.H. Voge, and G.M. Good. Catalytic and thermal cracking of pure hydrocarbons—Mechanisms of reaction. *Ind. Eng. Chem.*, 41(11), 2573–2584, 1949.

10 Molecular Lithography Using DNA Nanostructures

Sumedh P. Surwade and Haitao Liu

CONTENTS

10.1 INTRODUCTION

10.1.1 NANOLITHOGRAPHY

The development of integrated circuits and microchips has led to a new paradigm of smart computing and Internet devices. Photolithography, the most commonly used lithography method in industry, has made possible the scaling of integrated circuits all the way from several microns in the 1970s to tens of nanometers in current manufacturing [1]. The number of transistors that can be placed on an integrated circuit has doubled approximately every two years, obeying Moore's law. This development has fueled continuous improvement in the tools used in the photolithography process. However, state-of-the-art photolithography process has reached a saturation point and faces technological and economic challenges to produce features less than 22 nm. Therefore over the last few years, alternative lithography processes that can produce features with sizes in the range of tens of nanometers have been developed.

Nanolithography is a lithography process that can produce nanometer-scale features on the surface. The technique of nanolithography can be divided into two types depending upon whether the pattern is transferred using a mask, referred to as masked lithography, or without a mask, referred to as maskless lithography. The forms of masked nanolithography include photolithography [2–8], nanoimprint lithography [9–16], and soft lithography [17–20]. Maskless nanolithography includes methods such as electron beam lithography [21–27], x-ray lithography [28,29], focused ion beam lithography [30–33], and scanning probe lithography [34–40] that fabricate patterns by serial writing without the use of a mask. Since this chapter focuses on the use of DNA nanostructures as the mask for lithography, the maskless lithography methods will not be discussed.

Among various masked nanolithography methods, soft lithography processes are gaining interest due to their lower cost, effectiveness, availability, and ease of use compared to the photolithography process. In a soft lithography process, elastomeric stamps, molds, or masks with patterns on their surface are used to generate micro- or nanostructure patterns [20]. The elastomeric stamp is a key component of the soft lithography process, and a variety of elastomers such as polydimethoxysilane, poly-urethanes, polyimides, cross-linked Novolac™ resins, and their chemically modified derivatives have been used by researchers to fabricate stamps [20,41–44]. The stamps are commonly fabricated by pouring the pre-polymer of the elastomer over a master having the desired pattern on its surface, followed by the curing (cross-linking of the polymer) and peeling off steps. The master is fabricated using photolithography, micromachining, or e-beam lithography.

The different types of soft lithography techniques include microcontact printing (μCP), replica molding (REM), microtransfer molding (μTM), micromolding in capillaries (MIMIC), solvent-assisted micromolding (SAMIM), phase-shift photolithography cast molding, embossing, and injection molding [20,45–59]. The μCP method is different from other printing methods in that it uses self-assembly to form micropatterns. A pattern with SAMs on the surface of the PDMS stamp is brought in contact with the substrate surface to form micropatterns of SAMs on the substrate surface.

The soft lithographic techniques offer advantages in several applications. For example, one of the major advantages of soft lithographic techniques is the ability to pattern a nonplanar surface. The other advantages include patterning solid materials other than photoresist, patterning materials in liquid phase, patterning functional molecules, and the formation of 3D microstructures and systems. Compared to photolithography, soft lithography is advantageous in applications where the capital cost of equipment is a concern or where precise alignment, uniformity, and continuity of the final pattern are not required. Examples of such applications include microelectrodes, microelectromechanical systems (MEMS), sensors, biosensors, microreactors, and plastic electronics. In these cases, soft lithography is economical and easy to implement compared to photolithography.

However, soft lithography is still in its infancy. The technique relies on a mask/mold for pattern transfer. The deformation and distortion of the elastomeric stamp needs to be very well understood and studied. Optimization of the properties of elastomer for faithful pattern transfer, especially for small size features, is important. Other factors, such as density of defects in the final pattern, high-resolution patterns, uniformity of the patterns, and compatibility with the techniques used in production of microelectronics, also need to be optimized and improved.

10.1.2 DNA NANOSTRUCTURES

A single-stranded DNA is a long polymeric chain made up of repeating units of nucleotides. The nucleotide contains both the backbone of the molecule and a base. The backbone is made of alternating phosphate and sugar residues, while the base can be one of the four molecules: adenine (A), cytosine (C), guanine (G), and thymine (T). These bases form the alphabet of DNA and encode information.

In living organisms, DNA exists as a pair of molecules called double-stranded DNA that are held together tightly in the shape of a double helix. In a double helix, A bonds preferentially to T, and C bonds preferentially to G. The DNA double helix is stabilized by hydrogen bonds and hydrophobic interactions between the bases. Since the hydrogen bonding is noncovalent, the double-stranded DNA can be broken and rejoined by controlling the temperature of the solution [60–62].

The DNA nanostructure is a complex arrangement of single-stranded DNA molecules that are partially hybridized along their subsegments. Fabrication of DNA nanostructures is mainly based on the self-assembly process. The idea is to design a stable motif with sticky ends first that can hybridize to form DNA nanostructures [60,61,63–66]. For example, four motifs with sticky ends are assembled to form a quadrilateral with additional sticky ends on the outside, further allowing the structure to form a two-dimensional (2D) lattice [60,65]. Some of the useful motif types found in DNA nanostructures are stem loop (also called hairpins), Holliday junction, and sticky ends. Stem loop is a single-stranded DNA that loops back to hybridize on itself and is often used to form patterns on DNA nanostructures. Sticky end is an unhybridized single-stranded DNA that protrudes from the end of a double helix and is often used to combine two DNA nanostructures via hybridization. Holliday junctions are formed by two parallel DNA helices with one strand of each DNA helix crossing over the other DNA helix. Holliday junctions are often used to hold together various parts of DNA nanostructures. To achieve complicated DNA assemblies, it is important to design stable motifs. The use of branched DNA junctions with sticky ends to construct 2D arrays often did not yield the desired stable nanostructure due to the flexible nature of branched DNA junctions. To address this challenge, Seeman and coworkers constructed branched complexes with greater rigidity called crossover tiles, such as double-crossover tiles (DX) [67,68]. Inspired by the DX motif, several other motifs were engineered, such as triple-crossover (TX) motif, paranemic crossover tiles (PX), three-, five-, and six-point star motifs, and T-junctions, to provide more options to fabricate DNA nanostructures. Examples of these structures are shown in Figure 10.1 [67–79].

Another breakthrough in the design of the DNA nanostructure was achieved when Rothemund developed the concept of DNA origami [81,82]. Unlike the conventional crossover strategy, which is a two-step process that involves self-assembly of single building blocks (or motifs) into large structures, DNA origami provides a simple and versatile one-pot method in which a large single strand of DNA is folded into the desired shape using numerous short single strands. A very complex and arbitrary shape can be made by this DNA origami approach. However, scaling up the size of DNA origami is a critical challenge. Recently, Zhao et al. developed a new strategy

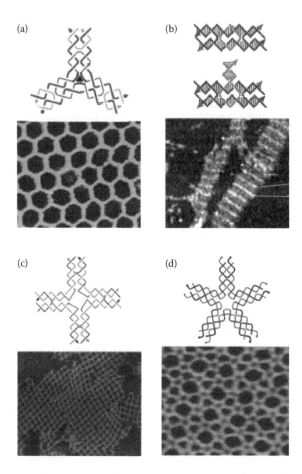

FIGURE 10.1 Two-dimensional (2D) DNA nanostructures. (a) 2D hexagonal lattices formed from three-point-star motifs [69]. (b) 2D periodic arrays formed from symmetric six-point-star motifs [71]. (c) 2D pseudohexagonal trigonal arrays formed from bulged junction triangle [80]. (d) 2D arrays assembled from six-helix bundle tube [76]. (Reprinted from He, Y., et al., *Journal of the American Chemical Society* 2005, 127, 12202; He, Y., et al., *Journal of the American Chemical Society* 2006, 128, 15978; Mathieu, F., et al., *Nano Letters* 2005, 5, 661; and Ding, B., et al., *Journal of the American Chemical Society* 2004, 126, 10230. With permission.)

to construct large-scale 2D origami [83]. They used rectangular-shaped DNA tiles instead of traditional staple strands that resulted in a DNA origami structure of a larger dimension than the structure obtained using staple strands.

Conceptually, there is no difference between 2D and 3D assembly. Still, 3D DNA nanostructures are significantly more challenging to design than 2D structures because of the limited rigidity of the DNA motifs. With the development of rigid and unique 3D motifs as well the assembling strategies, a variety of 3D DNA nanostructures are fabricated and reported in the literature [61,63,66,84–98].

10.2 DNA NANOSTRUCTURES AS A SELF-ASSEMBLED TEMPLATE

Assembly of interconnected nanoscale components in a controlled fashion is one of the key challenges in nanotechnology. The top-down lithographic patterning method is approaching its feature-size limitations, and so researchers are exploring bottom-up lithographic patterning methods as an alternative. DNA nanostructures are promising bottom-up alternatives because of their ease with which predesigned shapes/patterns can be obtained and their rich surface chemistry that can be used for precise positioning of nanoscale materials using a variety of covalent and noncovalent interactions. The most common way to use DNA nanostructures as templates for self-assembly is to pattern the DNA on the substrate and then use it to bind, nucleate, or mask other materials.

10.2.1 NANOSCALE PATTERNING OF METAL

DNA templated synthesis of metallic nanowires involves adsorption of positively charged metal ions on the negatively charged DNA backbone, followed by reduction of the adsorbed metal ions, resulting in metal nanowires grown on DNA templates. There are several reports on the synthesis of a variety of metallic nanowires, such as gold, platinum, silver, palladium, and nickel, using DNA as templates [79,99–108]. However, the metallization process results in nanowires that are at least 10 times thicker than the DNA templates. Also, the metallic nanowires obtained are linear structures, and features with specific shapes and patterns that are normally required for technological applications are challenging. To circumvent this issue, Deng and Mao reported the use of DNA as a mask to fabricate nanoscale gold patterns on the surface [109]. The method involved deposition of thick metal films over DNA nanostructures on mica to create imprinted metal replicas of the DNA features (negative tone pattern) with the nanometer-scale resolution. In another such technique, Becerril and Woolley reported the use of DNA molecules aligned on the substrate as shadow mask templates to obtain features with a spatial resolution in the sub-10 nm range [110]. The method involved alignment of DNA nanostructures on the surface, followed by vapor deposition of metal film at a certain angle relative to the surface, which resulted in films with nanometer-sized gaps corresponding to the DNA shadow mask. Anisotropic etching of these features resulted in transfer of these patterns into the substrate in the form of trenches.

10.2.2 ALIGNMENT OF MOLECULES, NANOCRYSTALS, AND CARBON NANOTUBES

DNA nanostructures have also been used as templates to assemble a variety of organic and inorganic moieties (Figure 10.2) [111,113–128]. The most common strategy is to incorporate functional moieties into DNA nanostructure architecture and use that as a binding site to organize nanoscale organic and inorganic moieties. For example, Zhang et al. demonstrated the incorporation of a single gold nanoparticle into a DNA nanostructure building block and used it for self-assembly of 2D nanoparticle arrays with well-defined periodic patterns and interparticle spacing [111]. Pinto et al. demonstrated

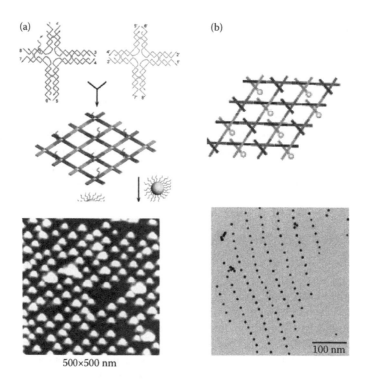

500×500 nm

FIGURE 10.2 Assembly of gold nanoparticles using (a) a square lattice [111] and (b) a DX-triangle tile [112]. (Reprinted from Zhang, J., et al., *Nano Letters* 2006, 6, 248, and Zheng, J., et al., *Nano Letters* 2006, 6, 1502. With permission.)

the periodic arrangement of nanoparticles of different sizes [114]. Yan et al. reported the possibility for designing complex assemblies of multicomponent nanomaterials by using peptides with specific affinity for particular inorganic materials. By coupling gold binding peptide to self-assembling DNA strands, they assembled gold nanoparticles suggesting the possibility of achieving mixed nanoparticle systems depending upon the peptide location within the self-assembling DNA nanostructure [79]. Hung et al. reported a combination of lithography and self-assembly to selectively deposit DNA origami structures hybridized with gold nanoparticles on lithographically patterned substrate resulting in a spatially ordered 2D assembly of 5 nm gold particle, demonstrating a versatile approach for large-scale nanoparticle patterning [129].

Carbon nanotubes have been studied extensively because of their unique mechanical, thermal, electronic, and optical properties. However, some of the major challenges in the widespread use of carbon nanotubes are their sorting, sizing, and arrangement to fabricate economical devices. DNA self-assembly offers the possibility for efficient arrangement of carbon nanotubes into composite structures. The early studies on assembly of carbon nanotubes using DNA relied on either a common moiety that would noncovalently link DNA and carbon nanotubes or covalent coupling of amino DNA to carboxylic groups on carbon nanotubes. During the last few years, a variety of methods were developed to assemble carbon nanotubes

[113,130–139]. For example, Keren et al. combined DNA self-assembly with other biomolecular recognition processes to construct a carbon nanotube field effect transistor [133]. In another report, Maune et al. used DNA origami templates mixed with DNA-functionalized nanotubes to align nanotubes into cross-junctions [113]. They synthesized rectangular origami templates with two lines of single-stranded DNA hooks in a cross-pattern; this origami served as a template for two different types of nanotubes, each functionalized noncovalently with a distinct DNA linker molecule. The mixing of DNA origami templates and functionalized nanotubes resulted in alignment of nanotubes into cross-junctions that also demonstrated stable field effect transistor-like behavior.

10.3 DNA NANOSTRUCTURES AS A CATALYTICALLY ACTIVE MASK FOR PATTERNING SIO$_2$

10.3.1 CONVENTIONAL METHODS FOR SIO$_2$ ETCHING

SiO$_2$ is the most commonly used dielectric material in silicon integrated circuits. The oxide growth on Si substrates can be divided into two categories: thermal and chemical vapor deposition (CVD). In the case of thermal oxide growth, bulk silicon is oxidized either in dry oxygen or in wet water vapor atmosphere at temperatures above 900°C, while in the case of the CVD process the oxide is formed by chemical reaction of silicon containing precursor vapor and oxygen. CVD oxides offer the advantage of lower deposition temperature than the thermal oxide process, and the deposition can be carried out either at atmospheric pressure (APCVD), low pressure (LPCVD), or by plasma activation (PECVD) [140,141].

The etching of SiO$_2$ can be carried in a wet, dry, or vapor phase process. In a typical wet etching process, liquid phase etchants are used where the wafer is immersed in the bath of etchants (e.g., hydrofluoric acid (HF) solution). In the case of dry etching, the wafer is exposed to plasma of some reactive ion (such as Cl and F) that can react with SiO$_2$. Dry etching requires expensive plasma instrumentation; it also has limitations on the etch rate. As a result, the wet etching method is most commonly used, as it offers the advantage of low cost, speed, and simplicity. However, one of the major challenges of using the wet etching process is the stiction problem [142–148]. Typically a wet etching process involves etching with a liquid etchant followed by a rinsing step. The rinse liquid eventually needs to be dried out. However, the capillary force at the menisci formed during the final drying step pulls the microstructures down to the substrate. The deformed structure will adhere to the surface due to strong van der Waals and electrostatic interactions. Various techniques have been developed to alleviate or remove the stiction problem, such as using low surface tension liquid or dimples to reduce contact area, or using a rough surface to reduce effective surface energy, or eliminating the formation of a liquid-gas interface through sublimation, plasma ashing, or supercritical drying [142,149–156].

The vapor phase etching is attractive in that it eliminates the elaborate rinsing and drying steps. Vapor phase etching can faithfully transfer lithographically defined photoresist patterns into the underlying layers with both isotropic and anisotropic etch methods. Compared to wet etching, vapor phase etching offers the advantage of better control of process variables, such as temperature, pressure, relative composition of gas phase etchants, and ease of automation. Importantly, vapor phase etching can be carried out under mild conditions that will not lift off or destroy the DNA-based templates.

10.3.2 MECHANISM OF SiO$_2$ ETCHING BY HF

The vapor phase etching of SiO$_2$ using HF gas to produce SiF$_4$ and H$_2$O is a thermodynamically favorable reaction [157]:

$$SiO_2 \text{ (s)} + 4HF \text{ (g)} \rightarrow SiF_4 \text{ (g)} + 2H_2O \text{ (g)} \quad \Delta G \text{ (298 K)} = -74.8 \text{ kJ/mol}$$

where ΔG is the change in Gibbs free energy. However, HF alone does not etch SiO$_2$, and so to lower the kinetic barrier, a polar molecule such as water is needed. It has been proposed that the adsorbed moisture on the oxide surface acts as a catalyst for the gas-solid reaction between HF and SiO$_2$.

Several experimental and theoretical studies have been reported proposing the mechanism of vapor phase etching of SiO$_2$ using HF and the significance of water in the etching reaction. For example, an in situ Fourier transform infrared (FTIR) spectroscopy study on the vapor phase etching suggested that the HF and H$_2$O molecules form a HF-H$_2$O complex that adsorbs on the SiO$_2$ surface more strongly than either molecule alone [158]. Water consequently acts as a carrier or a medium for HF to stick on SiO$_2$ and etch the surface [157]. A first principles molecular modeling study suggested that a concerted attack of HF and H$_2$O molecules on the Si-O bond of surface silanol (SiOH) groups is the preferred etching pathway [160]. According to the model for gas phase etching of SiO$_2$ proposed by Lee et al., HF vapor molecules and water molecules first adsorb physically onto the oxide [142,160].

$$HF \text{ (g)} \leftrightarrow HF \text{ (ads)}$$

$$H2O \text{ (g)} \leftrightarrow H_2O \text{ (ads)}$$

An ionization reaction between the adsorbed HF and water results in a HF^{2-} ion.

$$2 HF \text{ (ads)} + H_2O \rightarrow HF_2^- \text{ (ads)} + H_3O^+ \text{ (ads)} \quad \textit{(deprotonation)}$$

The ionized HF then reacts with the oxide by sequential substitution reactions to produce water and other by-products, as shown in the reaction below [157]:

$$3 HF_2^- + 3 H_3O^+ + SiO_2 \rightarrow 2 HF + SiF_4 \text{ (ads)} + 5 H_2O \text{ (ads)} \quad \textit{(etching)}$$

The by-products are removed from the oxide surface by desorption [142,160]:

$$SiF_4 \text{ (ads)} \leftrightarrow SiF_4 \text{ (g)}$$

$$H_2O \text{ (ads)} \leftrightarrow H_2O \text{ (g)}$$

Thus irrespective of the variety of proposed mechanisms in the literature, all the mechanisms suggest the adsorption of water on the surface of the oxide playing an important role in controlling the etching kinetics. Therefore if the water adsorption on the surface is controlled with nanometer-scale resolution, it will be possible to control the etching of SiO_2 at the same length scale. Since the etching reaction is autocatalytic, a small spatial variation in the initial concentration of surface-adsorbed water would be amplified by the reaction and will have a significant impact on the etching kinetics in the long run.

10.3.3 DNA-MEDIATED ETCHING OF SiO$_2$

Several reasons make DNA nanostructures good candidates to control the adsorption of moisture and therefore to pattern the SiO_2 surface. First, DNA nanotechnology offers the advantage of synthesizing a predetermined shape or pattern at the nanometer scale. Second, the chemical functionalities on the surface of DNA, such as phosphates, nitrogen, and oxygen, can hydrogen bond with water, and thus the moisture around the surface of DNA can be effectively controlled. Third, the etching is carried out in the vapor phase, eliminating the possibility of DNA being lifted off from the surface.

Both the DNA and the SiO_2 surface will adsorb water; however, the amount of surface-adsorbed water on the clean SiO_2 surface and the SiO_2 surface underneath the DNA will depend on several factors, such as relative humidity, temperature, and partial pressure of water. Therefore by controlling the etching conditions, it is possible to etch/pattern the SiO_2 surface at the nanometer-scale resolution. The resolution of the etching/pattern will be determined by the size of the DNA nanostructure.

We recently discovered that triangular DNA nanostructures deposited on the SiO_2 surface when exposed to HF vapor under high-humidity conditions (~50% relative humidity) result in triangular-shaped trenches on the substrate surface (Figure 10.3a). This is due to high adsorption of water around DNA under high-humidity conditions that results in faster etching kinetics underneath DNA, resulting in a trench (negative tone etching). Under low-humidity conditions, the etching rate of SiO_2 under DNA is less compared to DNA-free surface, resulting in DNA acting as a mask protecting the surface underneath it (positive tone etching). As shown in Figure 10.3a, triangular trenches in the range of 1–2 nm are obtained under high-humidity conditions while triangular ridges in the range of 2–3 nm are obtained under low-humidity conditions

The resolution limit of this technique was probed using a single double-stranded DNA (λ-DNA) as a template. As shown in Figure 10.3b, λ-DNA aligned on the SiO_2 substrate, when exposed to HF vapors under high moisture conditions, resulted in trenches of width ranging from 10 to 30 nm on the SiO_2 surface. Such a wide range of trench width is attributed to bundling of DNA with a single DNA strand producing features as small as 10 nm, while bundles produce features as large as 30 nm. Consistently, when λ-DNA aligned on the SiO_2 substrate is exposed to HF vapors under low moisture conditions, this results in ridges on the SiO_2 surface, indicating that even a single strand of DNA can slow the etching of SiO_2 underneath it acting as a mask.

Kinetics study on the etching reaction under low moisture conditions further supports the hypothesis (Figure 10.4). In this case, the aligned λ-DNA on SiO_2 substrates

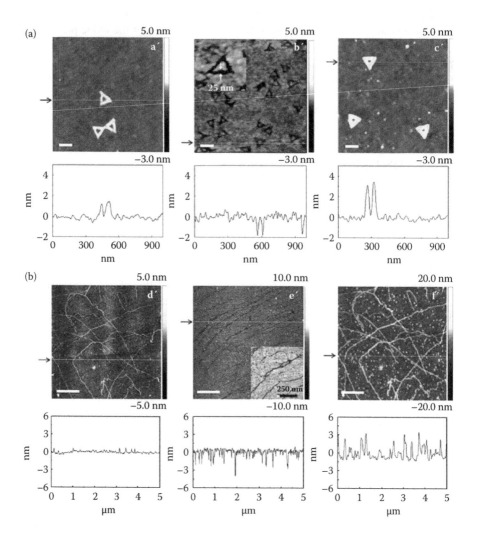

FIGURE 10.3 AFM images and cross section of (a) DNA nanostructures. (a′) DNA origami triangles self-assembled on SiO_2 wafer. (b′) Triangular trenches produced upon exposure of (a′) to high relative humidity air and HF vapor; the inset shows a high-magnification AFM image of a triangular trench with a width of 25 nm. (c′) Triangular ridges produced upon exposure of (a′) to low relative humidity air and HF vapor. Arrows indicate the lines along which the cross sections were determined. Scale bars represent 100 nm. (b) Single-stranded DNA. (d′) λ-DNA aligned on the SiO_2 substrate. (e′) Trenches produced after exposure of (d′) to high relative humidity air and HF vapor; the inset shows a high-magnification image of the trench. (f′) Ridges produced after exposure of (d′) to low relative humidity air and HF vapor. It should be noted that this AFM image was obtained at exactly the same location as the one in (d′). Arrows indicate lines along which the cross sections were determined. Scale bars represent 1 μm. (Reproduced from Surwade, S.P., et al., *Journal of the American Chemical Society* 2011, 133, 11868. With permission.)

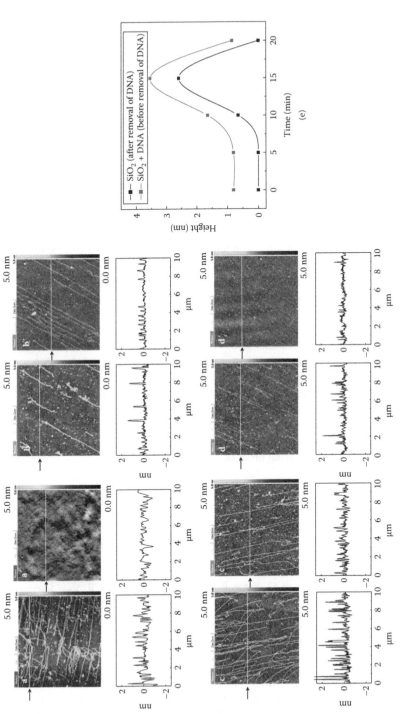

FIGURE 10.4 Kinetics study on etching rate. AFM images of SiO$_2$ wafer with single-stranded DNA after etching under relatively low-humidity air and HF vapor for (a) 5 min, (a′) sample (a) after piranha wash; (b) 10 min, (b′) sample (b) after piranha wash; (c) 15 min, (c′) sample (c) after piranha wash; and (d) 20 min, (d′) sample (d) after piranha wash. (e) Plot showing the temporal evolution of the height of the ridges obtained under relatively low-humidity air and HF vapor. (Reproduced from Surwade, S.P., et al., *Journal of the American Chemical Society* 2011, 133, 11868. With permission.)

was etched for 5, 10, 15, and 20 min under low moisture conditions, and the heights of the ridge features were measured using an atomic force microscope (AFM) before and after the removal of λ-DNA. The difference in the two measurements (~0.7 nm) indicates the presence of the DNA throughout the reaction. It is important to note, however, that AFM cannot detect if the chemical structure of DNA is still intact. The etching reaction observes an initiation or induction period (~5 min), where no etching is observed, followed by a fast etching step, indicated by the rapid buildup of the ridge height (~5–15 min). As the reaction progresses further, enough water is produced by the reaction to saturate the surface, leading to a decrease in ridge height.

Substrate temperature also plays an important role in determining the etching kinetics. The etching rate is observed to decrease with increasing substrate temperature, possibly due to a decrease in water adsorption on the SiO_2 surface at high temperatures.

10.4 APPLICATIONS AND FUTURE DIRECTIONS

DNA nanostructures hold great promise as a template for high-resolution nanofabrication. A lithography is essentially a process that can make arbitrary shaped patterns at arbitrary locations. In this regard, the DNA nanostructure is far superior to almost any other self-assembled material systems in that it can produce arbitrary-shaped patterns with 2–3 nm of resolution. However, its applications in lithography are greatly limited by its stability, both chemical and mechanical. The vapor phase etching scheme circumvented these stability issues by avoiding using liquid and plasma etchants. However, improving the chemical and mechanical stability of the DNA nanostructure will certainly make it compatible with a wide range of existing nanofabrication processes. In addition to being useful as a practical lithography process, ways to fabricate a large-scale DNA nanostructure and its deterministic positioning on the substrate need to be developed. Last but not least, the fidelity and reproducibility of DNA self-assembly needs to be understood and controlled before it can be used in any real-world manufacturing process.

REFERENCES

1. Ito, T., Okazaki, S. *Nature* 2000, 406, 1027.
2. Bruning, J.H. *Proceedings of SPIE* 2007, 6520, 652004/1.
3. Brunner, T.A. *Journal of Vacuum Science and Technology B* 2003, 21, 2632.
4. Flagello, D.G. *Proceedings of SPIE* 2008, 6827, 68271N/1.
5. Flagello, D.G., Arnold, B. *Proceedings of SPIE* 2006, 6327, 63270D/1.
6. Jeong, H.J., Markle, D.A., Owen, G., Pease, F., Grenville, A., von Bunau, R. *Solid State Technology* 1994, 37, 39.
7. Rai-Choudhury, P. *Handbook of Microlithography, Micromachining, and Microfabrication,* Vol 1. Bellingham, WA: SPIE Optical Engineering Press, 1997.
8 Renwick, S.P., Williamson, D., Suzuki, K., Murakami, K. *Optics and Photonics News* 2007, 18, 34.
9. Guo, L.J. *Advanced Materials* 2007, 19, 495.
10. Hu, Z., Jonas, A.M. *Soft Matter* 2010, 6, 21.
11. Kehagias, N., Hu, W., Reboud, V., Lu, N., Dong, B., Chi, L., Fuchs, H., Genua, A., Alduncin, J.A., Pomposo, J.A., Mecerreyes, D., Torres, C.M.S. *Physica Status Solidi C* 2008, 5, 3571.

12. Ofir, Y., Moran, I.W., Subramani, C., Carter, K.R., Rotello, V.M. *Advanced Materials* 2010, 22, 3608.
13. Reboud, V., Kehoe, T., Kehagias, N., Torres, C.M. S. *Nanotechnology* 2010, 8, 165.
14. Schift, H. *Journal of Vacuum Science and Technology B* 2008, 26, 458.
15. Schift, H., Kristensen, A. *Handbook of Nanotechnology*. 2nd ed. Berlin: Springer, 2007, p. 239.
16. Tiginyanu, I., Ursaki, V., Popa, V. *Nanocoatings and Ultra-Thin Films*, Philadelphia, PA: Woodhead, 2011, p. 280.
17. Duan, X., Reinhoudt, D.N., Huskens, J. Soft Lithography for Patterning Self-Assembling Systems. In Samorí P., Cacialli F., *Functional Supramolecular Architectures*, 2011, vol. 1. Weinheim, Germany: Wiley-VCH, 2011.
18. Rogers, J.A., Nuzzo, R.G. *Materials Today*, 2005, 8, 50
19. Whitesides, G.M., Ostuni, E., Takayama, S., Jiang, X., Ingber, D.E. *Annual Review of Biomedical Engineering* 2001, 3, 335.
20. Xia, Y., Whitesides, G.M. *Annual Review of Materials Science* 1998, 28, 153.
21. Hahmann, P., Fortagne, O. *Microelectronic Engineering* 2009, 86, 438.
22. Liu, M., Ji, Z., Shang, L. *Nanotechnology* 2010, 8, 3.
23. Pfeiffer, H.C. *Proceedings of SPIE* 2010, 7823, 782316/1.
24. Wnuk, J.D., Rosenberg, S.G., Gorham, J.M., van Dorp, W.F., Hagen, C.W., Fairbrother, D.H. *Surface Science* 2011, 605, 257.
25. Zhou, Z. Electron beam lithography. In Yao N., Wang, Z. L., *Handbook of Microscopy for Nanotechnology* (pp. 287–322), Boston: Kluwer Academic Publishers, 2005.
26. Vieu, C., Carcenac, F., Pepin, A., Chen, Y., Mejias, M., Lebib, A., Manin-Ferlazzo, L., Couraud, L., Launois, H. *Applied Surface Science* 2000, 164, 111.
27. Tseng, A.A., Chen, K., Chen, C.D., Ma, K.J. *IEEE Transactions on Electronics Packaging Manufacturing* 2003, 26, 141.
28. Silverman, J.P. *Journal of Vacuum Science and Technology B* 1998, 16, 3137.
29. Smith, H.I., Flanders, D.C. *Journal of Vacuum Science and Technology* 1980, 17, 533.
30. Bischoff, L. *Nuclear Instruments and Methods in Physics Research B* 2008, 266, 1846.
31. Gierak, J. *Semiconductor Science and Technology* 2009, 24, 043001/1.
32. Volkert, C.A., Minor, A.M. *MRS Bulletin* 2007, 32, 389.
33. Reyntjens, S., Puers, R. *Journal of Micromechanics and Microengineering* 2001, 11, 287.
34. Garcia, R., Martinez, R.V., Martinez, J. *Chemical Society Reviews* 2006, 35, 29.
35. Kraemer, S., Fuierer, R.R., Gorman, C.B. *Chemical Reviews* 2003, 103, 4367.
36. Liu, G.-Y., Xu, S., Qian, Y. *Accounts of Chemical Research* 2000, 33, 457.
37. Tseng, A.A., Notargiacomo, A., Chen, T.P. *Journal of Vacuum Science and Technology B* 2005, 23, 877.
38. Shim, W., Braunschweig, A.B., Liao, X., Chai, J., Lim, J.K., Zheng, G., Mirkin, C.A. *Nature* 2011, 469, 516.
39. Ginger, D.S., Zhang, H., Mirkin, C.A. *Angewandte Chemie, International Edition* 2004, 43, 30.
40. Lenhert, S., Fuchs, H., Mirkin, C.A. *Nanotechnology* 2009, 6, 171.
41. Choi, K.M. *NSTI Nanotech 2007, Nanotechnology Conference and Trade Show*, Santa Clara, CA, May 20–24, 2007, pp. 1, 348.
42. Choi, K.M., Rogers, J.A. *Materials Research Society Symposium Proceedings* 2003, 788, 491.
43. Rolland, J., Hagberg, E.C., Denison, G.M., Carter, K.R., De Simone, J.M. *Angewandte Chemie, International Edition* 2004, 43, 5796.
44. Kumar, A., Whitesides, G.M. *Applied Physics Letters* 1993, 63, 2002.
45. Kaufmann, T., Ravoo, B.J. *Polymer Chemistry* 2010, 1, 371.
46. Luttge, R. *Journal of Physics D* 2009, 42, 123001/1.

47. Brehmer, M., Conrad, L., Funk, L. *Journal of Dispersion Science and Technology* 2003, 24, 291.
48. Mallouk, T.E. *Science* 2001, 291, 443.
49. Xia, Y., Whitesides, G.M. *Polymeric Materials Science and Engineering* 1997, 77, 596.
50. Zhao, X.-M., Xia, Y., Whitesides, G.M. *Journal of Materials Chemistry* 1997, 7, 1069.
51. Cavallini, M., Albonetti, C., Biscarini, F. *Advanced Materials* 2009, 21, 1043.
52. Kim, E., Xia, Y., Whitesides, G.M. *Nature* 1995, 376, 581.
53. Qin, D., Xia, Y., Whitesides, G.M. *Nature Protocols* 2010, 5, 491.
54. Ye, X., Duan, Y., Ding, Y. *Nano-Micro Letters* 2011, 3, 249.
55. Terris, B.D., Mamin, H.J., Best, M.E., Logan, J.A., Rugar, D., Rishton, S.A. *Applied Physics Letters* 1996, 69, 4262.
56. Chou, S.Y., Krauss, P.R., Renstrom, P.J. *Applied Physics Letters* 1995, 67, 3114.
57. Masuda, H., Fukuda, K. *Science* 1995, 268, 1466.
58. Becker, H., Gartner, C. *Analytical and Bioanalytical Chemistry* 2008, 390, 89.
59. Sollier, E., Murray, C., Maoddi, P., Di Carlo, D. *Lab on a Chip* 2011, 11, 3752.
60. Seeman, N.C., Lukeman, P.S. *Reports on Progress in Physics* 2005, 68, 237.
61. Yang, D., Campolongo, M.J., Tran, T.N.N., Ruiz, R.C.H., Kahn, J.S., Luo, D. *Wiley Interdisciplinary Reviews* 2010, 2, 648.
62. Reif, J.H., Chandran, H., Gopalkrishnan, N., LaBean, H., Self-ssembled DNA nano-structures and DNA devices, In Cabrini, S., Kawata, S. *Nanofabrication Handbook,* Taylor & Francis, Boca Raton, FL, 2012.
63. Lin, C., Liu, Y., Yan, H. *Biochemistry* 2009, 48, 1663.
64. Seeman, N.C., Wang, H., Yang, X., Liu, F., Mao, C., Sun, W., Wenzler, L., Shen, Z., Sha, R., Yan, H., Wong, M.H., Sa-Ardyen, P., Liu, B., Qiu, H., Li, X., Qi, J., Du, S.M., Zhang, Y., Mueller, J.E., Fu, T.-J., Wang, Y., Chen, J. *Nanotechnology* 1998, 9, 257.
65. Seeman, N.C. *Nature* 2003, 421, 427.
66. Aldaye, F.A., Sleiman, H.F. *Pure and Applied Chemistry* 2009, 81, 2157.
67. Fu, T.J., Seeman, N.C. *Biochemistry* 1993, 32, 3211.
68. Winfree, E., Liu, F., Wenzler, L.A., Seeman, N.C. *Nature* 1998, 394, 539.
69. He, Y., Chen, Y., Liu, H., Ribbe, A.E., Mao, C. *Journal of the American Chemical Society* 2005, 127, 12202.
70. He, Y., Tian, Y., Chen, Y., Deng, Z., Ribbe, A.E., Mao, C. *Angewandte Chemie, International Edition* 2005, 44, 6694.
71. He, Y., Tian, Y., Ribbe, A.E., Mao, C. *Journal of the American Chemical Society* 2006, 128, 15978.
72. Ke, Y., Lindsay, S., Chang, Y., Liu, Y., Yan, H. *Science* 2008, 319, 180.
73. LaBean, T.H., Yan, H., Kopatsch, J., Liu, F., Winfree, E., Reif, J.H., Seeman, N.C. *Journal of the American Chemical Society* 2000, 122, 1848.
74. Liu, D., Wang, M., Deng, Z., Walulu, R., Mao, C. *Journal of the American Chemical Society* 2004, 126, 2324.
75. Mao, C., Sun, W., Seeman, N.C. *Journal of the American Chemical Society* 1999, 121, 5437.
76. Mathieu, F., Liao, S., Kopatsch, J., Wang, T., Mao, C., Seeman, N.C. *Nano Letters* 2005, 5, 661.
77. Park, S.H., Barish, R., Li, H., Reif, J.H., Finkelstein, G., Yan, H., LaBean, T.H. *Nano Letters* 2005, 5, 693.
78. Reishus, D., Shaw, B., Brun, Y., Chelyapov, N., Adleman, L. *Journal of the American Chemical Society* 2005, 127, 17590.
79. Yan, H., Park, S.H., Finkelstein, G., Reif, J.H., LaBean, T.H. *Science* 2003, 301, 1882.
80. Ding, B., Sha, R., Seeman, N.C. *Journal of the American Chemical Society* 2004, 126, 10230.
81. Rothemund, P.W.K., Papadakis, N., Winfree, E. *PLoS Biology* 2004, 2, 2041.
82. Rothemund, P.W.K. *Nature* 2006, 440, 297.

83. Zhao, Z., Yan, H., Liu, Y. *Angewandte Chemie, International Edition* 2010, 49, 1414.
84. Andersen, E.S., Dong, M., Nielsen, M.M., Jahn, K., Subramani, R., Mamdouh, W., Golas, M.M., Sander, B., Stark, H., Oliveira, C.L.P., Pedersen, J.S., Birkedal, V., Besenbacher, F., Gothelf, K.V., Kjems, J. *Nature* 2009, 459, 73.
85. Han, D., Pal, S., Nangreave, J., Deng, Z., Liu, Y., Yan, H. *Science* 2011, 332, 342.
86. Lo Pik, K., Metera Kimberly, L., Sleiman Hanadi, F. *Current Opinion in Chemical Biology* 2010, 14, 597.
87. Shen, X., Song, C., Wang, J., Shi, D., Wang, Z., Liu, N., Ding, B. *Journal of the American Chemical Society* 2012, 134, 146.
88. Zhang, C., He, Y., Su, M., Ko, S.H., Ye, T., Leng, Y., Sun, X., Ribbe, A.E., Jiang, W., Mao, C. *Faraday Discussions* 2009, 143, 221.
89. Lo, P.K., Metera, K.L., Sleiman, H.F. *Current Opinion in Chemical Biology* 2010, 14, 597.
90. Aldaye, F.A., Lo, P.K., Karam, P., McLaughlin, C.K., Cosa, G., Sleiman, H.F. *Nature Nanotechnology* 2009, 4, 349.
91. Aldaye, F.A., Sleiman, H.F. *Journal of the American Chemical Society* 2007, 129, 13376.
92. Dietz, H., Douglas, S.M., Shih, W.M. *Science* 2009, 325, 725.
93. Douglas, S.M., Dietz, H., Liedl, T., Hogberg, B., Graf, F., Shih, W.M. *Nature* 2009, 459, 414.
94. Goodman, R.P., Heilemann, M., Doose, S., Erben, C.M., Kapanidis, A.N., Turberfield, A.J. *Nature Nanotechnology* 2008, 3, 93.
95. Goodman, R.P., Schaap, I.A. T., Tardin, C.F., Erben, C.M., Berry, R.M., Schmidt, C.F., Turberfield, A.J. *Science* 2005, 310, 1661.
96. He, Y., Ye, T., Su, M., Zhang, C., Ribbe, A.E., Jiang, W., Mao, C. *Nature* 2008, 452, 198.
97. Ke, Y., Sharma, J., Liu, M., Jahn, K., Liu, Y., Yan, H. *Nano Letters* 2009, 9, 2445.
98. Shih, W.M., Quispe, J.D., Joyce, G.F. *Nature* 2004, 427, 618.
99. Houlton, A., Watson, S.M.D. *Annual Reports on the Progress of Chemistry A* 2011, 107, 21.
100. Braun, E., Eichen, Y., Sivan, U., Ben-Yoseph, G. *Nature* 1998, 391, 775.
101. Becerril, H.A., Stoltenberg, R.M., Wheeler, D.R., Davis, R.C., Harb, J.N., Woolley, A.T. *Journal of the American Chemical Society* 2005, 127, 2828.
102. Richter, J., Seidel, R., Kirsch, R., Mertig, M., Pompe, W., Plaschke, J., Schackert, H.K. *Advanced Materials* 2000, 12, 507.
103. Monson, C.F., Woolley, A.T. *Nano Letters* 2003, 3, 359.
104. Watson, S.M.D., Wright, N.G., Horrocks, B.R., Houlton, A. *Langmuir* 2010, 26, 2068.
105. Becerril, H.A., Ludtke, P., Willardson, B.M., Woolley, A.T. *Langmuir* 2006, 22, 10140.
106. Mertig, M., Ciacchi, L.C., Seidel, R., Pompe, W., De Vita, A. *Nano Letters* 2002, 2, 841.
107. Mohammadzadegan, R., Mohabatkar, H., Sheikhi, M.H., Safavi, A., Khajouee, M.B. *Physica E* 2008, 41, 142.
108. Nakao, H., Gad, M., Sugiyama, S., Otobe, K., Ohtani, T. *Journal of the American Chemical Society* 2003, 125, 7162.
109. Deng, Z., Mao, C. *Angewandte Chemie, International Edition* 2004, 43, 4068.
110. Becerril, H.A., Woolley, A.T. *Small* 2007, 3, 1534.
111. Zhang, J., Liu, Y., Ke, Y., Yan, H. *Nano Letters* 2006, 6, 248.
112. Zheng, J., Constantinou, P.E., Micheel, C., Alivisatos, A.P., Kiehl, R.A., Seeman, N.C. *Nano Letters* 2006, 6, 1502.
113. Maune, H.T., Han, S.-P., Barish, R.D., Bockrath, M., Goddard, W.A., III, Rothemund, P.W.K., Winfree, E. *Nature Nanotechnology* 2010, 5, 61.
114. Pinto, Y.Y., Le, J.D., Seeman, N.C., Musier-Forsyth, K., Taton, T.A., Kiehl, R.A. *Nano Letters* 2005, 5, 2399.
115. Sharma, J., Chhabra, R., Liu, Y., Ke, Y., Yan, H. *Angewandte Chemie, International Edition* 2006, 45, 730.
116. Sharma, J., Ke, Y., Lin, C., Chhabra, R., Wang, Q., Nangreave, J., Liu, Y., Yan, H. *Angewandte Chemie, International Edition* 2008, 47, 5157.

117. Xiao, S., Liu, F., Rosen, A.E., Hainfeld, J.F., Seeman, N.C., Musier-Forsyth, K., Kiehl, R.A. *Journal of Nanoparticle Research* 2002, 4, 313.
118. Zheng, J., Birktoft, J.J., Chen, Y., Wang, T., Sha, R., Constantinou, P.E., Ginell, S.L., Mao, C., Seeman, N.C. *Nature* 2009, 461, 74.
119. Alivisatos, A.P., Johnsson, K.P., Peng, X., Wilson, T.E., Loweth, C.J., Bruchez, M.P., Jr., Schultz, P.G. *Nature* 1996, 382, 609.
120. Cheng, W., Park, N., Walter. M.T., Hartman, M.R., Luo, D. *Nature Nanotechnology* 2008, 3, 682.
121. Loweth, C.J., Caldwell, W.B., Peng, X., Alivisatos, A.P., Schultz, P.G. *Angewandte Chemie, International Edition* 1999, 38, 1808.
122. Mastroianni, A.J., Claridge, S.A., Alivisatos, A.P. *Journal of the American Chemical Society* 2009, 131, 8455.
123. Maye, M.M., Kumara, M.T., Nykypanchuk, D., Sherman, W.B., Gang, O. *Nature Nanotechnology* 2010, 5, 116.
124. Maye, M.M., Nykypanchuk, D., van der Lelie, D., Gang, O. *Small* 2007, 3, 1678.
125. Mirkin, C.A., Letsinger, R.L., Mucic, R.C., Storhoff, J.J. *Nature* 1996, 382, 607.
126. Mucic, R.C., Storhoff, J.J., Mirkin, C.A., Letsinger, R.L. *Journal of the American Chemical Society* 1998, 120, 12674.
127. Nykypanchuk, D., Maye, M.M., van der Lelie, D., Gang, O. *Nature* 2008, 451, 549.
128. Park, S.Y., Lytton-Jean, A.K.R., Lee, B., Weigand, S., Schatz, G.C., Mirkin, C.A. *Nature* 2008, 451, 553.
129. Hung, A.M., Micheel, C.M., Bozano, L.D., Osterbur, L.W., Wallraff, G.M., Cha, J.N. *Nature Nanotechnology* 2010, 5, 121.
130. Dwyer, C., Guthold, M., Falvo, M., Washburn, S., Superfine, R., Erie, D. *Nanotechnology* 2002, 13, 601.
131. Dwyer, C., Johri, V., Cheung, M., Patwardhan, J., Lebeck, A., Sorin, D. *Nanotechnology* 2004, 15, 1240.
132. Hazani, M., Shvarts, D., Peled, D., Sidorov, V., Naaman, R. *Applied Physics Letters* 2004, 85, 5025.
133. Keren, K., Berman, R.S., Buchstab, E., Sivan, U., Braun, E. *Science* 2003, 302, 1380.
134. Li, S., He, P., Dong, J., Guo, Z., Dai, L. *Journal of the American Chemical Society* 2005, 127, 14.
135. Taft, B.J., Lazareck, A.D., Withey, G.D., Yin, A., Xu, J.M., Kelley, S.O. *Journal of the American Chemical Society* 2004, 126, 12750.
136. Xin, H., Woolley, A.T. *Nanotechnology* 2005, 16, 2238.
137. Ekani-Nkodo, A., Kumar, A., Fygenson Deborah, K. *Physical Review Letters* 2004, 93, 268301.
138. Han, S.-P., Maune, H.T., Barish, R.D., Bockrath, M., Goddard, W.A. *Nano Letters* 2012, 12, 1129.
139. Xu, P.F., Noh, H., Lee, J.H., Cha, J.N. *Physical Chemistry Chemical Physics* 2011, 13, 10004.
140. Barron, A.R. *Advanced Materials for Optics and Electronics* 1996, 6, 101.
141. Buehler, J., Steiner, F.P., Baltes, H. *Journal of Micromechanics and Microengineering* 1997, 7, R1.
142 Lee, Y.-I., Park, K.-H., Lee, J., Lee, C.-S., Yoo, H.J., Kim, C.-J., Yoon, Y.-S. *Journal of Microelectromechanical Systems* 1997, 6, 226.
143. Winters, H.F. *Journal of Applied Physics* 1978, 49, 5165.
144. Zhao, Y.P., Drotar, J.T., Wang, G.C., Lu, T.M. *Physical Review Letters* 2001, 87, 136102/1.
145. Van Spengen, W.M., Puers, R., De Wolf, I. *Journal of Adhesion Science and Technology* 2003, 17, 563.
146. Agache, V., Quevy, E., Collard, D., Buchaillot, L. *Applied Physics Letters* 2001, 79, 3869.
147. Maboudian, R., Howe, R.T. *Journal of Vacuum Science and Technology B* 1997, 15, 1.
148. Mastrangelo, C.H. *Tribology Letters* 1997, 3, 223.

149. Abe, T., Messner, W.C., Reed, M.L. *Journal of Microelectromechanical Systems* 1995, 4, 66.
150. Kim, J.Y., Kim, C.-J. 10th Proceedings of IEEE Annual International Workshop on Micro Electro Mechanical Systems: An Investigation of Micro Structures, Sensors, Actuators, Machines and Robots, Nagoya, January 26–31, 1997, p. 442.
151. Kobayashi, D., Kim, C.J., Fujita, H. *Japanese Journal of Applied Physics* 1993, 32, L1642.
152. Mastrangelo, C.H., Hsu, C.H. *Journal of Microelectromechanical Systems* 1993, 2, 33.
153. Fushinobu, K., Phinney, L.M., Tien, N.C. *International Journal of Heat and Mass Transfer* 1996, 39, 3181.
154. Hurst, K.M., Roberts, C.B., Ashurst, W.R. *Nanotechnology* 2009, 20, 185303.
155. Tien, N.C., Jeong, S., Phinney, L.M., Fushinobu, K., Bokor, J. *Applied Physics Letters* 1996, 68, 197.
156. Yee, Y., Chun, K., Lee, J.D., Kim, C.-J. *Sensors and Actuators A* 1996, A52, 145.
157. Reinhardt, K.A., Kern, W., Eds. *Handbook of Silicon Wafer Cleaning Technology.* 2nd ed. Norwich, NY: William Andrew Publishing, 2008.
158. Montano-Miranda, G., Muscat, A.J. *Diffusion and Defect Data—Solid State Data B* 2003, 92, 207.
159. Kang, J.K., Musgrave, C.B. *Journal of Chemical Physics* 2002, 116, 275.
160. Lee, C.S., Baek, J.T., Yoo, H.J., Woo, S.I. *Journal of the Electrochemical Society* 1996, 143, 1099.
161. Surwade, S.P., Zhao, S.-C., Liu, H.-T. *Journal of the American Chemical Society* 2011, 133, 11868.

11 CMOS-Compatible Nanowire Biosensors

Thanh C. Nguyen, Wanzhi Qiu, Matteo Altissimo, Paul G. Spizzirri, Laurens H. Willems van Beveren, and Efstratios Skafidas

CONTENTS

11.1 INTRODUCTION

Seralogical point-of-care diagnostic tests that are able to detect disease-causing viruses, bacteria, antigens, or prions are required for clinical diagnosis. Beyond blood serum, the development of robust and inexpensive protocols that can operate in widely varying environments (e.g., biological and physical) can prove even more challenging. Underpinning many conventional detection protocols that include enzyme-linked immunosorbent assay (ELISA) techniques is the use of biological antibodies due to their unique selectivity to target antigens. Being able to directly determine the binding status of these efficient molecular detectors through some form of electronic readout has become an active research area for lab-on-a-chip (LOC) applications.

Silicon nanowire devices, which are effectively low- (or one-) dimensional resistive semiconductor channels, display conductance sensitivity to surface charges (and their resulting electric fields) due to their high surface area-to-volume ratio. When an external charge is in close proximity to the surface of the nanowire, it modifies the charge carrier distribution in the channel, which in turn substantially affects the electrical properties: conductance, quantum capacitance, and kinetic inductance. Since it is also known that biological molecules can exhibit a net charge due to ionized groups at pH values far from their isoelectric point, biomolecules attached to the surface of nanowires should be capable of influencing the electronic response of such a device.

Nanowire biosensors designed using this principle, namely, measuring changes in conductance, have been reported recently [1–12]. Gengfeng et al. recently reported a direct, real-time electrical detection of a single influenza A virus [12], while Hahm, Lieber, and Gao have demonstrated the use of a PNA probe to detect a particular DNA sequence with a limit of detection of 10 fM [1,9]. The methods reported in the literature are based upon direct measurement of the change in DC conductance of the nanowire due to the attachment of a molecule of interest to the functionalized surface. Unfortunately, measurements of small changes in conductance can be difficult and require sensitive, low-noise amplifiers and high-resolution, analog-to-digital converters. Very low-noise, low-frequency, and high-gain amplifiers are difficult to implement on small-geometry complementary metal-oxide-semiconductor (CMOS) processes because of the inherently high value of the low-frequency flicker noise [13].

Recent results on nanowire detection are presented here that indicate that silicon nanowires exhibit a frequency-dependent transfer function that resembles that of a high-pass filter. In order to better understand this response, we describe in this chapter methods to model and simulate the frequency response of a three-dimensional silicon nanowire (SiNW) field effect transistor (FET) biosensor. We will show using these models that as biomolecules with a higher net charge attach to the nanowire, they displace more charge carriers within the nanowire channel, causing the corner frequency (i.e., the location of the 3 dB point of the transfer function) of the filter to decrease along with the conductance, quantum capacitance, and kinetic inductance. This property of silicon nanowires, which was first shown in [14], is further developed in this work to build a low-cost CMOS, frequency-based detection system.

In addition to simulating the device response, we also consider factors that affect the high-pass filter behavior of these nanostructures in addition to methods that could be used to help quantify the amount of the antigen present in the sample. Competitive versus captive antigen binding scenarios can also change the analysis paradigm from that of single positive detection to quantitative analysis, depending upon the clinical need. Simulations in this work have shown, however, that competitive binding, which results in a high number of antibody/antigen attachments per unit time, gives rise to the frequency response of the device, and that the location of the corner frequency of the high-pass filter varies with the average number of these attachments. Therefore practical realization of such a sensor requires the development of readout circuitry capable of processing the SiNW FET frequency-dependent characteristic in real time. Lastly, several CMOS circuit designs that could be used to determine the amount of charge attached to the nanowire are presented and discussed.

11.2 DEVICE CONSTRUCTION

Although many fabrication techniques exist for fabricating silicon nanowires, most are not compatible with planar CMOS fabrication processes. The top-down method [1], however, which results in the device being fabricated on a thin device layer atop a silicon-on-insulator (SOI) wafer, is compatible with CMOS fabrication processes.

The silicon dioxide insulator layer (buried oxide) is approximately 150 nm thick and provides both electrical isolation and a reduction in parasitic device capacitance. The top silicon device layer, which has a thickness of 50 nm, is phosphorus doped to a concentration of $1.0 \; e^{15} \, cm^{-3}$ to produce a semiconducting nanowire using low-energy (<30 keV) ion implantation techniques. Photoresist, negative tone electron-beam resist, and electron-beam lithography are used to fashion the nanowire (NW) structures on the top of the SOI, and the isotropic reactive ion etching technique is finally used to remove the nonmasked areas, leaving a fin structure as shown in Figure 11.1. The two contact ends of the nanowire must be further doped to a higher concentration of phosphorus, creating local n^+ regions capable of producing near-ohmic contact to the metallic bonding pad on top. The next step involves performing a rapid thermal anneal (950°C for 5 s) to activate the phosphorus dopants. Finally, metallic contact pads are deposited at the two ends of the nanowire to form the ohmic source and drain contacts, and these are spike annealed in forming gas. Antibody conjugation to the device surface is performed using procedures described in later sections in order to produce the chemical gate. Figure 11.2 illustrates the final nanowire design along with the simulated doping profiles.

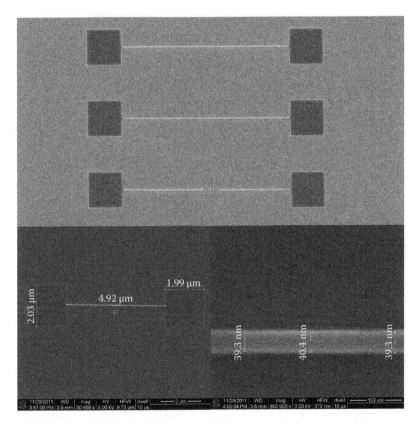

FIGURE 11.1 Silicon-on-insulator (SOI)-based nanowires (40 nm in width and ~5 um long) fabricated using e-beam lithography followed by RIE etching.

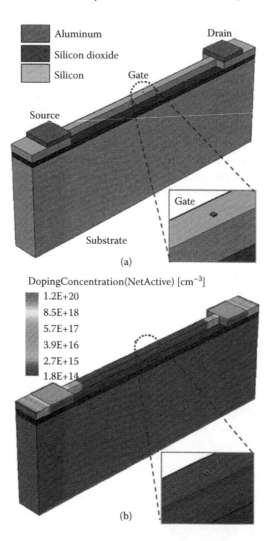

FIGURE 11.2 (a) SiNW FET device structure material view. (The insets show a magnified view of the gate on the channel surface.) (b) Doping concentration view.

11.3 DC CHARACTERISTICS OF AN N-TYPE SINW FET

The sensitivity of the silicon nanowire is defined to be the ratio between the change and initial conductance $S = \Delta G/G_0$ [2]. When the drain-to-source voltage is constant, the sensitivity can be written as $S = |I - I_0|/I_0$, where I_0 is the current at zero charge and $|I - I_0|$ is the current change upon the attachment of charge to the nanowire surface. Figure 11.3(c) shows the sensitivity versus substrate voltage with surface charge varied from −5 e to +5 e, where e is the fundamental unit of charge. It can be seen from Figure 11.3(b) and (c) that negative charges on the surface of an n-type silicon nanowire exhibit greater sensitivity than when positive charges are attached to the same surface. This result is in agreement with the nanowire field effect transistor

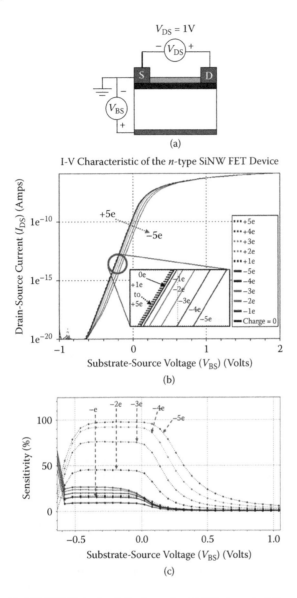

FIGURE 11.3 (a) SiNW FET device DC schematic. (b) I-V characteristics as a function of gate net charge. Substrate-source voltage is swept from –1 V to 2 V while drain-source bias is 1 V. (c) Device sensitivity as a function of gate charge.

MATLAB®* simulation of Nair and Alam [18], whose simulations were performed in air. In the same work, it was suggested that the device sensitivity should be reduced when the NW is in aqueous solution because of the high dielectric constant of water, which will reduce the depletion depth into the NW body [18].

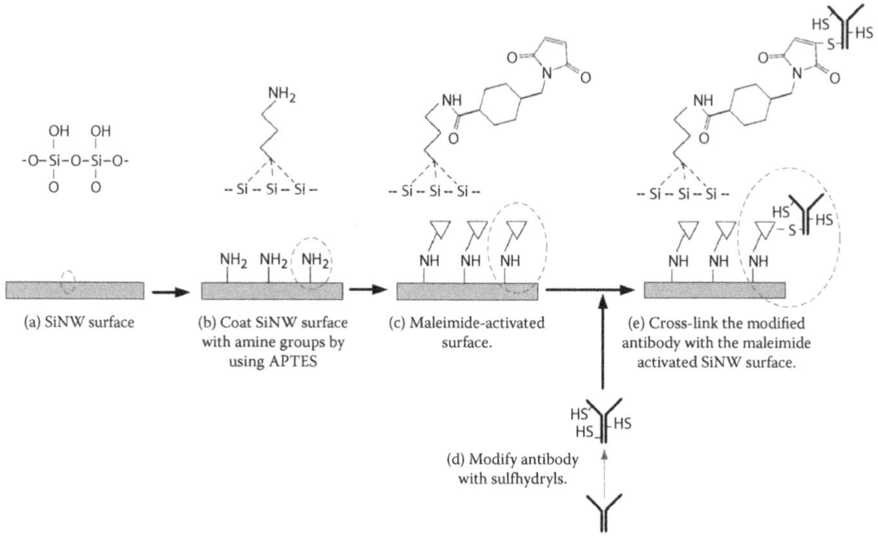

FIGURE 11.4 Steps involved in nanowire functionalization. Processing of the nanowire surface in order to conjugate antibodies that function as chemical gates.

11.4 NANOWIRE FUNCTIONALIZATION PROTOCOL

Protocols for functionalizing the nanowire can be found in [1,2,4–14,16,17]. In this work, the protocol for functionalization is shown in Figure 11.4 and is based on the method described in [15]. The functionalization of the nanowire surface is achieved by first thoroughly washing the device with a mixture of acetone and ethanol (1:1 v/v). After drying, the nanowire is silylated using an aminosilane reagent (3-amino-propyltriethoxysilane (APTES)). This step results in a nanowire surface that is coated with amine groups. A cross-linker (Sulfo-SMCC) is then applied to the silylated surface and activated with maleimide. Sulfhydryl groups are made available on an antibody by using Traut's reagent, which reacts with primary amines ($-NH_2$) present on the side chain of lysine residues of antibodies, creating sulfhydryl groups upon reaction with the maleimide-activated surface. The next step is to immediately cover the maleimide-activated surface with the modified antibody solution, which is then incubated for 2–4 h at room temperature for the antibody to covalently attach to the nanowire surface. The surface is then thoroughly rinsed with coupling buffer phosphate-buffered saline (PBS). After this step, the nanowire and antibody system is ready for use as a detection assay.

11.5 FREQUENCY-DEPENDENT METHOD FOR BIOMOLECULE DETECTION

The DC change in conduction has been described for biomolecule sensing purposes [1]. Although the relative change in conduction can be large, the absolute conductance is small, requiring large amplification of the signal received across the nanowire. The resulting signal can be noisy and difficult to determine accurately.

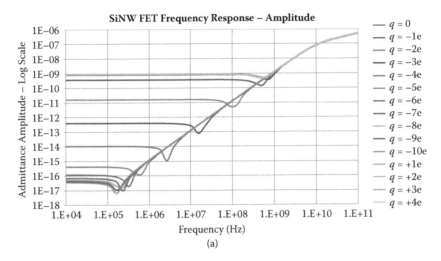

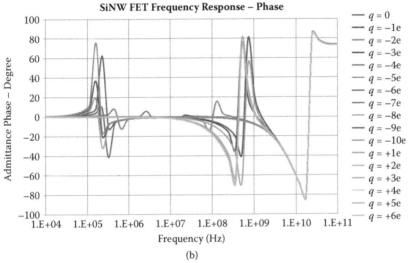

FIGURE 11.5 SiNW FET frequency response as a function of gate charge. The nanowire has dimensions of 10 nm (W) × 10 nm (H). (a) Amplitude plot data of admittance. (b) Phase plot of admittance.

In [14], an AC signal analysis method was employed to detect antigen-antibody bonding with the subsequent determination of antigen concentration. From the frequency-based approach described, a signal was used to interrogate the device by sweeping from 10 kHz to many gigahertz while keeping the gate and source-drain voltage constant.

Figure 11.5 illustrates that both the amplitude (a) and phase (b) of the received signal versus frequency curves depend on the amount of charge attached to the nanowire. Furthermore, it can be illustrated that the corner frequencies in the amplitude plot correspond to those in the phase plot where the change in phase sign occurs. The actual corner frequency at zero charge can be determined once the nanowire device is fabricated. Then, when negative elementary charges attach to the nanowire surface, this will result in a significant difference in the nanowire's frequency response amplitude plot. We propose, therefore, that during the antigen detection process, by recognizing a shift in the corner frequency, the amount of negative charge on the nanowire surface can be determined once the nanowire is calibrated. The target antigen that is specifically bound by the antibody receptor on the modified surface, if detected, carries this charge.

To further study the frequency domain behaviors of silicon nanowire devices, AC signal analysis versus the nanowire device dimensions was performed. Sizes of 10 × 10 nm, 30 × 30 nm, and 50 × 50 nm (width by height) nanowires were compared with other parameters, such as length, doping profile, and aluminum pads, which were kept identical. Figure 11.6 illustrates the frequency response of three SiNW FET devices with different dimensions (i.e., width by height). It can be seen that an increase in channel width and height leads to an increase in device admittance, although the nanowire's high-pass filter behavior is retained. In addition, the change in the channel dimension also results in a change of the corner frequency of the transfer function. It can be seen from Figure 11.7 that while the charge varies from -1 e to -10 e, as the cross-sectional area of a nanowire increases, we observe less distinction in the corner frequencies, making it more difficult to distinguish between the amounts of attached charge.

Sensitivity in these devices is ultimately dependent on how easy it is to recognize the difference in measuring signal during detection [18]. In traditional DC methods, detection is based on observing the change in conductance. This is difficult, especially when the absolute change is too small due to a low concentration of the target molecules, and requires sensitive low-noise amplifiers and high-resolution analog-to-digital converters, which is not ideal for low-cost and highly integrated systems. The frequency-dependent method of antigen detection, the task of discriminating the corner frequencies, seems to be closely located, as in Figure 11.6(c), and is straightforward as the difference is on the order of 100 MHz. As can be observed in Figure 11.7, this task becomes less complicated as the dimension of the nanowire channel shrinks. As a result, it can be concluded that there is an inversely proportional relationship between the device's sensitivity and its channel dimension.

Figure 11.7(a) also illustrates how the corner frequency varies as a function of attached charge and nanowire dimensions. Overall, with the same amount of charge, larger nanowire dimensions exhibit higher corner frequencies. It is also clear that at the lower end (i.e., smaller number of attached charges), corner frequencies linearly drop and diverge further as nanowire dimension gets smaller, making them significantly easier to be distinguished. However, as more surface charge is applied, the corner frequencies saturate. Moreover, the frequency saturation regions, which are the regions in Figure 11.6 where more charge does not result in corner frequency change, depend on nanowire channel size. Figure 11.7(b) also depicts a graph that shows the slope of the high-pass filter for an n-type nanowire device as described in the simulations. This graph demonstrates that regardless of the device width and height, these nanowire devices have approximately the same slopes.

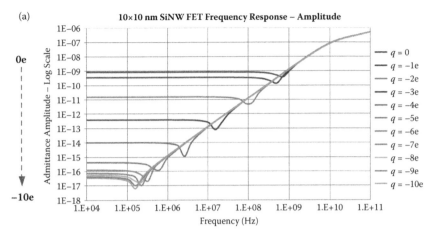

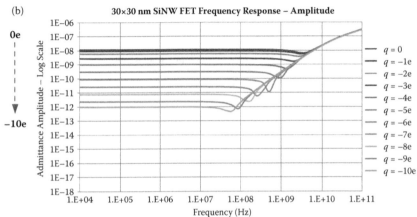

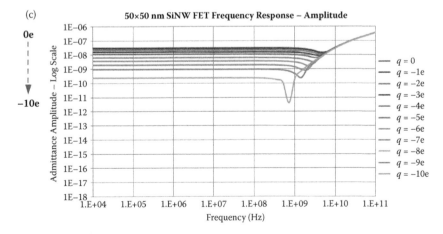

FIGURE 11.6 SiNW FET frequency response as a function of gate charge and nanowire channel dimension: (a) 10×10 nm nanowire, (b) 30×30 nm nanowire, and (c) 50×50 nm nanowire. In these simulations, the nanowire length is 2.2 µm.

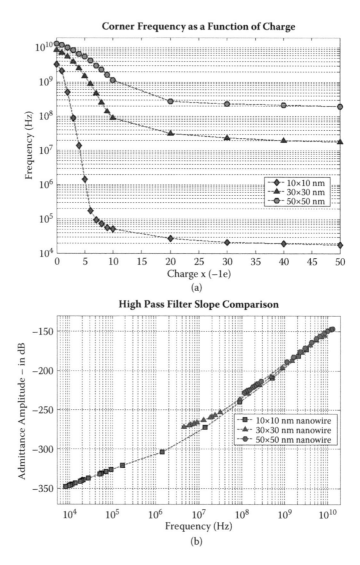

FIGURE 11.7 (a) Corner frequencies as a function of charge attached and nanowire dimension showing how the corner frequency drops at low charge but saturates as the charge increases. (b) Comparison of nanowire devices' high-pass filter slope.

11.6 DETECTION CIRCUIT

The circuit proposed in Figure 11.8 can be used to determine the amount of charge that is attached to the nanowire. The circuit shown transmits an up-converted I/Q signal through the nanowire, which itself introduces a frequency-dependent phase change depending on the amount of charge attached to the nanowire. The receiving circuit amplifies this signal and down-converts it to a low frequency that is then sampled by two analog-to-digital converters.

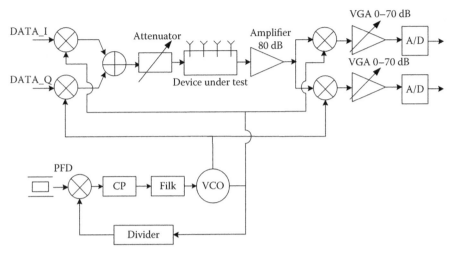

FIGURE 11.8 A proposed detection circuit for the frequency-dependent method of biomolecule detection.

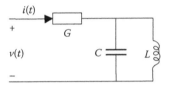

FIGURE 11.9 Equivalent representation of the nanowire system.

This signal is then used to determine the phase change of the received signal relative to the transmitted signal. The charge is then determined by estimating the frequency at which there is a 45° phase change. This approach has the advantage of permitting robust identification of the attached charge when compared to methods trying to determine changes of conductance at DC, as it makes multiple measurements reducing the noise effect, but also avoids the need for accurate knowledge about the gain of the reception amplifiers. Here a ratio is used that is robust to low noise amplifier (LNA) amplification error. Calibration of the system is performed be applying a sample of known contribution. This would be done at the time of manufacture.

11.7 DETERMINATION OF CHANGE OF NANOWIRE PARAMETERS

The high-pass nature of the frequency responses revealed in Figures 11.5 and 11.6 suggests that the system can be represented by the circuit depicted in Figure 11.9. As a result, system identification techniques [19] can be utilized to identify the associated parameters (i.e., the conductance G, capacitance C, and inductance L) and, consequently, the corner frequency or phase change of the system. Together with the known frequency response properties obtained experimentally or through simulation, this result enables one to determine the amount of gate charge, leading to the detection of the antigen attached to the nanowire.

The system identification is based on measured current samples $\{i(n)\}$ in response to a certain excitation voltage sequence $\{v(n)\}$. According to Figure 11.8, the transfer function of the system in the Z-domain is given by

$$\frac{I(Z)}{V(Z)} = \frac{b_0 + b_2 Z^{-2}}{1 + a_1 Z^{-1} + a_2 Z^{-2}} \tag{11.1}$$

where
$$b_0 = G$$
$$b_2 = G/(LC)$$
$$a_1 = G/C$$
$$a_2 = 1/(LC)$$

The corresponding time domain expression of (11.1) is given by

$$i(n) = -a_1 i(n-1) - a_2 i(n-2) + b_0 v(n) + b_2 v(n-2)$$

For an N-sample input sequence $\{v(n), n = 0, 1, ..., N - 1\}$, the current (output) samples $\{i(n), n = 2, 3, ..., N - 1\}$ can be expressed in the following vector form:

$$y = Hb$$

where

$$y = \left[i(2) i(3) ... i(N-1) \right]^T$$

$$H = \begin{pmatrix} -i(1) & -i(0) & v(2) & v(0) \\ -1(2) & -i(1) & v(3) & v(1) \\ \vdots & \vdots & \vdots & \vdots \\ -i(N-2) & -i(N-3) & v(N-1) & v(N-2) \end{pmatrix}$$

$$b = \left[a_1 \ a_2 \ b_0 \ b_2 \right]^T$$

The least-squares [20] estimation of vector **b** is accomplished via

$$\hat{b} = H^+ y$$

where ( )+ denotes the Moore-Penrose pseudoinverse. Note that by setting the initial condition $i(0) = i(1) = 0$, both H and **y** are known matrices. The estimation of vector **b** is accomplished via the following constraint least-squares minimization [20]:

$$\hat{b} = \arg \min_b \left\| H\mathbf{b} - y \right\|_2, \text{ subject to } b(4) = b(2) * b(3)$$

11.8 CONCLUSIONS

In this chapter, silicon nanowires that are compatible with CMOS fabrication processes have been described. It has been shown that these nanowires can be functionalized by conjugating monoclonal antibodies to their surface in order to build sensitive biochemical sensors. It has also been shown that by using frequency-based signals, all the necessary components to interrogate these nanowires can be built on low-cost CMOS processes.

ACKNOWLEDGMENTS

The authors would like to acknowledge the generous support of National ICT Australia and the Centre for Neural Engineering at the University of Melbourne. This work was performed in part at the Melbourne Centre for Nanofabrication, an initiative partly funded by the Commonwealth of Australia and the Victorian Government.

REFERENCES

1. Z. Gao, A. Agarwal, A.D. Trigg, et al. Silicon nanowire arrays for label-free detection of DNA. *Analytical Chemistry*, 79(9), 3291–3297, 2007.
2. C.-C. Wu, F.-H. Ko, Y.-S. Yang, et al. Label-free biosensing of a gene mutation using a silicon nanowire field-effect transistor. *Biosensors and Bioelectronics*, 25(4), 820–825, 2009.
3. X.T. Vu, R. Ghosh Moulick, J.F. Eschermann, et al. Fabrication and application of silicon nanowire transistor arrays for biomolecular detection. *Sensors and Actuators B: Chemical*, 144(2), 354–360, 2010.
4. C.-Y.H. Chih-Heng Lin, C.-H. Hung, Y.-R. Lo, C.-C. Lee, C.-J. Su, H.-C. Lin, F.-H. Ko, T.-Y. Huang, and Y.-S. Yang. Ultrasensitive detection of dopamine using a polysilicon nanowire field-effect transistor. *Chemical Communications*, 5749–5751, 2008.
5. G.-J. Zhang, Z.H.H. Luo, M.J. Huang, et al. Morpholino-functionalized silicon nanowire biosensor for sequence-specific label-free detection of DNA. *Biosensors and Bioelectronics*, 25(11), 2447–2453, 2010.
6. J.H. Chua, R.-E. Chee, A. Agarwal, et al. Label-free electrical detection of cardiac biomarker with complementary metal-oxide semiconductor-compatible silicon nanowire sensor arrays. *Analytical Chemistry*, 81(15), 6266–6271, 2009.
7. C.-Y. Hsiao, C.-H. Lin, C.-H. Hung, et al. Novel poly-silicon nanowire field effect transistor for biosensing application. *Biosensors and Bioelectronics*, 24(5), 1223–1229, 2009.
8. G.-J. Zhang, L. Zhang, M.J. Huang, et al. Silicon nanowire biosensor for highly sensitive and rapid detection of Dengue virus. *Sensors and Actuators B: Chemical*, 146(1), 138–144, 2010.
9. J.-I. Hahm and C.M. Lieber. Direct ultrasensitive electrical detection of DNA and DNA sequence variations using nanowire nanosensors. *Nano Letters*, 4(1), 51–54, 2003.
10. G.-J. Zhang, J.H. Chua, R.-E. Chee, et al. Label-free direct detection of MiRNAs with silicon nanowire biosensors. *Biosensors and Bioelectronics*, 24(8), 2504–2508, 2009.
11. Z. Gengfeng, F. Patolsky, C. Yi, et al. Multiplexed electrical detection of cancer markers with nanowire sensor arrays. *Nature Biotechnology*, 23(10), 1294–1301, 2005.
12. F. Patolsky, G. Zheng, O. Hayden, et al. Electrical detection of single viruses. *Proceedings of the National Academy of Sciences of the United States of America*, 101(39), 14017–14022, 2004.

13. J. Chang, A. Abidi, and C. Viswanathan. Flicker noise in CMOS transistors from sub-threshold to strong inversion at various temperatures. *IEEE Transactions on Electron Devices*, 41(11), 1965–1971.

14. T. Nguyen, W. Qiu, and E. Skafidas. Functionalised nanowire based antigen detection scheme using frequency based signals. *IEEE Transactions on Biomedical Engineering*, 59(1), 213–218, 2012.

15. C. Niemeyer (Ed.). Bioconjugation protocols. In *Methods in Molecular Biology*. Vol. 283. Humana Press, Clifton, NJ.

16. C. Yi, W. Qingqiao, P. Hongkun, et al. Nanowire nanosensors for highly sensitive and selective detection of biological and chemical species. *Science*, 293(5533), 1289, 2001.

17. E. Stern, J.F. Klemic, D.A. Routenberg, et al. Label-free immunodetection with CMOS-compatible semiconducting nanowires. *Nature*, 445(7127), 519–522, 2007.

18. P.R. Nair and M.A. Alam. Design considerations of silicon nanowire biosensors. *IEEE Transactions on Electron Devices*, 54(12), 3400–3408, 2007.

19. L. Ljung. *System Identification: Theory for the User*. Prentice Hall, Upper Saddle River, NJ, 1999.

20. S. Boyd and L. Vandenberghe. *Convex Optimization*. Cambridge University Press, Cambridge, 2004.

12 Trace Explosive Sensor Based on Titanium Oxide-B Nanowires

Danling Wang and Antao Chen

CONTENTS

12.1 INTRODUCTION

Nearly every terrorist attack involves explosives. To protect society from the increasing terrorist threat, effective techniques to detect high-explosive materials commonly used in terrorist attacks, such as TNT (2,4,6-trinitrotoluene), RDX (cyclotrimethylene trinitramine), PETN (pentaerythritol tetranitrate), and picric acid (2,4,6-trinitrophenol), at trace levels in luggage, vehicles, mail, aircraft, and soils, are very necessary. However, trace explosives are known to be very difficult to detect due to several factors, including the physical state of the sample to be detected (solid, liquid, and gas), very low vapor pressure of explosives [1], complicating or inhibiting the detection due to degradation by-products of explosives, and interferences leading to false signal due to other chemicals in the environment and the lack of selectivity of the detection techniques. In recent years, governments and industries have made a lot of effort in improving existing detecting technologies as well as developing new methods that could allow high sensitivity, selectivity, small size, and low cost.

Techniques to detect concealed explosives can be classified into two categories [2]: bulk detection and trace detection. Bulk detection is typically based on x-ray and gamma ray imaging techniques [3]. Considering extremely low vapor pressure of some explosives, a detection device needs to be very sensitive. There are a wide variety of analytical techniques for trace explosive detection that have been developed in recent years. They include ion mobility spectrometry [4], gas chromatography–mass spectrometry [5], and optical techniques such as fluorescing quenching, Raman, and laser breakdown spectroscopy [6,7]. However, detection systems based on those techniques are usually bulky, expensive, and with high-power consumption. These factors limit their applications in explosives detection [8–11]. Trained animals like dogs are an extremely sensitive sniffer to detect specific explosives, but these animals only produce qualitative alarms rather than quantitative data, and animals require a lot of effort to maintain. There are fundamental limitations to those techniques in significantly reducing the size, weight, power consumption, and quantification of detecting trace explosives. The development of nanotechnology potentially provides a feasible solution for building substantially smaller, highly sensitive, and selective trace explosive detectors through the use of a variety of nanostructured materials. Nanomaterials also have a high surface-to-volume ratio. This favors the adsorption of gases on the sensor and can increase the sensitivity and response time of the device because the interaction between the analytes and the sensing material is stronger [12–16]. It was found that in a chip-size device with greatly reduced size, weight, and power consumption, both the sensitivity and response speed of the sensor based on the nanostructured materials have surpassed those of current technologies [17–19]. Nanostructures of metal-oxide semiconductors such as In_2O_3 and SnO_2 are already widely used as a base material for commercial gas sensors for the detection of toxic (e.g., CO) or dangerous (e.g., CH_4) gases, owing to the lower production costs, high sensitivity, and long-term stability [20]. The material characteristics and size effects of metal oxides have been well explored for sensor applications [21].

Compared to other wide-band-gap metal oxides, such as SnO_2, Ga_2O_3, ZnO, and WO_2, titania (TiO_2) exhibits unique chemical and electrical characteristics, including superior photocatalytic properties and excellent chemical stability [22]. TiO_2-B is

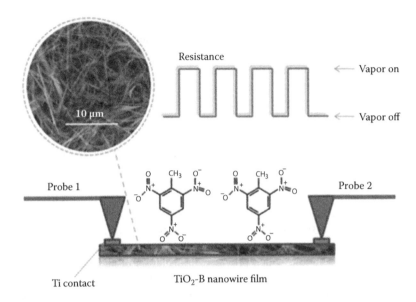

FIGURE 12.1 Schematic of a test sample for the chemiresistive response to nitroaromatic and nitroamino explosives. (Reproduced from D.L. Wang et al., *IEEE Sensors Journal*, 1352–1358, 2011. Copyright © 2011 IEEE. With permission.)

one of the crystal polymorphs of TiO_2 [23]. It proves to be the least dense polymorph of TiO_2, which has a relatively open structure with significant voids and continuous channels compared to the other titanium dioxide polymorphs [24–26]. TiO_2-B as a good functional material promising for sensor applications was first suggested in Wang et al.'s paper [27]. Recently, our group reported that the TiO_2-B nanowires exhibit sensitive and rapid chemiresistive responses to subtrace concentrations of nitroaromatic and nitroamine explosives (Figure 12.1) [14]. Detection limits below sub-ppb levels and response times below a second were observed. This chapter will present a summary of recent developments of TiO_2-B nanowires as explosive sensors. The mechanism of TiO_2-B nanowires to effectively detect explosive vapors will be discussed, and in the final section we will point out new potential improvements that can be developed to further increase the sensing properties of TiO_2-B nanowires as explosive sensors.

12.2 MATERIAL SYNTHESIS AND DEVICE FABRICATION

This section describes nanowire synthesis, test sample fabrication, equipment, and procedures to test nanowire samples for chemiresistive responses.

12.2.1 MATERIAL SYNTHESIS AND CHARACTERIZATION

The TiO_2-B nanotubes were synthesized with a hydrothermal method [28–32]. Typically, a suspension containing 0.5 g of commercial anatase TiO_2 powder (J.T. Baker Chemical Co.) dispersed in 20 ml of 10 M NaOH aqueous solution was prepared as precursor. After vigorous stirring for 5 ~ 10 min, the suspension was transferred to a Teflon vessel and placed in a hermetically sealed autoclave and heated at 180°C for 32 h. The precipitate produced was washed with 0.1 M HCl several times and then with deionized (DI) water until the pH value reached 7. This treatment removes Na^+ ions remaining in the titanate nanoproducts and results in the formation of H_2TiO_3 nanotubes. A following post-treatment at 450°C for 1 h was carried out to promote the phase transformation from H_2TiO_3 to TiO_2-B [28]. The length of the nanowires could be controlled by choosing the ultrasonic treatment time to the suspension solution of TiO_2 nanoparticles, as described in literature [30]. The diameter of the nanowires was determined by the temperature during hydrothermal growth.

A Joel JSM-7000F scanning electron microscope (SEM) was used to character-ize the morphology of the $TiO_2(B)$ nanowires. An x-ray diffraction (XRD) pattern of as-synthesized TiO_2 nanowires was obtained with a Bruker F8 Focus Power XRD to confirm the $TiO_2(B)$ crystal structure. Fourier transfer infrared (FTIR) spectra were taken using a Bruker Vector 33 spectrophotometer in the 500–4000 cm^{-1} range with a 0.6 cm^{-1} resolution, which provided information of chemical bonds on the TiO_2 nanowires' surface. The morphology of the nanowire films was studied with scanning electron microscopy (SEM) in Figure 12.2, and the films have a highly porous structure made of a three-dimensional (3D) mesh of randomly orientated and interconnected nanowires. The length of the nanowires ranges from 1.1 to 2.2 μm, and the diameter of the nanowires is 40–100 nm, adjustable through synthesis conditions. Electric contact pads made of titanium were deposited onto the thin film

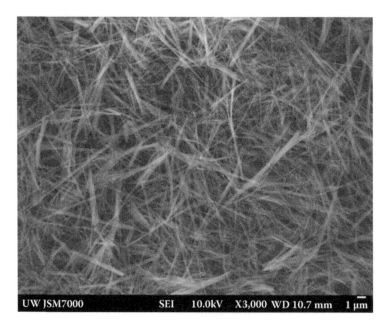

FIGURE 12.2 SEM image of the interconnected 3D mesh structure of a TiO_2-B thin film. (Reproduced from D.L. Wang et al., *IEEE Sensors Journal*, 1352–1358, 2011. Copyright © 2011 IEEE. With permission.)

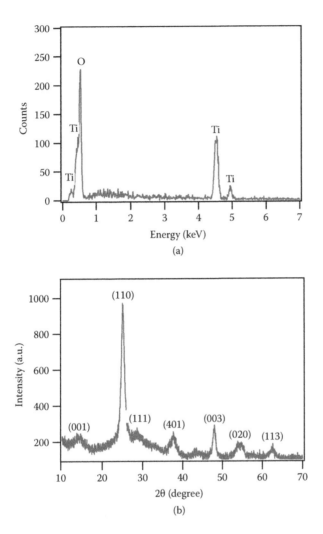

FIGURE 12.3 (a) EDS spectrum and (b) XRD pattern of synthesized TiO$_2$-B nanowires. θ is the x-ray diffraction angle. The XRD pattern matches the TiO$_2$-B pattern in literature [23, 31]. (Reproduced from D.L. Wang et al., *IEEE Sensors Journal*, 1352–1358, 2011. Copyright © 2011 IEEE. With permission.)

by sputtering. Titanium forms a good ohmic contact of low junction resistance with TiO$_2$(B). Shown in Figure 12.3 are an energy-dispersive spectrum (EDS) and x-ray diffraction (XRD) of the nanowire film, revealing the composition of titanium oxide and the crystal structure of TiO$_2$-B.

12.2.2 DEVICE FABRICATION

As-synthesized nanowires were dispersed in ethanol to form a suspension solution. This solution was then drop-casted on glass substrates and heated at 70°C to attain a thin film of nanowires about 10 μm in film thickness. To fabricate a sensor, as shown in Figure 12.1, patterned titanium electrodes were deposited through a shadow mask over the nanowire film by sputtering. The titanium electrodes have a circular shape and are 4 mm in diameter. The spacing between contacts is 1 cm for the convenience of shadow mask patterning and probing during testing. The nominal thickness of the titanium electrodes is 200 nm.

12.2.3 SENSOR TESTING

Vapors of equilibrium concentrations at room temperature were generated using glass beads coated with various explosive analytes (Inert Products, LLC) in a vapor generator based on the method described in [33]. The vapor concentrations were confirmed by HP 5797 gas chromatography–mass spectrometry (GC-MS). After changing an analyte, all the tubings in the downstream of the vapor generator were replaced and the vapor generator was operated for several days before the vapor was used for sensor testing. This is to ensure that impurities and water in the generated vapor are purged, and the adsorption and desorption of analyte molecules on the inside walls of the tubing are at equilibrium. As vapor left the generator with the heated analyte, the temperature of the vapor gradually dropped to room temperature and excess explosives condensed on the walls of the tubing. At the exit end of the tubing, air that contained saturated concentration of analyte vapor at room temperature was obtained.

To test the sample for response to explosive vapor, a solenoid valve alternated the gas flow to the sample between air that contained saturated analyte vapor and pure air without the vapor. The cycle time of the valve was a few seconds. The sensitivity and response time of the sensor were determined from the change of the resistance between the electrodes with a Keithley 617 electrometer while the valve that controlled the gas flow to the test sample was cycled. The resistance measurements were made at room temperature and in ambient air.

12.2.4 SURFACE MODIFICATIONS OF SENSOR SAMPLES

Hydrogen and oxygen plasma treatments of nanowires were carried out in a microwave plasma cleaner/etcher (Plasma-preen). The plasma treatments were made at pressure of 1 Torr and radio frequency (RF) power of 300 W for 5 min, when the effect of plasma treatments became saturated.

12.3 RESULTS AND DISCUSSION

12.3.1 SENSITIVITY AND RESPONSE TIME

The sensing performance of nanowire film samples was tested in ambient air, except for tests of temperature effects. The response time τ ($1/e$ time constant) is obtained by fitting the resistance change to an exponential function. The chemiresistive response is defined as $S = (R_v - R_0)/R_0$, where R_v is the resistance when the sample is exposed to the vapor, and R_0 is the resistance of the fresh sample before it is exposed to any

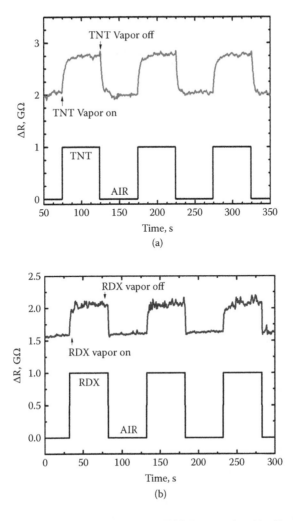

FIGURE 12.4 Typical resistance change of a TiO2-B nanowire thin film in response to vapors. (a) 1 ppb of 2,4,6-trinitrotoluene (TNT). (b) 5 ppt of 1,3,5-trinitroperhydro-1,3,5-triazine (RDX) at ambient conditions. (Reproduced from D.L. Wang et al., *IEEE Sensors Journal*, 1352–1358, 2011. Copyright © 2011 IEEE. With permission.)

explosive vapor [34]. The resistance of the nanowires increases to R_v upon exposure to explosive vapor. After the sample is returned to fresh air, its resistance decreases and reaches a stabilized value R_s, which is usually 0 to 100% higher than R_0 and is likely due to incomplete desorption of explosive molecules from the surface of nanowires. Subsequent switching cycles between vapor of the same concentration and pure air make the resistance vary between R_v and R_s, as shown in Figure 12.4. The recovery time of the resistance after the explosive vapor is replaced by pure air is almost the same as the response time to the vapor. The incomplete desorption mentioned above can be eliminated by heating the sample at 80 to 100°C for several minutes, and the resistance of the sample returns to R_0.

TABLE 12.1

Response of a TiO_2-B Nanowire Thin Film to Explosives

Symbol	Equilibrium Vapor Concentration	Percentage Response $(100 \times (R_v - R_0)/R_0)$	Response Time (s)	Molar Mass (g/mol)
Nitrotoluene (NT)	130 ppm	55	0.57	132
2,4-Dinitrotoluene (DNT)	100 ppb	58	0.64	182.13
2,4,6-Trinitrotoluene (TNT)	5 ppb	57	1.67	227.13
RDX	5 ppt	50	2.35	222.12

Note: Response of a TiO_2-B nanowire thin-film sample to room temperature equilibrium vapor of common high explosives. ppm = parts per million, ppb = parts per billion; ppt = parts per trillion.

Table 12.1 lists the percentage resistance change and response times of the TiO_2-B nanowire thin film for vapor of representative explosive compounds at equilibrium concentrations, where the values of the equilibrium vapor concentration are referenced from literature [36]. Significant and fast changes in resistance have been observed with all major explosives, including cyclotrimethylene trinitramine (RDX), which has extremely low vapor pressure. The concentrations of TNT and RDX vapors were analyzed by HP 5797 GC-MS spectrum and compared with the results in literature [14,35]. Note that molecules of smaller mass produce faster response, because the response time is limited by the diffusion of the vapor molecules through the nanowire thin film, and smaller molecules permeate the 3D matrix of nanowires faster.

12.3.2 SPECIFICITY AND STABILITY

Since the detection of explosives based on TiO_2-B nanowires was conducted at room temperature, the interaction between explosive molecules and TiO_2-B nanowires should be quite different from previous semiconductor gas sensors, which typically operate at above 400°C. Explosive molecules are unlikely to be capable of producing significant and fast chemical reduction-oxidation reactions at room temperature to produce sensitivity below the ppt level (Figure 12.4) and a fast response on the order of a second. The synthetic TiO_2-B nanowires have a relatively higher charge carrier transfer ability than anatase TiO_2 [27], less compact structure, and higher level of oxygen vacancies due to Ti^{4+} ions. The Ti^{4+} ions can be coordinated by hydroxyl groups and form hydroxyl-terminated surfaces [37–39]. The surface hydroxyl groups can trap electrons and facilitate adsorption of explosive compounds via their nitro groups. This indicates that the hydroxyl groups on the TiO_2-B nanowires' surface play an important role in determining the explosive gas-sensing properties at room temperature, and further study is described in the following section. Previous studies have shown that titanium oxide is an n-type semiconductor due to oxygen vacancies [40,41]. It is also known that nitroaromatic compounds and most high explosive compounds are highly electronegative, meaning that they tend to attract electrons from other molecules through charge transfer interactions from nitro groups in explosives to hydroxyl groups on TiO_2-B nanowires' surface. When explosive molecules adsorb on the surface of n-type semiconductor nanowires, the explosive molecules can trap charge carriers via surface hydroxyl groups and create a carrier depletion region near the TiO_2-B nanowires' surface. This could explain the increase of the resistance when the nanowires are exposed to explosive vapors.

In order to determine whether the response is due to charge transport within individual wires or across junctions between connecting wires, test samples of nanowires of the same diameter but different lengths were fabricated and their test results compared. Test results showed that the length of the wires has no significant effect on the chemiresistive response. However, shorter wires respond to the vapor at a slower rate. The slower response can be attributed to the denser packing of shorter nanowires and the consequent slower permeation of the thin film by vapor. Since the test samples have the same separation between the electrodes, the average path length for an electron to travel from one electrode to the other electrode is largely independent of the length of nanowires. However, the average number of junctions between connecting nanowires in the path of the electron is strongly dependent on the length of individual wires. In the film made of shorter wires, electrons need to pass through a greater number of junctions. The fact that the response is not affected by the length of nanowires indicates that the junction between nanowires does not play a significant role in the sensing process. In addition, different metal electrodes, including gold, aluminum copper, and titanium, have been used. Different types of metals did not change sensitivity and response time, indicating that the interface between the metal electrodes and nanowires did not play a significant role in the chemiresistive response. Test samples with titanium electrodes exhibited a more linear current-voltage (I-V) relationship, characteristic of a good ohmic contact between the metal electrode and nanowires. Based on these observations, it can be concluded that the electrical response is dominated by the charge transport within individual nanowires.

The surface depletion being the origin of the chemiresistive response is confirmed by the effect of the wire diameter on the response. Samples made of films with different wire diameters were prepared [42] and tested. The same film thickness and electrode spacing were used for all test samples. Wires of 50 ± 10 nm in diameter produced a response of 30%, higher than the 22% response observed for wires of 100 ± 20 nm in diameter. This observation supports the surface depletion hypothesis because thinner wires are more susceptible to surface depletion [43]. The response time is found to be largely independent of the diameter of nanowires.

Because the band-gap of TiO_2 is much greater than the thermal energy, the chemiresistive response of TiO_2-B nanowires to explosive trace vapors is found to be reliable and insensitive to temperature. There is only a few percent decrease in (R_v-R_s), and a slightly faster response at 75°C over 25°C. Such stability is the key to reliable sensors for practical applications. Also, as indicated in Figure 12.5, although the resistant baseline has a drift from 0.6 GΩ to 1.5 GΩ, the resistance change between R_v and R_s is found to be highly consistent over a long-term test of 15,000 switching cycles. The material demonstrated the exceptional stability of sensing performance. The sample was at ambient temperature throughout the test.

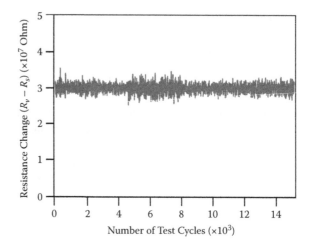

FIGURE 12.5 Resistance change R_v-R_s over 15,000 continuous test cycles between vapor of 100 ppb of DNT and pure air. Each test cycle consists of 6 s of vapor and 6 s of air. (Reproduced from D.L. Wang et al., *IEEE Sensors Journal*, 1352–1358, 2011. Copyright © 2011 IEEE. With permission.)

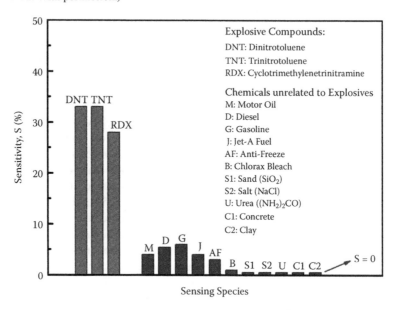

FIGURE 12.6 The sensing response of TiO_2-B nanowires to different chemical compounds. (Reproduced from D.L. Wang et al., *IEEE Sensors Journal*, 1352–1358, 2011. Copyright © 2011 IEEE. With permission.)

On the other hand, the nanowires are found to have good specificity (seen in Figure 12.6) and are not sensitive to chemicals that are unrelated to explosives that cause false positives to other explosive detectors, such as inert chemicals that have high nitrogen content (for example, urea).

12.4 UNDERSTANDING THE BASIC MECHANISM OF THE CHEMIRESISTIVE RESPONSE

According to the above results, the achievement of TiO_2-B nanowires in detecting explosives is related to two common characteristics of the nitroaromatic and nitroamine explosive compounds: (1) high electronegativity due to strong oxidizing chemical groups and (2) strong tendency to adsorb on the surfaces of objects exposed to their vapor. Therefore to effectively detect these explosive compounds, it is desirable that the sensing material has a high surface-to-volume ratio as well as a large surface area for the adsorption of explosive molecules, and moreover, it must be able to facilitate a strong surface charge transfer interaction with the molecules. TiO_2-B nanowires meet both conditions. Also, the Ti^{4+} ions in TiO_2-B can be coordinated by hydroxyl groups and tend to have hydroxyl-terminated surfaces [38]. The surface hydroxyl groups could effectively interact with nitro groups in explosives at room temperature and facilitate TiO_2-B surface physisorption and charge transfer interactions. In order to further study the role of surface hydroxyl groups in the chemiresistive response of explosives, a series of surface treatments to change the density of surface hydroxyl groups on TiO_2-B nanowires have been carried out, and the effects of each surface treatment on the chemiresistive sensing response have been characterized. On the other hand, the surface hydroxyl groups mean that TiO_2-B nanowires have a polar surface [38,44–47]. A stronger dipole-dipole interaction between TiO_2-B nanowires and polar analyte molecules can be another way to understand the sensing mechanism of TiO_2-B nanowires being the explosives sensor.

12.4.1 THE ROLE OF SURFACE HYDROXYL GROUPS ON TiO_2-B NANOWIRES TO RESPONSE EXPLOSIVES

As we mentioned above, the resistant baseline has a drift from 0.6 to 1.5 G in the long-term scan, indicating that TiO_2-B nanowires' chemiresistive responses to explosive vapors are affected by the humidity of the environment to which the nanowires are exposed [14]. This indicates that the hydroxyl groups on the surfaces of nanowires participate in the chemiresistive response [48], as it is known that humidity strongly affects the density of hydroxyl groups on the surface of metal oxides. In this section, a series of surface treatments to change the density of surface hydroxyl groups on TiO_2-B nanowires have been carried out in order to show how the density of hydroxyl groups on the TiO_2-B nanowires' surface influences its chemiresistive response.

12.4.1.1 Varying the Density of Surface Hydroxyl Groups through Plasma Treatment

The surface hydroxyl groups on the TiO_2-B nanowires and the charge transfer via the surface hydroxyl groups and nitro group in explosives were initially observed via a FTIR spectrum. Figure 12.7 is FTIR spectra of the TiO2 -B nanowires before and after exposure to 2,4-dinitrotulene (DNT) vapor. Three main bands were detected in FTIR, and they were consistent with the bands described by Vargas and Nunez [39]. Compared to the FTIR spectrum of the control sample, the broad band at 3370 cm^{-1}

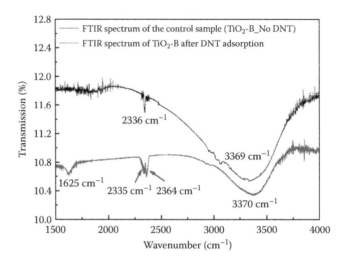

FIGURE 12.7 FITR spectrum of TiO₂-B nanowires after DNT exposure. The background was recorded by as-synthesized TiO₂-B nanowires. (Reproduced from D.L. Wang et al., *Journal of Materials Chemistry*, 7369–7373, 2011. Copyright © 2011 The Royal Society of Chemistry. With permission.)

is due to surface hydroxyl groups (Ti-OH) in the TiO₂-B thin film [45,46]. This band is evidence that TiO₂-B nanowires' surface is terminated with hydroxyl groups. The new band at 1625 cm⁻¹ is attributed to the complex between nitro groups of DNT bonded with hydroxyl groups at the TiO₂-B surface (-O-Ti-OH-N-O-) [39]. This band did not exist in neat thin films (samples that had not been exposed to DNT). The new band at 1625 cm⁻¹ after adsorption of DNT vapor onto TiO₂-B indicates a charge transfer pathway from the TiO₂-B surface to nitro groups in explosives via hydroxyl groups. The bands at 2335 and 2364 cm⁻¹ are from carbon dioxide and aromatic groups of DNT. Figure 12.8 shows the sensing responses of nanowires to explosive vapor, 2,4,6-trinitrotoluene (TNT), after oxygen and hydrogen plasma treatments, respectively. A higher response of TiO₂-B nanowires to TNT after an oxygen plasma treatment was observed compares to the response of the untreated TiO₂-B nanowires and the nanowires after a surface hydrogen plasma treatment. It is noticed that the hydrogen plasma treatment did not change the sensitivity significantly. This is probably due to the competing effects of surface cleaning and reduction of hydroxyl groups.

The water wettability tests also confirmed that plasma treatments change the contact angles of water on the TiO₂ surface, and therefore change the density of surface hydroxyl groups. Figure 12.9 shows the contact angles on the TiO₂-B nanowires' surfaces after different surface plasma treatments. It is evident that the contact angle increases when TiO₂-B nanowires are treated with hydrogen plasma and decreases after oxygen plasma treatment, indicating a more hydrophilic TiO₂-B nanowire surface after an oxygen plasma treatment. These results suggest that TiO₂-B nanowires' surface with an oxygen plasma treatment has a higher density of surface hydroxyl groups. This is because an oxygen-rich nanowire surface created by oxygen plasma could enhance the dissociative adsorption of water molecules in air and increase the density

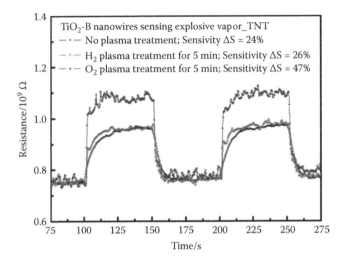

FIGURE 12.8 Effects of surface plasma treatments on the sensitivities of TiO$_2$-B nanowires to TNT vapor. (Reproduced from D.L. Wang et al., *Journal of Materials Chemistry*, 7369–7373, 2011. Copyright © 2011 The Royal Society of Chemistry. With permission.)

FIGURE 12.9 Contact angles of TiO$_2$-B nanowire surfaces and water. (a) As-fabricated TiO$_2$-B nanowire surface. (b) The nanowire surface after oxygen plasma treatment. (c) The nanowire surface after hydrogen plasma treatment. (Reproduced from D.L. Wang et al., *Journal of Materials Chemistry*, 7369–7373, 2011. Copyright © 2011 The Royal Society of Chemistry. With permission.)

of the hydroxyl groups on the TiO$_2$-B surface [48,49]. The plasma-induced hydroxyl groups (OH$^-$), as additional surface defects, can trap electrons and facilitate the charge transfer between the titania nanowires and explosive compounds further [50].

12.4.1.2 Varying the Density of Surface Hydroxyl Groups through Self-Assembled Monolayer (SAM) and Water Treatment

The importance of surface hydroxyl groups to the chemiresistive response of TiO$_2$-B nanowire was further confirmed through surface functionalization. Since surface functionalization of TiO$_2$-B nanowires with a self-assembled monolayer of certain hydrophobic acids can reduce the density of surface hydroxyl groups, this treatment should result in lower sensitivity of TiO$_2$-B nanowires to explosives if surface hydroxyl groups contribute to the sensitivity. Surface modification was carried out

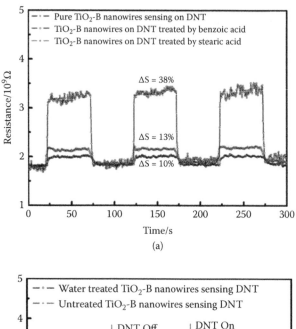

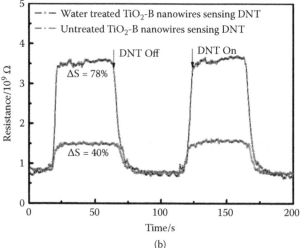

FIGURE 12.10 Sensitivities of TiO$_2$-B nanowires to (a) DNT with surface modification and (b) DNT before and after water treatment of the surfaces. (Reproduced from D.L. Wang et al., *Journal of Materials Chemistry*, 7369–7373, 2011. Copyright © 2011 The Royal Society of Chemistry. With permission.)

by immersing the nanowire thin films into stearic and benzoic acids, respectively, and rinsing with de-ionized water before drying the thin films in an oven. As indicated in Figure 12.10(a), as-fabricated TiO$_2$-B nanowires exhibited a sensitivity of 38% to DNT. After surface modification by stearic acid, the sensitivity of the same sample dropped to 10%. A similar decrease in sensitivity was also observed after the nanowire surfaces were modified with a self-assembled hydrophobic monolayer of benzoic acid. To the contrary, sensing responses of a TiO_2-B nanowire thin film with water treatment were designed to increase the density of surface hydroxyl groups. Water treatment was achieved by immersing the test sample into de-ionized water and letting it become completely dry in the air. As shown in Figure 12.10(b), the sensitivity of TiO_2-B nanowire thin film to DNT vapor after the water treatment was almost two times higher than that before the water treatment. This increase in sensitivity is most likely due to water-treated TiO_2-B nanowires having a higher density of surface hydroxyl groups of bonded water molecules [50,51], and hydroxyl groups enhance the sensing properties of the TiO_2-B nanowires.

According to the above results, it can be concluded that surface hydroxyl groups on TiO_2-B nanowires indeed enhance the charge transfer interactions on the nanowire surfaces and increase the chemiresistive response to nitro-explosive compounds. By functionalizing the surface with a high density of hydroxyl groups strongly bonded to TiO_2-B, one can achieve a high chemiresistive response to nitro-explosives and low sensitivity to air humidity.

12.4.2 The Relationship between the Dipolar Strength of the Analyte Molecule and Speed of Sensor Response

In order to improve the performance of the sensor further, a series of experiments were designed to study how the molecular polarity and electron deficiency influence the charge transfer process between nitro-containing explosives and TiO_2-B. Depending on their molecular structures with functional groups, two types of analytes have been chosen. The first is positional isomers of dinitrobenzene (DNB); these positional isomers have varying dipole moments but are similar in electron deficiency. The other series of compounds are nitroanilines (NAs) and nitrotoluenes (NTs), such as 4-nitroaniline and 4-nitrotoluene. Although NAs and NTs have similar molecular structures except for one functional group, their molecular polarities are quite different because the electron-rich amino group ($-NH_2$) in the nitroanilines is a stronger electron donor than the methyl group ($-CH_3$) in nitrotolune.

12.4.2.1 Chemiresistive Response to Different Positional Isomers of Nitro-Compounds

Table 12.2 lists the response time and sensitivity of TiO_2-B nanowires to the positional isomers of dinitrobenzene (DNB). The relative position of the two nitro groups of dinitrobenzene determines the polarity of the molecule, which can be estimated theoretically by the density functional theory (DFT) [52]. As is evident in Table 12.2, the response time and the dipole moment of the analyte molecules are strongly correlated. Molecules with higher dipole moments exhibit faster responses. For example, 1,2-dinitrobenzene has the most asymmetric structure in terms of electron-withdrawing group position, and therefore it has the highest dipole moment, and produces the shortest response time of 1.54 s. In contrast, 1,4-dinitrobenzene has a symmetrical molecular structure, and its dipole moment is zero. As a result, its response time is the slowest (6.31 s).

TABLE 12.2

Sensing Response to Positional Isomers of Dinitrobenzene

Analyte Molecule	Molecular Structure	μ (Debye)	LUMO (eV)	P at 20°C (Torr) [53]*	τ (s)
1,4-DNB	(structure: benzene with NO$_2$ at positions 1 and 4)	0	−2.46	3.95×10^{-3}	6.31
1,3-DNB	(structure: benzene with NO$_2$ at positions 1 and 3)	4.22	−3.14	2.06×10^{-3}	3.15
1,2-DNB	(structure: benzene with NO$_2$ at positions 1 and 2)	6.67	−3.03	1.03×10^{-3}	1.54

Note: The dipole moments and LUMO levels were calculated using the DFT method [52].

* P = vapor pressure; τ = response time.

Another observation is that the response time is not completely correlated with the calculated LUMO level.

12.4.2.2 Chemiresistive Response to Nitroanilines and Corresponding Nitrotoluenes

Another factor to affect the molecular dipole moment μ is the type of functional groups in analytes. In Table 12.3 a series of nitroanilines and corresponding nitrotoluenes are compared. Nitroanilines have an amino group (-NH$_2$) and a nitro group, while in nitrotoluenes the amino group is replaced by a methyl group (-CH$_3$). Both amino and methyl groups are electron donating; however, the amino group (-NH$_2$) is more electropositive than the methyl group (-CH$_3$). The nitro group is electronegative; therefore the dipole moment of a nitroaniline is greater than the dipole moment of its corresponding nitrotoluene [54,55]. Once again, it is observed that the response times of more polar NAs are faster than their corresponding NTs, consistent with the trend shown in Table 12.2.

The fact that the response time is dominated by the polarity of the sensing molecules supports previous arguments that TiO$_2$-B nanowires have a polar surface. Since a polar surface facilitates a stronger dipole-dipole interaction between surface dipoles and more polar analyte molecules, it can facilitate surface adsorption and result in a faster response.

TABLE 12.3

Comparison of Chemiresistive Responses to Nitroanilines and Corresponding Nitrotoluenes

Analyte Molecule	Molecular Structure	μ (Debye)	LUMO (eV)	P at 20°C (Torr) [56,57]*	T (s)
4-NA		7.12	−1.96	1.50×10^{-3}	0.42
4-NT		5.21	−3.14	2.06×10^{-3}	0.73
3-NA		5.65	−2.24	0.96×10^{-4}	3.18
3-NT		4.89	−2.36	2.05×10^{-1}	5.68
2-NA		4.74	−2.17	3.00×10^{-3}	6.48
2-NT		4.33	−2.31	1.88×10^{-1}	6.85

* P = vapor pressure; T = response time.

12.4.3 THE RELATIONSHIP BETWEEN ELECTRONEGATIVITY OF THE ANALYTE MOLECULE AND THE LEVEL OF SENSOR RESPONSE

In order to study the effect of electron deficiency of analytes, we have measured and compared chemiresistive responses of the TiO_2-B nanowire films with nitrotoluenes with an increasing number of nitro groups, 4-nitrotoluene (NT), 2,4-dinitrotoluene (DNT), and 2,4,6-trinitrotoluene (TNT). By comparing the responses of analytes functionalized with electropositive amino groups and electronegative nitro groups, the surface charge transfer interaction, and the consequent depletion of surface electron charge carrier density, is proposed as the origin of the chemiresistive effect of TiO_2-B nanowires. The electron deficiency of nitroaromatic compounds varies with the number of nitro groups in a molecule. For example, in the nitrotoluene family, the electron deficiency increases (indicated by decreases in their LUMO levels) from 4-nitrotoluene (NT) to 2,4-dinitrotoluene (DNT) and 2,4,6-trinitrotoluene (TNT). TNT has the most number of electron-withdrawing nitro groups, and it leads to the lowest LUMO level. TNT is more electron deficient than both NT and DNT and tends to cause a stronger charge transfer interaction with the TiO_2-B nanowires.

12.4.3.1 Sensing Response to Nitrotoluenes of Different Numbers of Nitro Groups

The results in Table 12.4 demonstrate this trend between LUMO levels and sensitivity. As shown in Table 12.4, although the vapor pressure of TNT is about 1000 times lower than that of DNT, TNT vapor can still produce the same level of sensitivity as DNT vapor of much higher concentration. This means that TNT can induce stronger charge transfer interactions with the TiO_2-B surface than DNT. Amazingly, the sensitivity to TNT is even higher than the sensitivity to NT despite the fact that the

TABLE 12.4

Comparison of Chemiresistive Responses to Nitroanilines and Corresponding Nitrotoluenes

Analyte Molecule	Molecular Structure	μ (Debye)	LUMO (eV)	P at 20°C (Torr) [56,58]*	S (%)	τ (s)
4-NT		5.21	−2.46	1.2×10^{-1}	16	0.73
2,4-DNT		4.85	−2.86	1.4×10^{-4}	33	1.94
2,4,6-TNT		1.00	−3.40	8.0×10^{-6}	33	3.31

* P = vapor pressure; τ = response time.

TABLE 12.5

Comparison of Chemiresistive Responses to Nitroanilines and Corresponding Nitrotoluenes

Analyte Molecule	Molecular Structure	μ (Debye)	LUMO (eV)	P at 20°C (Torr) [57,59]*	S (%)	τ (s)
Aniline	NH$_2$	3.5	1.44	0.60	–23	26.7
NB		3.98	–0.36	0.15	25	2.83
4-NA	NH$_2$ ... NO$_2$	7.12	–1.96	1.5×10^{-3}	8	0.42

* P = vapor pressure; τ = response time.

vapor pressure of NT is almost 10^5 higher than TNT. From these results, it is clear that the sensitivity is strongly dependent on the number of nitro groups and thus electron deficiency of analyte.

12.4.3.2 Comparison between Aniline and Nitrobenzene

A comparison between the effect of aniline, nitrobenzene (NB), and 4-nitroaniline (4-NA) on the nanowires in Table 12.5 reveals the mechanistic details of the chemiresistive effect at the interface due to nitro-compounds. As described before, aniline is an electron-rich compound (with a LUMO level of +1.44 eV). Replacing the amino group with an electron-withdrawing nitro group results in electron-deficient nitrobenzene (with a LUMO level of –0.36 eV). The electron-rich aniline causes a *decrease* of resistance ($S = -23\%$). In contrast, the electron-deficient nitrobenzene causes an *increase* of resistance ($S = +25\%$). In the case that there is both an amino and a nitro group linked to the aromatic ring, as in 4-nitroaniline, the resistance still increases. Also, the response time (26.7 s) of aniline is much longer than that of nitrobenzene (2.83 s), although they have the similar dipole moment. All those suggest the interaction between the nitro group and TiO$_2$-B prevails over the interaction between the amino group and TiO$_2$-B. The trend can be explained by the orientation of permanent dipole moments of analytes relative to the hydroxylated surface dipoles of TiO$_2$-B nanowires [42]. On the other hand, like most wide-band-gap metal-oxide semiconductors, TiO$_2$-B is an n-type semiconductor with electrons as the majority charge carrier. When electron-deficient nitrobenzene is adsorbed onto TiO_2-B nanowires, its electron-withdrawing nitro groups likely create localized surface states that trap TiO_2-B to cause surface depletion of electron carriers and result in the decreasing conductivity and faster response. The observations from this study show that the rate of surface adsorption is dominated by the molecular polarity, and the degree of the charge transfer interaction is dominated by the electron deficiency of analytes.

12.5 CONCLUSIONS

In summary, a large and fast increase of the electrical resistivity of TiO_2-B nanowires in response to subtrace vapors of nitroaromatic and nitroamine explosive compounds at room temperature has been observed. Experimental results indicate that the response originates from a depletion of electron carriers by the surface states produced by adsorbed molecules of electronegative explosive compounds via hydroxyl groups on TiO_2-B nanowires' surface. The results of FTIR spectrum, surface plasma treatments, and surface modification have consistently shown that surface hydroxyl groups on TiO_2-B nanowires are important functional groups in modulating the sensor properties of TiO_2-B nanowires. A stronger response was observed after increasing the density of surface hydroxyl groups. These observations indicate that surface hydroxyl groups provide a major charge transfer pathway between nitro groups in nitroaromatic explosives and TiO_2-B nanowires and cause the chemiresistive response. Also, how the molecular polarity and electron deficiency influence the charge transfer process between nitroaromatic explosives and TiO_2-B has been studied. The results suggest that the sensing response time is dominated by the dipole moment of the analyte molecules. The sensitivity is dominated by the electron deficiency of analytes. Charge transfer interaction between the analyte and TiO_2-B is the major factor of the chemiresistive effect in TiO_2-B nanowires, and electron-deficient and electron-rich chemical compounds cause opposite chemiresistive change. In nitro-containing compounds, nitro groups have strong interactions with TiO_2-B, and the interaction between nitro groups and TiO_2-B prevails over that of amino groups.

TiO_2-B nanowires can be chemically synthesized with a low-cost and high-yield hydrothermal method, and explosive sensors based on TiO_2-B nanowires are compatible with standard microelectronic manufacturing processes. For these reasons this technology is suitable for the mass production of low-cost microelectronic sensors and has excellent potential for practical application.

ACKNOWLEDGMENTS

This work was supported by Office of Naval Research Grants N00014–05–1-0843 and N00014–09–1-0706, and NSF Center on Materials and Devices for Information Technology Research (CMDITR) Grant DMR-0120967. Danling Wang was supported in part by University of Washington WRF-APL Fellowship.

REFERENCES

1. B.C. Dionne, D.P. Roundbehler, E.K. Achter, J.R. Hobbs, and D.H. Fine. Vapor pressure of explosives. *Journal of Energetic Materials*, 4, 447–472, 1986.
2. A.M. Jimenez and M.J. Navas. Chemiluminescence detection systems for the analysis of explosives. *Journal of Hazardous Material*, 106, 1–8, 2004.
3. J.E. McFee and A.A. Faust. Nuclear methods for explosive detection. *SPIE, Newsroom*, 2011. DOI: 10.1117/2.1201108.003761.
4. K.M. Roscioli, E. Davis, W.F. Siems, A. Mariano, W.S. Su, S.K. Guharay, and H.H. Hill. Modular ion mobility spectrometer for explosives detection using corona ionization. *Analytical Chemistry*, 83, 5965–5971, 2011.
5. R. Batlle, H. Carlsson, P. Tollback, A. Colmsjo, and C. Crescenzi. Enhanced detection of nitroaromatic explosive vapors combining solid-phase extraction-air sampling, supercritical fluid extraction, and large-volume injection-GC. *Analytical Chemistry*, 75, 3137–3144, 2003.
6. J.S. Yang and T.M. Swager. Porous shape persistent fluorescent polymer films: An approach to TNT sensory materials. *Journal of American Chemical Society*, 120, 5321–5322, 1998.
7. J.S. Yang and T.M. Swager. Fluorescent porous polymer films as TNT chemosensors: Electronic and structural effects. *Journal of American Chemical Society*, 120, 11864–11873, 1998.
8. Q.L. Fang, J.L. Geng, B.H. Liu, D.M. Gao, F. Li, Z.Y. Wang, G.J. Guan, and Z.P. Zhang. Inverted opal fluorescent film chemosensor for the detection of explosive nitroaromatic vapors through fluorescence resonance energy transfer. *Chemistry—A European Journal*, 15, 11507–11514, 2009.
9. F. Rock, N. Barsan, and U. Weimar. Electronic nose: Current status and future trends. *Chemical Reviews*, 108, 705–725, 2008.
10. R.G. Ewing and C.J. Miller. Detection of volatile vapors emitted from explosives with a handheld ion mobility spectrometer. *Field Analytical Chemistry and Technology*, 5, 215–221, 2001.
11. H.B. Liu, Y.Q. Chen, G.J. Bastiaans, and X.C. Zhang. Detection and identification of explosive RDX by THz diffuse reflection spectroscopy. *Optical Express*, 14, 415–423, 2006.
12. A.D. Aguilar, E.S. Forzani, M. Leright, F. Tsow, A. Cagan, R.A. Iglesias, L.A. Nagahara, I. Amlani, R. Tsui, and N.J. Tao. A hybrid nanosensor for TNT vapor detection. *Nano Letters*, 10, 380–384, 2010.
13. Y.H. Gui, C.S. Xie, J.Q. Xu, and G.Q. Wang. Detection and discrimination of low concentration explosives using MOS nanoparticle sensors. *Journal of Hazardous Material*, 164, 1030–1035, 2009.
14. D.L. Wang, Q.F. Zhang, G.Z. Cao, and A.T. Chen. Room-temperature chemiresistive effect of TiO_2-B nanowires to nitroaromatic and nitroamine explosives. *IEEE Sensors Journal*, 11, 1352–1358, 2011.
15. O.K. Varghese and C.A. Grimes. Metal oxide nanoarchitectures for environmental sensing. *Journal of Nanoscience and Nanotechnology*, 3, 277–293, 2003.
16. A. Rothschild and Y. Komem. The effect of grain size on the sensitivity of nanocrystalline metal-oxide gas sensors. *Journal of Applied Physics*, 95, 6374–6380, 2004.
17. K. Shiraishi, T. Sanji, and M. Tanaka. Trace detection of explosive particulates with a phosphole oxide. *ACS Applied Materials and Interfaces*, 1, 1379–1382, 2009.
18. S.J. Toal and W.C. Trogler. Polymer sensors for nitroaromatic explosives detection. *Journal of Materials Chemistry*, 16, 2871–2883, 2006.
19. J.S. Yang and T.M. Swager. Porous shape persistent fluorescent polymer films: An approach to TNT sensory materials. *Journal of the American Chemical Society*, 120, 5321–5322, 1998.

20. G. Korotcenkov. Metal oxides for solid-state gas sensors: What determines our choice? *Materials Science and Engineering B*, 139, 1–23, 2007.
21. N. Barsan, D. Koziej, and U. Weimar. Metal oxide-based gas sensor research: How to? *Sensors and Actuators B*, 121, 18–35, 2007.
22. F. Millot, M.G. Blanchin, R. Tetot, J.F. Marucco, B. Poumellec, C. Picard, and B. Touzelin. High temperature nonstoichiometric rutile TiO_{2-x}. *Progress in Solid State Chemistry*, 17, 263–293, 1987.
23. R. Marchand, L. Brohan, and M. Tournoux. $TiO_2(B)$: A new form of titanium-dioxide and the potassium octatitanate $K_2Ti_8O_{17}$. *Materials Research Bulletin*, 15, 1129–1133, 1980.
24. G. Nuspl, K. Yoshizawa, and T. Yamabe. Lithium intercalation in TiO_2 modifications. *Journal of Materials Chemistry*, 7, 2529–2536, 1997.
25. A.R. Armstrong, G. Armstrong, J. Canales, and P.G. Bruce. TiO_2-B nanowires. *Angewandte Chemie International Edition*, 43, 2286–2288, 2004.
26. T.P. Feist and P.K. Davies. The soft chemical synthesis of $TiO_2(B)$ from layered titanates. *Journal of Solid State Chemistry*, 101, 275–295, 1992.
27. G. Wang, Q. Wang, W. Lu, and J.H. Li. Photoelectrichemical study on charge transfer properties of TiO_2-B nanowires with an application as humidity sensors. *Journal of Physical Chemistry B*, 110, 22029–22034, 2006.
28. J. Jitputti, S. Pavasupree, Y. Suzuki, et al. Synthesis of TiO_2 nanotubes and its photocatalytic activity for H-2 evolution. *Japanese Journal of Applied Physics*, 47, 751–756, 2008.
29. S. Pavasupree, Y. Suzuki, S. Yoshikawa, and R. Kawahata. Synthesis of titanate, TiO_2 (B), and anatase TiO_2 nanofibers from natural rutile sand. *Journal of Solid State Chemistry*, 128, 3110–3116, 2005.
30. N. Viriya-Empikul, N. Sano, T. Charinpanitkul, T. Kikuchi, and W. Tanthapanichakoon. A step towards length control of titanate nanotubes using hydrothermal reaction with sonication pretreatment. *Nanotechnology*, 19, 035601, 2008.
31. Y.Q. Wang, G.Q. Hu, X.F. Duan, H.L. Sun, and Q.K. Xue. Microstructure and formation mechanism of titanium dioxide nanotubes. *Chemical Physics Letters*, 365, 427–431, 2002.
32. R. Yoshida, Y. Suzuki, and S. Yoshikawa. Syntheses of $TiO_2(B)$ nanowires and TiO_2 anatase nanowires by hydrothermal and post-heat treatments. *Journal of Solid State Chemistry*, 178, 2179–2185, 2005.
33. P.A. Pella. Generator for producing trace vapor concentrations of 2,4,6-trinitrotoluene, 2,4-dinitrotoluene, and ethylene-glycol dinitrate for calibrating explosives vapor detectors. *Analytical Chemistry*, 48, 1632–1637, 1976.
34. S. Ahlers, G. Muller, and T. Doll. A rate equation approach to the gas sensitivity of thin film metal oxide materials. *Sensors and Actuators B—Chemical*, 107, 587–599, 2005.
35. M.-W. Ahn, K.-S. Park, J.-H. Heo, J.-G. Park, and D.-W. Kim. Gas sensing properties of defect-controlled ZnO-nanowire gas sensor. *Applied Physics Letters*, 93, 263103–263106, 2008.
36. NRC, Committee on the Review of Existing and Potential Standoff Explosives Detection Techniques. *Existing and Potential Standoff Explosives Detection Techniques*. National Academies Press, Washington, DC, 2004.
37. G. Munuera, V. Rivesarnau, and A. Saucedo. Photo-adsorption and photo-desorption of oxygen on highly hydroxylated TiO_2 surfaces. 1. Role of hydroxyl-groups in photo-adsorption. *Journal of the Chemical Society—Faraday Transactions I*, 75, 736–747, 1979.
38. R. Marchand, L. Brohan, and M. Tournoux. $TiO_2(B)$: A new form of titanium-dioxide and the potassium octatitanate $K_2Ti_8O_{17}$. *Materials Research Bulletin*, 15, 1129–1133, 1980.

39. R. Vargas and O. Nunez. Hydrogen bond interactions at the TiO_2 surface: Their contribution to the pH dependent photo-catalytic degradation of p-nitrophenol. *Journal of Molecular Catalysis A—Chemical*, 300, 65–71, 2009.

40. M.D. Earle. The electrical conductivity of titanium dioxide. *Physical Review*, 61, 56–62, 1942.

41. M. Gratzel. Photoelectrochemical cells. *Nature*, 414, 338–344, 2001.

42. M.H. Seo, M. Yuasa, T. Kida, J.-S. Huh, K. Shimanoe, and N. Yamazoe. Gas sensing characteristics and porosity control of nanostructured films composed of TiO_2 nanotubes. *Sensors and Actuators B—Chemical*, 137, 513–520, 2009.

43. A. Rothschild and Y. Komem. The effect of grain size on the sensitivity of nanocrystalline metal-oxide gas sensors. *Journal of Applied Physics*, 95, 6374–6380, 2004.

44. O.A. El Seoud, A.R. Ramadan, B.M. Sato, and P.A.R. Pires. Surface properties of calcinated titanum dioxide probed by solvatochromic indicators: Relevance to catalytic application. *Journal of Physical Chemistry C*, 114, 10436–10443, 2010.

45. K.S. Finnie, D.J. Cassidy, J.R. Bartlett, and J.L. Woolfrey. IR spectroscopy of surface water and hydroxyl species on nanocrystalline TiO_2 films. *Langmuir*, 17, 816–820, 2001.

46. A.S. Vuk, R. Jese, B. Orel, and G. Drazic. The effect of surface hydroxyl groups on the adsorption properties of nanocrystalline TiO_2 films. *International Journal of Photoenergy*, 7, 163–168, 2005.

47. R. Muller, H.K. Kammler, K. Wegner, and S.E. Prtsinis. OH surface density of SiO_2 and TiO_2 by thermogravimetric analysis. *Langmuir*, 19, 160–165, 2003.

48. D.L. Wang, A.T. Chen, S.-H. Jang, H.-L. Yip, and A.K.-Y. Jen. Sensitivity of titania(B) nanowires to nitroaromatic and nitroamino explosives at room temperature via surface hydroxyl groups. *Journal of Materials Chemistry*, 21, 7369–7373, 2011.

49. B. Meyer, D. Marx, O. Dulub, U. Diebold, M. Kunat, D. Langenberg, and C. Woll. Partial dissociation of water leads to stable superstructures on the surface of zinc oxide. *Angewandte Chemie International Edition*, 43, 6641–6645, 2004.

50. H.W. Ra, R. Khan, J.T. Kim, B.R. Kang, K.H. Bai, and Y.H. Im. Effects of surface modification of the individual ZnO nanowire with oxygen plasma treatment. *Materials Letters*, 63, 2516–2519, 2009.

51. J.M. Pan, B.L. Maschhoff, U. Diebold, and T.E. Madey. Interaction of water, oxygen, and hydrogen with $TiO_2(110)$ surfaces having different defect densities. *Journal of Vacuum Sicence and Technology A*, 10, 2470–2476, 1992.

52. R.G. Parr. Density functional theory. *Annual Review of Physical Chemistry*, 34, 631–656, 1983.

53. D. Ferro, V. Piacente, R. Gigli, and G. D'Ascenzo. Determination of the vapour pressures of *o-*, *m-*, and *p*-dinitrobenzene by the torsion-effusion method. *Journal of Chemical Thermodynamics*, 8, 1137–1143, 1976.

54. J.E. Abbott, X.Z. Peng, and W. Kong. Symmetry properties of electronically excited states of nitroaromatic compounds. *Journal of Chemical Physics*, 117, 8670–8675, 2002.

55. T. Tanaka, A. Nakajima, A. Watanabe, T. Ohno, and Y. Ozaki. Surface-enhanced Raman scattering spectroscopy and density functional theory calculation studies on adsorption of o-, m-, and p-nitroaniline on silver and gold colloid. *Journal of Molecular Structure*, 661–662, 2003.

56. K. Aim. Saturated vapor pressure measurements on isomeric monoitrotoluenes at temperatures between 380 and 460 K. *Journal of Chemical and Engineering Data*, 39, 591–594, 1994.

57. http://actrav.itcilo.org/actrav-english/telearn/osh/ic/100016.htm.

58. H. Carlsson, G. Robertsson, and A. Colmsjo. Response mechanisms of thermionic detectors with enhanced nitrogen selectivity. *Analytical Chemistry*, 73, 5698–5703, 2001.

59. A. Kawski, B. Kuklinski, and P. Bojarski. Dipole moment of aniline in the excited S_1 state from thermochromic effect on electronic spectra. *Chemical Physics Letters*, 415, 251–255, 2005.

13 Properties of Different Types of Protective Layers on Silver Metallic Nanoparticles for Ink-Jet Printing Technique

Andrzej Mościcki, Anita Smolarek,
Jan Felba, and Tomasz Fałat

CONTENTS

13.1 INTRODUCTION

One of the fastest developing fields of nanotechnology is obtaining metals or chemical compounds on the nanometer scale. As it turns out, the vast majority of materials reveal their completely new properties when reduced to such small sizes. This allows us not only to better explore the properties of materials but also to use them in quite unexpected applications.

One such material of great significance in electronic applications is undoubtedly silver. One of the best conductors of electricity and heat, and used in microelectronics for a long time, silver, among other applications as a filler for electrically conductive adhesives (ECAs), is used in assembling electronic elements. The possibility of splitting silver into nanosized particles has enabled the development of a new group of materials, i.e., electrically conductive inks (ECIs). These inks are applied using mainly roll-to-roll (R2R) methods or ink-jet printers, and therefore their viscosity must be very low and they must have a stable form in both static (storage) and dynamic (printing) states, i.e., the characteristics closest to those of a homogeneous liquid.

These requirements are of particular importance in ink-jet printing since this technology is based on pulse ink dispersing. Each pulse causes a small droplet of ink (with a volume of up to several picoliters—10–12 L) to be dispensed with a very high acceleration, reaching 100,000 g. The manner of dispensing being so dynamic, the homogeneity of liquid (ink) is a necessary condition for the components of liquid not to be separated in the process of printing.

It is widely known that in order for the dispersion of silver to be stable in time, each metal particle is covered with an organic protective layer. It is the type and properties of these layers that the subsequent properties of inks and their operating parameters mainly depend on.

13.2   REVIEW OF METHODS OF OBTAINING NANOSILVER

At present, there are several methods of obtaining silver powders with nanometric dispersion. However, the common feature of all the methods is that at a definite moment the produced nanopowders are coated with an organic protective layer.

Among the most popular are the following technologies:

- DC plasma sputtering of silver particles
- Metal vapor deposition
- Chemical methods—through reduction of silver compounds
- Thermal decomposition of silver fatty acids under inert atmosphere

In each of the foregoing methods, a chemical compound is administered in an appropriate manner to produce on the surface of the forming metal particles a thin organic coating that effectively prevents coagulation of particles and their aggregation into bigger structures (Figure 13.1).

13.3   SIGNIFICANCE OF PROTECTIVE COATING

It is only possible to obtain single particles of metallic silver by coating them with protective stabilizing substances in the process of production. The stabilizer is adsorbed or bound to the surface of silver and forms a protective layer, thus preventing agglomeration with other particles. Only unprotected particles tend to aggregate and form compact microscale structures. The protective layer is popularly called a surfactant.

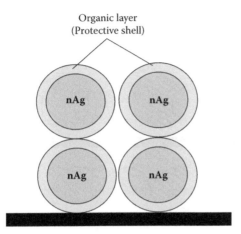

FIGURE 13.1   Organic layer (protective shell).

The selection of protective coating is one of the most important parameters in the synthesis of nanosilver. Besides protecting against sedimentation, the surfactant significantly influences the morphology of nanoparticles, including their sizes, distribution, or even their shape. Zhang et al. [1] have analyzed the influence of the amount of stabilizer on the grain size distribution of nanosilver. It has been shown that as the molar ratio of surfactant to silver salts grows, smaller particles with a more regular and spherical shape are formed.

There is also evidence that under certain conditions of reaction the stabilizer can play the role of reducer. Slistan-Grijalva et al. have demonstrated that the stabilizer reduces silver ions to metallic silver if the solution is heated to 100°C and the reaction takes place in darkness for 1 h [2]. This method is often accepted since stabilizers are more environmentally friendly than reducers. Furthermore, during such chemical reduction there are no undesired reaction substrates (reducer residue) that need to be washed out [3,4]. Thus it is very important for the protective coating to be reactive and ensure the solubility of metal nanoparticles in a medium. It is only thanks to solubility that there is the possibility of producing stable ink with a very good dispersion of metal nanoparticles and low resistance.

In order for printed structures to conduct electricity, they must be heated at a temperature that enables effective removal (as a result of thermal decomposition, desorption, or evaporation) of the isolating protective layer. The coating separates silver nanoparticles and prevents their contact, thereby causing lack of electrical connection between them. The removal of surfactant is accompanied by a relatively small decrease in resistance of printed layer; however, it is only the process of sintering the filler nanoparticles that enables us to achieve a sufficiently high level of electrical conductivity (Figure 13.2).

Thus it is very important that the protective coating should be easily removed from the surface of silver in the sinterization process at a possibly low temperature, which offers a possibility of printing on a flexible polymer foil substrate. So, the protective coating, on the one hand, plays the role of stabilizer for filler nanoparticles, while, on the other hand, it impedes obtaining conductive structures. Magdassi et al.

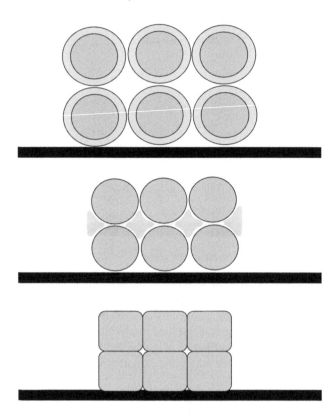

FIGURE 13.2   The mechanism of removing protective coating from nanosilver surface.

[5] have demonstrated in their paper that it also functions as an adhesion promoter. Silver nanoparticles were obtained with a coating that acted as an adhesive, ensuring good adhesion of inks to typical glass and flexible substrates.

13.4   MECHANISM OF ACTION OF COATINGS

When particles are of very small sizes—up to 100 nm—van der Waals forces and Brownian motion have great influence, while gravitational forces are not too important. Van der Waals forces are very weak and their range is very limited; however, Brownian motion makes nanoparticles collide, and then as a result of action of van der Waals forces, aggregates can be formed.

Thus one of the important functions of protective coatings is to prevent aggregation of metal nanoparticles, and this function is usually classified in two categories: electrostatic or steric stabilization.

13.4.1   ELECTROSTATIC (CHARGED) STABILIZATION

Electrostatic stabilization is achieved by producing a surface charge on the surface of nanoparticles.

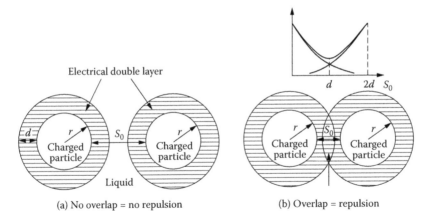

FIGURE 13.3 Mechanism of electrostatic stabilization. (From G. Cao, *Nanostructures and Nanomaterials*, Imperial College Press, London, 2004.)

The mechanisms of this process may be various, e.g.:

- Preferential adsorption of ions of one sign
- Isomorphic substitution of ions
- Accumulation or transfer of electrons
- Dissociation of surface charged particles
- Physical adsorption of charged particles

A charge formed around a particle is known as the zeta potential (ζ) and strictly depends on the pH of the nanoparticle environment. According to Magdassi et al.'s study, the most stable AgNP colloids have negative potentials (–33) at a pH range of 6–8. Tests were made for both nonstabilized silver particles and those stabilized with sodium citrate. A layer of counterions, a so-called double layer, is formed around charged particles to neutralize the charge of particles.

Distant particles do not interact with each other because van der Waals forces have a short range, and the formed double layer neutralizes the electrical charge of particles (Figure 13.3a). When particles are so close that their double layers partially overlap, there emerges the resultant electrostatic repulsive force (Figure 13.3b).

The examples of electrostatic stabilizers are the compounds containing such functional groups as sulfo, carboxyl, and amino, including citrates, sodium dodecyl sulfate (SDS), amines, amides, saccharides, fatty acids, surface active agents, and many others (Figure 13.4).

13.4.2 STERIC (POLYMER) STABILIZATION

Steric stabilization consists in attaching polymer chains to the surface of particles; it results in the fact that because of spatial (steric) action the particles cannot come close to each other and remain dispersed (suspended) in liquid.

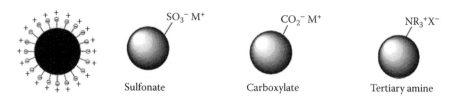

FIGURE 13.4 Electrostatic (charged) stabilization by the same protective coating. (From www.cabot-corp.com. With permission.)

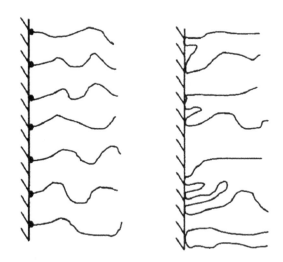

FIGURE 13.5 Mechanism of steric (polymer) stabilization.

Polymer chains may be chemically attached to the surface of the particle (Figure 13.5 left) or physically adsorbed on its surface (Figure 13.5 right). The mechanism of steric stabilization (also called polymer stabilization) depends on the degree of coating of the surface of particles with macroparticles, as well as on the type of solvent. When the surface of particles is densely covered by polymer chains, the formed layers prevent the particles from approaching each other. When the degree of coating is relatively low, stabilization relies on the solvent. In a so-called good solvent, whose particles interact with the polymer, the chains of macroparticles are straight, while in a weak solvent the chains are curled up (Figure 13.6).

Typical polymers used for protection against agglomeration include poly(vinyl pyrrolidone) (PVP), poly(ethylene glycol) (PEG), poly(methacrylic acid) (PMAA), poly(methyl methacrylate) (PMMA), poly(vinyl alcohol) (PVA), and others.

Steric stabilization can be combined with electrostatic stabilization:

- Polymer chains are attached to the surface of an electrically charged particle.
- Polyelectrolyte chains are attached to the surface of an electrically neutral particle.

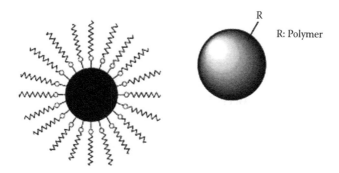

FIGURE 13.6 Steric (polymer) stabilization by the same protective coating. (From www. cabot-corp.com. With permission.)

13.5 EXAMPLES OF COATING EXAMINATION (EXPERIMENTAL)

The basic problem while producing nanocomposites (for instance, silver nanocomposites) where nanoparticles should be uniformly dispersed and stable in the medium is to ensure their proper stabilization and to split the aggregated metal clusters into individual nanoparticles. The experiments were performed for colloids with nanosilver with various types of coatings:

- Carboxyl-coated nanosilver (Ag1)
- Amino-coated nanosilver (Ag2)
- Polymer-coated nanosilver (Ag3)

The first two coatings provide the electrostatic stabilization of nanoparticles in the medium, while the polymer layer stabilizes nanosilver sterically (Figure 13.7).

The first objective of the experiment was to characterize nanoparticles. The obtained nanoparticles with different protective coatings required a full description in order to assess their sizes and shapes (SEM), size distribution (Malvern), and optical properties (UV-Vis). In the second stage of testing, attention was mainly focused on quantitative analysis of the protective coatings used, their chemical composition (energy-dispersive x-ray spectroscopy, or EDX), and the removal dynamics. The tests aimed at assessing the behavior of coating at the sintering (synthesizing) temperature of printed structures.

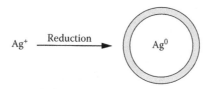

FIGURE 13.7 The preparation of silver nanoparticles with different protective coatings via reduction.

Three types of nanosilver were used for the tests: carboxyl-coated Ag1, amino-coated Ag2, and polymer-coated Ag3, which were the final products of three different types of synthesis reaction.

13.6 CHARACTERIZATION OF THE TESTED SILVER NANOPARTICLES

13.6.1 SEM INVESTIGATIONS

The morphology of the tested silver samples was determined using scanning electron microscopy (SEM). The electron microscope uses a beam of electrons for imaging and enables us to examine the structure of matter at the atomic level. The greater the energy of electrons, the shorter their wave and the higher the microscope resolution.

The exemplary photographs from numerous measurements made in various enlargements are shown in Figure 13.8.

The methodology of tests consisted of applying a sample in the dry form to a substrate placed in vacuum. The sample of Ag1 was in the form of powder, while imaging the morphology with a microscope and Ag2 and Ag3 silver required the application of a colloid droplet, and next the evaporation of solvent. The obtained results clearly indicate that all nanoparticles are spherical and have a uniform structure of an average size of 50–70 nm for amino-coated and polymer-coated Ag, and ca. 80–100 nm for carboxyl-coated nanoparticles.

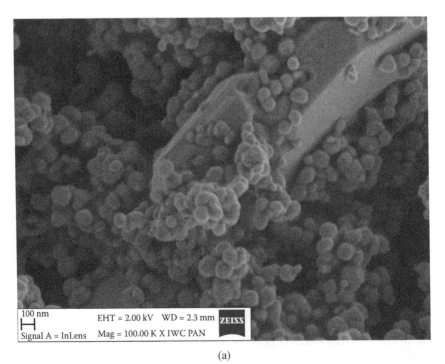

100 nm
EHT = 2.00 kV   WD = 2.3 mm   ZEISS
Signal A = InLens   Mag = 100.00 K X IWC PAN

(a)

FIGURE 13.8   SEM pictures of nanosilver with different protective coatings: (a) Ag1.

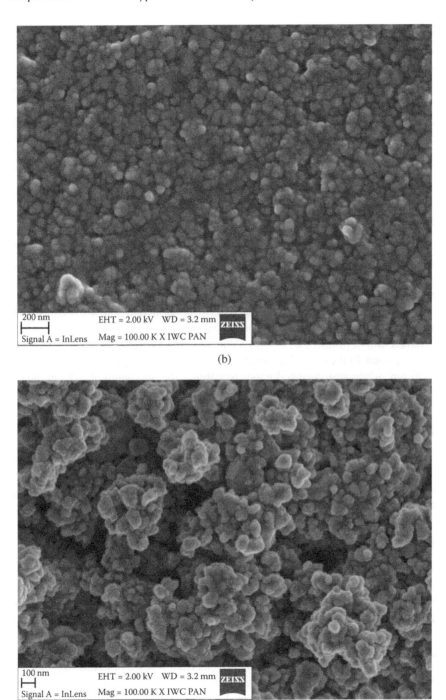

(b)

(c)

FIGURE 13.8 *(Continued)* SEM pictures of nanosilver with different coatings: (b) Ag2, and (c) Ag3.

13.6.2 MALVERN INVESTIGATIONS

In order to investigate grain size distribution of samples, tests were performed on a zetasizer, which made it possible to measure the size of particles within the range between 0.6 nm and 6 microns. Examples of the obtained results are shown in Figure 13.9.

The presented diagrams of grain size distribution in the function of their quantity show that the diameters of tested particles were ca. 100, 50, and 70 nm, respectively, for Ag1, Ag2, and Ag3. The measurements were characterized by a very narrow distribution range and very high accuracy and repeatability. Despite using different coating sizes, the samples show similar grain size distributions.

13.6.3 UV-VIS ANALYSIS

The UV-Vis spectroscopic investigations were performed using a spectrometer with a wavelength range of 190–1100 nm for Ag1, Ag2, and Ag3 samples. For all the samples, absorption spectra were obtained with the absorption maximum in a wavelength range of 415–430 nm (Figure 13.10).

The presented diagrams clearly demonstrate the formation of nanocrystalline silver in each sample, regardless of a protective coating used. All the peaks are smooth and have a similar profile. The absorption maximum is shifted depending on the type of coating used. In the case of carboxyl-coated nanosilver, it appears at a wavelength of 430 nm, while amino-coated nanosilver shows the maximum at 415 nm, and a shift toward longer waves (420 nm) can again be observed with a polymer coating.

13.7   EXAMINATION OF PROTECTIVE COATINGS OF NANOSILVER

Testing the properties of protective coating formed on the surface of nanosilver in the process of its production is key for further applications of nAg as a filler for electrically conductive inks. In order to determine the influence of the size of coatings as well as the dynamics of their removal, further tests were made.

13.7.1   QUANTITATIVE ANALYSIS OF PROTECTIVE COATING

To determine the interdependence of removing the protective coating surrounding the silver as the function of sintering temperature, thermal tests were made. To this end, it is necessary to begin with determining the total amount of coatings. Therefore a quantitative analysis of sample weight loss in a thermal process was performed. The simple control of weight loss at a temperature of 500°C for 1 h constituted a significant criterion allowing us to assess the content of each type of coating in the tested samples (Figure 13.11). Analyzing the thermal process of nanosilver, one can assume that under the test conditions any organic substances are removed, and after the heating process pure metallic silver is left. Percentage weight loss corresponds to the quantity by weight of protective coating.

It is possible to determine the total amount of protective coating by performing the test as described above. As can be seen, the tested silvers contain only a small

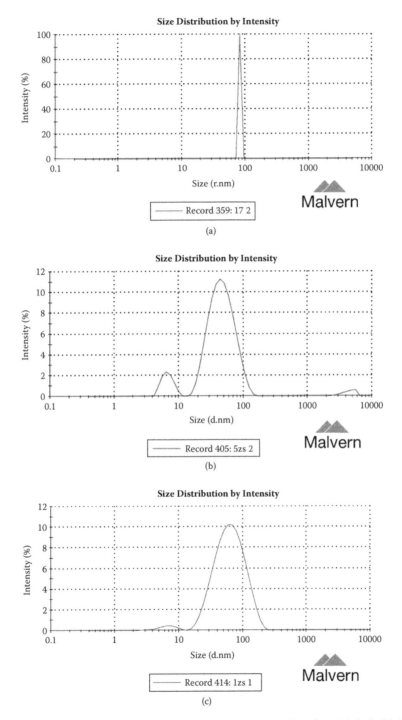

FIGURE 13.9 Distribution of nanosilver particles by Malvern Zetasizer: (a) Ag1, (b) Ag2, and (c) Ag3.

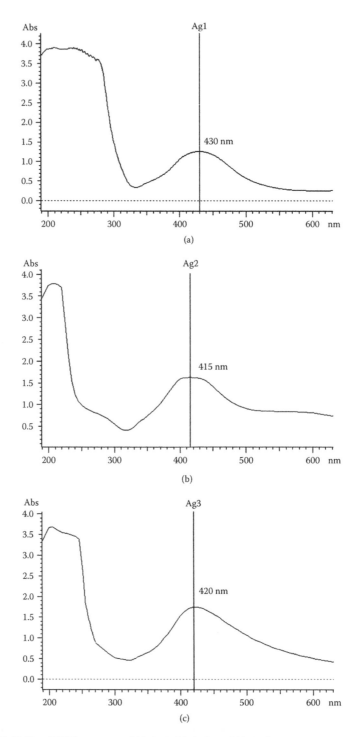

FIGURE 13.10   UV-Vis spectra of (a) Ag1, (b) Ag2, and (c) Ag3.

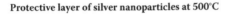

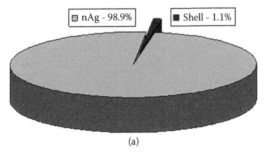

(a)

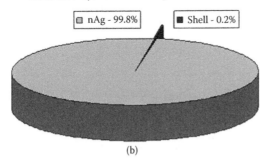

(b)

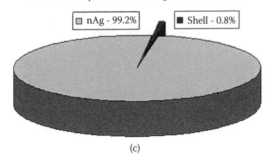

(c)

FIGURE 13.11 The result of quantitative analysis for (a) Ag1, (b) Ag2, and (c) Ag3.

amount of protective coating—less than 1.5%. The maximum amount of protective coating was determined for Ag1. In the remaining cases, 0.8% of polymer coating was for Ag3, and the smallest amount of 0.2% was assessed for amino coating (Ag2).

13.7.2 THE DYNAMICS OF REMOVAL OF PROTECTIVE COATINGS

Having knowledge on the total amount of coatings, it is possible to perform tests of weight loss of silver samples having various coatings in the function of time at specified temperatures planned for their removal. Figure 13.12 presents the dependence

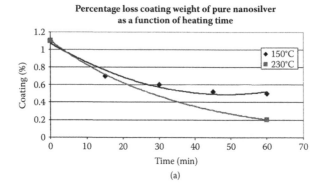

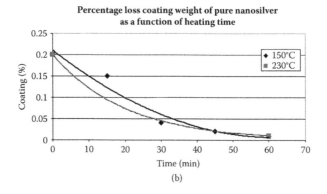

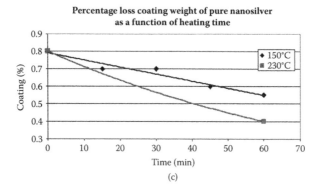

FIGURE 13.12 The dynamics of removal of protective coatings at temperatures of 150 and 230°C in the function of time (a) Ag1, (b) Ag2, and (c) Ag3.

of protective coating weight loss in the function of time (15, 30, 45 min, and 1 h) at assumed temperatures of silver sintering.

An analysis of the sintering process of tested samples enables the observation that the decomposition of coatings takes place already at 150°C. The obtained results indicate that the weight loss at any temperature is time dependent. As can be seen in the figures, the total content of protective material for respective samples amounted to 1.1% Ag1, 0.2% Ag2, and 0.8% Ag3, and was determined in the diagram as an initial value (time 0 min). The diagram of changes of the coating amount at both 150°C and a higher temperature has the same profile. However, a greater weight loss was observed at a temperature of 230°C and for a longer sintering time.

To sum up the results of studies concerning the dynamics of removing the coating from the surface of silver during the thermal sintering process in the presented cases, it can be stated that this is an individual process for each of the protecting materials used. In the case of Ag1, there was 0.5 and 0.3% of the carboxyl coating left, respectively, at temperatures of 150 and 230°C. Moreover, Figure 13.12c shows that not the entire polymer material was decomposed. It is only in the case of amino-coated silver that there is a possibility of removing nearly the whole amount of protective material, even at a temperature of 150°C.

13.7.3 EDX ANALYSIS

To additionally verify the methodology of the tests applied, EDX spectral analysis was performed for the Ag3 sample. The measurement consisted of using x-radiation generated by an electron beam for examining the chemical composition of a sample on a microanalyzer. The collaboration of EDX with a scanning electron microscope ensures, moreover, a qualitative analysis of chemical composition of the tested surface in a small area and a very short time, while the registration and analysis of radiation enable qualitative and quantitative measurements of chemical composition of the tested sample in these micro areas.

In these tests, the polymer protective coating is manifested as carbon, which constitutes ca. 1.1% of the whole volume of product (Figure 13.13a), which is confirmed by previous quantitative tests. The carbon is removed in the thermal process (Figure 13.13b), in accordance with the dynamics of coating removal (Figure 13.11c).

The tests performed and their results, shown in Figures 13.10 and 13.11, allow drawing an initial conclusion that for silver nanoparticles with different coatings, the effect of steric (polymer) type stabilization prevails over other types since:

- It is a more efficient stabilization, even in the case of very concentrated suspensions, e.g., inks with a concentration of conductive nanoparticles of 20–40%.
- It can be used for multicomponent systems.
- It is a thermodynamic method.
- The agglomerates formed can be split again into single nanoparticles.

On the other hand, electrostatic stabilization is of limited practical significance because:

- It is effective only for diluted suspensions.
- It may not be applied in multicomponent systems since different particles will produce different surface charges and different double layers.
- It is a kinetic method.
- If agglomerates are formed, it is practically impossible to split them.

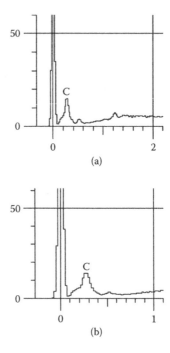

FIGURE 13.13   Test result before the sintering process (a) and after the process (b).

Therefore the polymer coating P has been chosen for further studies concerning production of a stable electrically conductive ink with a low thermal processing temperature and nAg content of ca. 20%.

13.7.4   CHARACTERIZATION OF THE TESTED POLYMER COATING P

The accepted amorphous polymer coating P is characterized by a small value of vitrification temperature Tg (ca. 60°C) [8], and its action as stabilizer depends to a great extent on its molecular mass. It is very well soluble in water and alcohol solvents, and therefore it is easy to wash out its unnecessary quantity immediately after the synthesis of nanoparticles. The rest of the coating is removed in the thermal process only, in order to obtain pure silver with perfect electrical properties. However, Lee et al. have demonstrated that even 5% of protective coating after the sintering of structure allows us to obtain very good electrical conductivity parameters of printed pathways [4].

13.8   EXAMPLE OF INK FOR INK-JET PRINTING TECHNOLOGY

The sample of conductive ink was based on obtained Ag nanoparticles with a polymer coating, using solvents mainly composed of ethanol and ethylene glycol, which are currently used in commercial ink. This composition is expected to be environmentally friendly. The base properties of produced ink formulation are presented in Table 13.1. Figure 13.14 shows several printed structures made on flex polymer polyethylene (PE) type foil.

TABLE 13.1
Ink Specifications

Number of components	One
Consistency	Very low viscous ink
Color	Dark green to gray
Percentage of silver filler	20–30%
Viscosity	5–6.5 m Pas
Thixotropy index (1/10)	~1.0
Surface tension value	~35 dynes/cm
Recommended curing and sintering conditions in convection oven	150°C, 60 min
Storage	2 months at room temperature (do not keep in temperatures below 5°C)

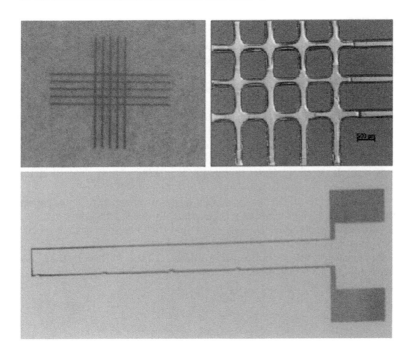

FIGURE 13.14 Example printout on the PE foil using tested inks at a sintering temperature of 150°C.

13.9 SUMMARY

During the production process of nanosilver, all particles are safely dispersed without an agglomeration possibility. This is because each particle is protected by a special kind of layer, which is electrically insulating. Therefore to obtain good electrical conductivity of printed structures, the additional energy—mostly thermal by heating—has to be delivered. In case of flexible electronics, where the ink-jet printing technique is commonly used, the temperature of the sintering process plays a very important role, because of the low thermal resistance of polymeric substrates. As it was presented above, it is possible to produce silver nanoparticles with coating layers of different materials, which are characterized by different parameters. The deep study about the different coatings of silver nanoparticles presented above gives the opportunity to modify the properties of inks for ink-jet printing technology that is based on nanosilver. The sintering temperature of the printed microstructure depends on the dynamics of removal of protective coatings. So the proper selection of the protective layer during the manufacturing process of silver nanoparticles may have a significant impact on the properties of microstructures performed by the ink-jet printing technique. Moreover, other properties, like viscosity, surface tension, particle size distribution, etc., which are also important from the ink-jet printing technique point of view, depend on the properties of the organic protective layer on silver nanoparticles.

REFERENCES

1. Z. Zhang, B. Zhao, L. Hu. PVP protective mechanism of ultrafine silver powder synthesized by chemical reduction processes. *Journal of Solid State Chemistry*, 121, 1996, 105–110.
2. A. Slistan-Grijalva, R. Herrera-Urbina, J.F. Rivas-Silvac, M. Ávalos-Borja, F.F. Castillón-Barraza, A. Posada-Amarillas. Synthesis of silver nanoparticles in a polyvinylpyrrolidone (PVP) paste, and their optical properties in a film and in ethylene glycol. *Materials Research Bulletin*, 43, 2008, 90–96.
3. S.L.-C. Hsu, R.-T. Wu. Synthesis of contamination-free silver nanoparticle suspensions for micro-interconnect. *Material Letters*, 61, 2007, 3719–3722.
4. H.-H. Lee, K.-S. Chou, K.-C. Huang. Inkjet printing of nanosized silver colloids. *Nanotechnology*, 16, 2005, 2436–2441.
5. S. Magdassi, A. Bassa, Y. Vinetsky, A. Kamyshny. Silver nanoparticles as pigments for water-based ink-jet inks. *Chemistry of Materials*, 15, 2003, 2208–2217.
6. G. Cao. *Nanostructures and Nanomaterials*. Imperial College Press, London, 2004.
7. www.cabot-corp.com.
8. G. Carotenuto, M. Valente, G. Sciumè, T. Valente, G. Pepe, A. Ruotolo, and L. Nicolais. Preparation and characterization of transparent/conductive nano-composites films. *Journal of Materials Science*, 41, 2006, 5587–5592.

14 Fabrication of Nanostructured Thin Films Using Microreactors

Chih-hung Chang, Brian K. Paul, and Si-Ok Ryu

CONTENTS

Among the many and diverse opportunities for embedding nanotechnology within industrial products is the opportunity to assemble nanostructured films from nanomaterial building blocks for many emerging clean energy technologies. Cheap, green, solution-phase oxide, sulfide, and phosphide nanosynthesis chemistries provide opportunities for buffer layers in solar cells [1,2], enhanced catalysis [3] in solar thermal chemical processing, superlubricity [4] in wind turbines, supercapacitance in grid storage batteries [5], and accelerated convection for industrial waste heat recovery [6], among many others. One promising approach is Microreactor-Assisted Nanomaterial Deposition (MAND™) (Figure 14.1), a continuous, liquid-phase alternative to high-temperature, high-vacuum vapor-phase thin-film production such as sputtering and chemical vapor deposition (CVD). MAND has been used by our team to produce high-performance nanoscale coatings on substrates up to 150 mm in dimension with substantially lower processing temperatures.

14.1 MICROREACTOR-ASSISTED NANOMATERIAL DEPOSITION (MAND)

MAND processes combine the merits of microreaction technology with solution-phase nanomaterial synthesis, purification, functionalization, and deposition. MAND architectures are a flexible and versatile nanomanufacturing platform for nanomaterials synthesis and deposition. It can be implemented in various ways for

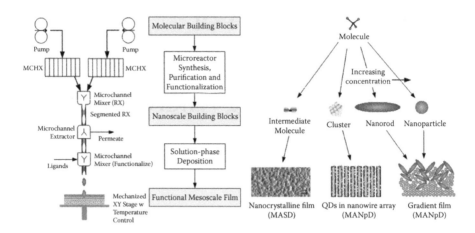

FIGURE 14.1 (Center) MAND architecture: Microreactor technology generates nanoscale building blocks (e.g., monodispersed, functionalized nanoparticle streams; reacting fluxes; etc.) for precise, economical solution-phase deposition. (Right) MASD and MANpD routes to nanostructured films. (Left) Schematic of a reactor for implementing MANpD.

the manufacturing of functional nanostructures. Microreactor-Assisted Solution Deposition (MASD) involves the use of microreactor technology to produce reactive fluxes of short-life, intermediate molecules for heterogeneous growth on a temperature-controlled substrate. Another variant of MAND is Microreactor-Assisted Nanoparticle Deposition (MANpD) involving the use of microreactor technology to implement real-time nucleation, growth, purification, and functionalization of nanoparticles (NPs) for deposition and assembly of NP films and structures. Figure 14.1 illustrates the process concept of MAND and various examples of nanostructured thin films fabricated by the MAND platform.

Current batch processes for NP production are uncoupled from functionalization and deposition processes that can lead to NP agglomeration. Key potential advantages of MANpD over batch synthesis and deposition of NPs include, among others: (1) shorter cycle times, (2) controlled agglomeration, (3) higher conversion of reactants, (4) purer yields, and (5) lower solvent usage. These advantages extend from the accelerated heat and mass transport possible within microchannels, allowing for rapid changes in reaction temperatures and concentrations and more uniform heating and mixing. At Oregon State University and elsewhere, microreactors have been found to yield reductions in the dispersity of nanocrystal size distributions [7] and increases in macromolecular yields [8]. Microreactors also offer to minimize the environmental impact of nanomanufacturing practices through solvent-free mixing, integrated separations, and reagent recycling [9]. Finally, the use of microreactor synthesis at the point of deposition has been demonstrated to eliminate particulate agglomeration/degradation during processing and storage [10].

We have developed several prototypes of continuous flow microreactor systems for the synthesis and deposition of nanomaterials in a number of applications.

14.2 MASD FOR THE DEPOSITION OF NANOCRYSTALLINE THIN FILMS

Chemical bath deposition (CBD) is an aqueous analogue of chemical vapor deposition (CVD). The constituent ions are dissolved in an aqueous solution, and the thin films are produced through a heterogeneous surface reaction. A fundamental understanding of CBD, however, is far less developed than that of CVD. This has limited the development and application of this growth technique. CBD is normally carried out as a batch process in a beaker and involves both heterogeneous and homogeneous reactions. Furthermore, the bath conditions change progressively as a function of time. It is known that CBD is capable of producing an epitaxial layer on a single crystal surface. Many compound semiconductors that are major candidates for solar energy utilization have been deposited by CBD, such as CdSe, Cu_2S, SnO, TiO_2, ZnO, ZnS, ZnSe, CdZnS, and $CuInSe_2$ [11]. Among these, CBD CdS deposition is the most studied CBD process due to its important role in fabricating CdTe and $CuInSe_2$ thin-film solar cells.

The fundamental aspects of CBD are similar to those of the CVD process. It involves mass transport of reactants, adsorption, surface diffusion, reaction, desorption, nucleation, and growth. The earlier studies suggested a colloidal-by-colloidal growth model. However, investigations by Ortega-Borges and Lincot [12], based on initial rate studies using a quartz crystal microbalance, suggested different growth kinetics. They identified three growth regimes: an induction period with no growth observed, a linear growth period, and finally, a colloidal growth period, followed by the depletion of reactants. They proposed a molecular level heterogeneous reaction mechanism given in Equations 14.1 to 14.3.

$$CD(NH_3)_4^{2+} + 2OH^- + Site \underset{k_{-1}}{\overset{k_1}{\rightleftharpoons}} Cd(OH)_{2\,ads} + 4NH_3 \qquad (14.1)$$

$$CD(OH)_{2\,ads} + SC(NH_2)_2 \xrightarrow{k_2} \left[CD\big(SC(NH_2)_2(OH)_2\big)\right]_{ads} \qquad (14.2)$$

$$\left[CD\big(SC(NH_2)_2(OH)_2\big)\right]_{ads} \xrightarrow{k_3} CdS + CN_2H_2 + 2H_2O + Site \qquad (14.3$$

This model has provided a good foundation for an understanding of the CBD process at the molecular level. It is well known that the particle formation plays an important role in the CBD process. Thus there is a need to find a method to decouple the homogeneous particle formation and deposition from the molecular level heterogeneous surface reaction. For this reason, we have developed the MASD process and implemented it first for CBD CdS, and investigated the fundamental reaction kinetics and growth mechanism using MASD as a tool.

A schematic diagram of a laboratory-scale continuous flow microreactor for MASD is shown in Figure 14.2. The reactant streams A and B were dispensed through two syringe pumps and allowed to mix in an interdigital micromixer [13].

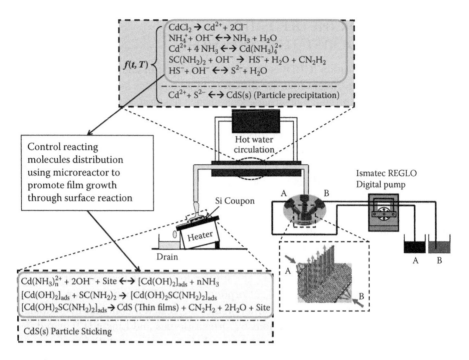

FIGURE 14.2 A schematic diagram of a continuous flow microreactor for CBD. (From Y.J. Chang et al., *Electrochem. Solid-State Lett.,* 12(7): H244–H247 (2009). With permission.)

Stream A consists of $CdCl_2$, NH_4OH, and NH_4Cl, and stream B consists of thiourea. A detailed schematic diagram of an interdigital micromixer is shown in the inset. Fluids A and B to be mixed are introduced into the mixing element as two counterflows and enter interdigital channels (~20–50 μm), then split into many interpenetrated substreams. The substreams leave the interdigital channel perpendicular to the direction of the feed flows, initially with a multilayered structure. Fast mixing through diffusion will soon follow due to small thickness of the individual layers. The resulting mixture from the micromixer then is passed through a temperature-controlled microchannel before it is delivered to a temperature-controlled substrate. The homogeneous chemistry of the reacting flux is controlled precisely by the inlet concentrations, temperature, and residence time, as illustrated in Figure 14.2.

Previous results indicated that for CBD CdS deposition, small particles were forming and growing even at the beginning of the deposition process, as supported by real-time dynamic light scattering measurements and transmission electron microscopy (TEM) characterization [12]. We have observed a similar result using the continuous flow microreactor. Experiments were carried out by preheating the precursor solutions (streams A and B) at 80°C. At this temperature, thiourea releases more sulfide ions through hydrolysis. Free sulfide ions react with free cadmium ions to form CdS particles at these operation conditions. The source chemicals were maintained at room temperature before they entered the micromixer in order to obtain a reacting flux without the homogeneous particle formation. The mixed reactants

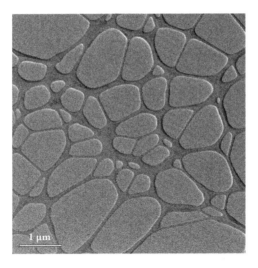

FIGURE 14.3   TEM evidence showing no nanoparticle formation in the chemical solution.

were maintained at 80°C using heat exchanging fluid from a constant temperature circulator. TEM samples were obtained by collecting drops of hot solution from the PEEK tube on the lacey carbon-coated TEM copper grid. TEM images (Figure 14.3) indicate that at very short residence times (e.g., 1 s), there was no evidence of particle formation on the surface of the grid under these processing conditions. This suggests the reaction was carried out within the induction timescale for homogeneous particle formation. Using this particle-free flux, study of the CdS deposition kinetics was simplified by focusing on the heterogeneous surface reaction through a molecule-by-molecule growth mechanism.

A series of CdS thin-film deposition experiments at different residence times (1, 3.5, 7, 35, and 70 s) were performed. The film thickness was determined by a surface profiler. Figure 14.4(a) shows the deposited CdS thin-film thickness versus the deposition time at different residence times. These growth rate results in Figure 14.4(b) clearly indicate that a lower CdS thin-film growth rate was obtained (~77 Å/min) when a 1 s residence time was used. The growth rate increases significantly (about four times higher) when a 3.5 s residence time was used compared to a 1 s time. The thin-film growth rate increases gradually from 3.5 s to 35 s residence time. However, when a 70 s residence time was applied, a decrease of the growth rate was observed (due to particle formation). This deposition result suggests that ions formed through the thiourea hydrolysis reaction are the dominant sulfide ion source responsible for the CdS deposition, rather than thiourea itself, which had been widely discussed in almost all of the previous literature. This finding could not be observed previously by a conventional CBD batch setup because all the reactant solutions were sequentially pulled into the reaction beaker and mixed all at once. These results also demonstrate the capability of MASD in controlling the reactant flux through varying the initial concentration and residence time for continuous growth of thin films. It is not possible to continuously grow high-quality films using the batch CBD process since the reactant flux is changing as a function of time and depletes in limited time.

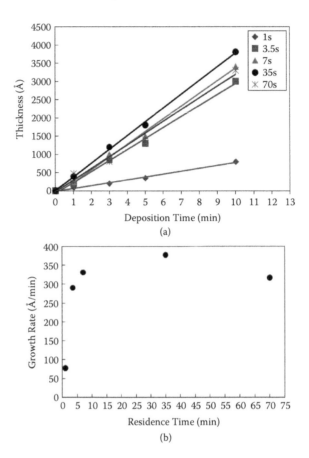

FIGURE 14.4 CdS thin-film thickness versus deposition time using flux at different residence times.

Using the MASD reactor, we obtained a particle-free flux that is capable of depositing high-quality CdS thin films. We were able to deposit smooth, dense, and highly oriented nanocrystalline CdS thin films in shorter time than typical CBD approaches [14]. Figure 14.5 shows the AFM images of the CBD-deposited CdS surfaces for a (a) continuous flow microreactor and (b) batch process. These images demonstrate the capability of using the continuous flow microreactor to obtain a dense and uniform deposition [15]. The x-ray diffraction (XRD) pattern given in Figure 14.5(c) clearly shows a highly (111)-oriented cubic CdS thin film produced by the continuous flow microreactor at low temperature.

Enhancement-mode CdS thin film transistors (TFTs) were fabricated using this continuous flow microreactor at low temperature (80–90°C) without any post-deposition annealing [16]. The schematic diagram of the transistor and the transfer characteristics are shown in Figure 14.6. An effective mobility, $\mu_{eff} \cong 1.5$ cm^2/V-s, and an on-off ratio of 10^5 were obtained from these devices. These values verify that the quality of films deposited by MASD at low temperature is suitable for electronic applications. This new approach could be adopted for low-temperature deposition of

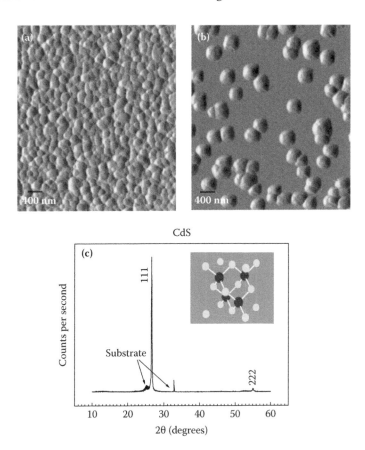

FIGURE 14.5 AFM images of the CBD-deposited CdS surfaces for (a) MASD, (b) batch process, and (c) XRD pattern for CdS film deposited by MASD. (From P.-H. Mugdur et al., *J. Electrochem. Soc.*, 154(9): D482–D488 (2007). With permission.)

other compound semiconductor thin films using chemical solution deposition with better control over processing chemistry and reduced waste production.

Using the continuous flow microreactor, we were able to deposit uniform, dense, and highly oriented CdS thin films in shorter time with excellent conformal coverage, even on 6-inch highly textured substrates using a particle-free solution (see Figure 14.7a). Figure 14.7b shows a cross-sectional transmission electron microscopy image of nanocrystalline CdS thin film deposited on a highly textured fluorine-doped tin oxide (FTO) substrate using our scale-up continuous flow microreactor [17].

McPeak and Baxter [18] demonstrated the deposition of dense arrays of well-aligned, single-crystal ZnO nanowires using a continuous flow microreactor (Figure 14.8). They utilized the spatial resolution of the microreactor to enable rapid and direct correlation of material properties to growth conditions. They observed that nanowire lengths decreased, morphology changed from pyramidal tops to flat tops, and the growth mechanism transitioned from two-dimensional nuclei to spiral growth.

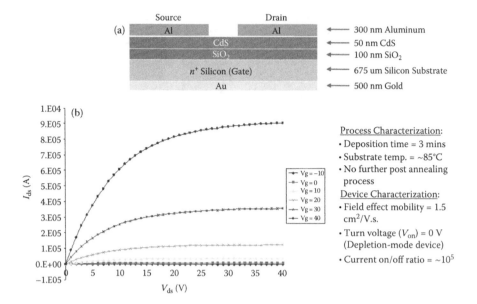

FIGURE 14.6 (a) CdS thin-film transistor fabricated by MASD at low temperature. (a) Schematic diagram of the CdS TFT. (b) Transfer characteristics (I_{ds}-V_{ds} at different V_g values) curve of the CdS TFT. (From Y.-J. Chang et al., *Electrochem. Solid-State Lett.*, 9(5): G174–G177 (2006). With permission.)

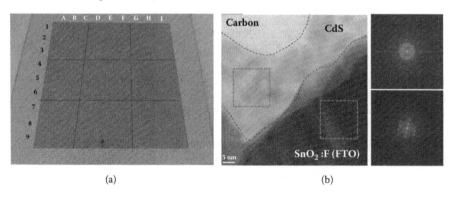

FIGURE 14.7 CdS thin film deposited by MASD on 6-inch FTO-coated glass fabricated by MASD at low temperature. (a) Optical image. (b) Cross-sectional high resolution transmission electron microscopy (HRTEM) image and electron diffraction pattern.

14.3 MAND FOR THE DEPOSITION OF NANOSTRUCTURED THIN FILMS

Another variant of MAND is Microreactor-Assisted Nanoparticle Deposition (MANpD) involving the use of microreactor technology to implement real-time nucleation, growth, purification, and functionalization of NPs for the deposition and assembly of NP films and structures. We have also used this approach to deposit

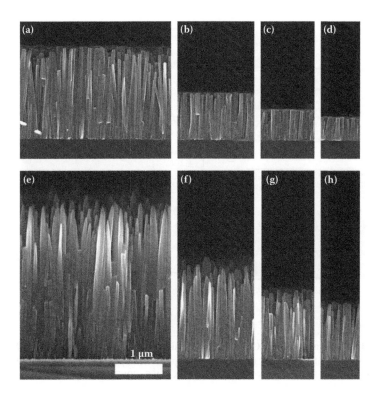

FIGURE 14.8 Cross-sectional SEM images of ZnO nanowires grown at flow rates of 0.72 ml/h (top row) and 2.88 ml/h (bottom row) with an equimolar inlet concentration of 0.025 M zinc nitrate and hexamethylenenetetramine (HMT). Images were taken at positions (a, e) 0, (b, f) 6, (c, g) 12, and (d, h) 18 mm downstream from the inlet. Scale bar applies to all panels. (From K.M. McPeak and J.B. Baxter, *Crystal Growth Des.*, 9: 4538–4545 (2009). With permission.)

various nanostructured thin films. This technique follows a thin-film growth mechanism based on nanoparticle formation and deposition. We have fabricated a nanoporous ZnO film [19] that is highly transparent using MANpD. Figure 14.9a shows a top-view SEM image of a ZnO film deposited by MANpD. The film consists of uniformly distributed nanoparticles and nanopores. The cross-sectional image shows a uniform nanoporous film with a thickness around 24 nm (see Figure 14.9b). The nanoporous ZnO thin film is highly transparent with an optical bandgap of 4.35 and 3.27 eV before and after thermal annealing, respectively (see Figure 14.9c).

A variety of nanostructures could be fabricated using MAND by changing the process parameters. For example, we were able to produce a variety of ZnO structures using our continuous flow microreactor by varying the concentration of NaOH [20]. In this experiment, 0.05 M zinc acetate dehydrate and 0.25 M ammonium acetate were mixed with the four different concentrations of NaOH varying from 0.005 M to 0.15 M. The SEM images of the synthesized ZnO at the four different NaOH concentrations are shown in Figure 14.10. It can be clearly observed that the concentration of NaOH exerted a great influence on the morphology of the synthesized ZnO. From Figure 14.10, several types of structures can be clearly seen from

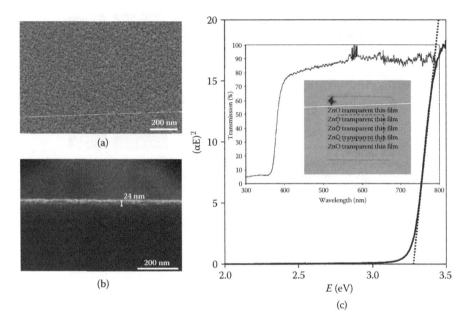

FIGURE 14.9 Nanoporous ZnO thin films deposited by MANpD. (a) Top-view SEM image. (b) Cross-sectional SEM image. (c) UV-Vis absorption and optical image (inset). (From S.-Y. Han et al., *Electrochem. Solid-State Lett.*, 10(1): K1–K5 (2007). With permission.)

the SEM images. The flower-like ZnO structure was fabricated on the substrate at a higher NaOH concentration, while the chrysanthemum-like ZnO structure was grown at a lower one. The petals of the flower-like ZnO structures became narrower in width, smaller in size, sharper in tip, and simpler in shape as the concentration of NaOH was decreased from 0.15 M to 0.05 M. Meanwhile, the synthesized ZnO has chrysanthemum-like ZnO structures if the synthesis of the ZnO microstructure was carried out at a concentration of NaOH less than 0.05 M. The synthesized chrysanthemum-like ZnO structure at 0.005 M NaOH has a lot of petals, and its petals are narrow and nanosized. These assembled ZnO structures have been deposited onto a substrate.

Figure 14.11(a) shows the SEM image of a flower-like ZnO structure, which was synthesized by delivering the mixed solution on the substrates for 1 min. Figure 14.10(b) shows the SEM image of a ZnO structure after 10 min of depositions. It can be seen clearly that the size of the flower-like ZnO structures increases as a function of deposition time. The conventional batch CBD ZnO process would produce films from large aggregates that exhibit coarse morphologies that dominate homogeneous reactions. The conventional batch CBD ZnO process could also form a mixture of heterogeneously grown ZnO columns or rods, along with homogeneously grown large aggregated particles. In addition, the growth rate of the solid phase on the substrates decreases rapidly as the time duration for the reaction is prolonged. The deposited ZnO structures continue to grow in the MAND method because a fresh and constant solution is continuously fed to the active sites of ZnO.

FIGURE 14.10 SEM images of flower-like ZnO structures synthesized with four different concentrations of NaOH at 90°C for 5 min: (a) 0.005 M, (b) 0.01 M, (c) 0.1 M, and (d) 0.15 M. (From J.Y. Jung, N.-K. Park, S.-Y. Han, G.B. Han, T.J. Lee, S.O. Ryu, C.-H. Chang, The growth of the flower-like ZnO structure using a continuous flow microreactor, *Curr. Appl. Phys.*, 8(6): 720–724. 2008. With permission.)

These flower-like nanostructures show enhanced pool boiling critical heat fluxes (CHFs) at reduced wall superheat [6]. In particular, we observed a pool boiling CHF of 82.5 W/cm^2 with water as the fluid for ZnO on Al versus a CHF of 23.2 W/cm^2 on a bare Al surface with a wall superheat reduction of 25–38°C. These CHF values on ZnO surfaces correspond to a heat transfer coefficient of ~23,000 W/m^{2}K.

14.4 DEPOSITION OF DENDRONS

Dendrimers are highly branched, nanometer-sized molecules with fascinatingly symmetrical fractal morphologies. The word *dendrimer* (coined by Tomalia et al. [21]) is derived from the Greek words *dendri* (branch, tree-like) and *meros* (part of). Dendrimers consist of a core unit, branching units, and end groups located on their peripheries. Their dendritic architecture presents great potential for a wide variety of applications. For example, they hold great promise as building blocks for complex supramolecular structures with specifically designed functions. They

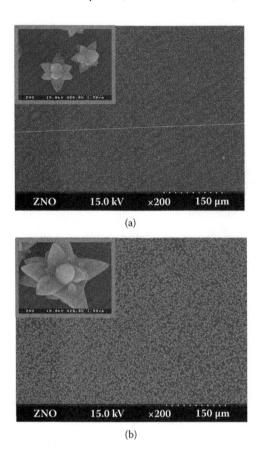

FIGURE 14.11 SEM images of flower-like ZnO structures synthesized at 90°C (0.1 M NaOH) (a) for 1 min and (b) for 10 min. (From J.Y. Jung, N.-K. Park, S.-Y. Han, G.B. Han, T.J. Lee, S.O. Ryu, C.-H. Chang, The growth of the flower-like ZnO structure using a continuous flow microreactor, *Curr. Appl. Phys.*, 8(6): 720–724. 2008. With permission.)

can be considered versatile nanoscale components for building nanoscale structures. Dendrimers can act as nanoscale carrier molecules in drug delivery, where nanoparticles and nanocapsules are gaining popularity. Structural variety, yielding molecules having differing optical, electrical, and chemical properties, makes dendrimers potentially even more versatile than the alternatives. The molecules can be assembled with startling precision, a necessity when the goal is construction of nanoscale structures or devices with sophisticated and complex functionality. Along with targeting tumor cells and drug delivery systems, dendrimers have shown promising results as tools for MRI [22–24] and gene transfer techniques. Also, dendrimer-based nanocomposites are being studied as possible antimicrobial agents to fight *Staphylococcus aureus*, *Pseudomonas aeruginosa*, and *Escherichia coli* [25]. Dendrimers have been shown to act as scavengers of metal ions, offering the potential to meet environmental cleanup needs. Their size allows them to be filtered out post-extraction using common ultrafiltration techniques.

The synthesis of dendrimer is precisely controlled, so a certain generation of dendrimer is characterized with a single size and molecular weight. This property distinguishes dendrimers from typical polymers.

The general synthetic strategy of dendrimers includes divergent and convergent approaches. The divergent approach, arising from the independent discovery in parallel works from Buhleier et al. [26] and Tomalia et al. [27], initiates growth at the core of the dendrimer and continues outward by several repetitions of activation and coupling steps. Convergent synthesis, first reported by Fréchet et al. in 1989 [28], initiates growth from the exterior of the molecule, and progresses inward by coupling focal point to each branch of the monomer. Usually, the divergent approach generates more side products with uncompleted addition of branches due to the steric hindrance of multifunctional reaction sites on the surface. This will result in difficulty for the separation of products. On the other hand, the convergent approach will provide a much purer product, because only one reaction site exists for each reactant molecule.

For dendrimers to realize their full potential, methods must be developed in the production of these macromolecules. We have demonstrated a convergent approach to the synthesis of polyamide dendrons and dendrimers, and the deposition of the G1 dendron on the aminosilanized glass surfaces, by using a continuous flow microreactor. The microreactor has proven to be an effective tool to synthesize the polyamide dendrons and dendrimers. The microreactor demonstrated several advantages over a conventional batch reactor. One of the most attractive advantages is that the reaction time was reduced tremendously from a few hours to seconds or minutes [29]. Therefore the continuous flow microreactor could potentially reduce the production cost, which will enhance the opportunities for the large-scale application of dendrimers in various fields. In addition, the required reaction conditions are easier to implement.

We have demonstrated an approach that used the microreactor to activate and deposit G1 dendrons directly on the silanized glass slides [30]. The process scheme is illustrated in Figure 14.12. A solution containing dendron G1 and another solution containing thionyl chloride were mixed through the micromixer. The activated G1 dendron from the microreactor was impinged onto a spinning silanized glass slide that was mounted on a heated rotating disk substrate holder. The entire deposition process using the microreactor took about 2 min, and the residence time in the outlet tubing was 17 s. Figure 14.13 shows the time-of-flight secondary ion mass spectrometry (TOF-SIMS) spectrum and corresponding ion images of the deposited dendrons.

14.5   CONCLUSION AND FUTURE DIRECTION

Microreactor-Assisted Nanomaterial Depostion (MAND) processes combine the merits of microreaction technology with solution-phase nanomaterial synthesis, purification, functionalization, and deposition. MAND architectures are a flexible and versatile nanomanufacturing platform for nanomaterial synthesis and deposition. Microreactors offer exciting opportunities for high levels of process control over nanomaterial synthesis via precise and rapid changes in reaction conditions.

Microreactor-Assisted Solution Deposition (MASD) involves the use of microreactor technology to produce reactive chemical fluxes of short-life, intermediate molecules for heterogeneous growth on a temperature-controlled substrate. MASD

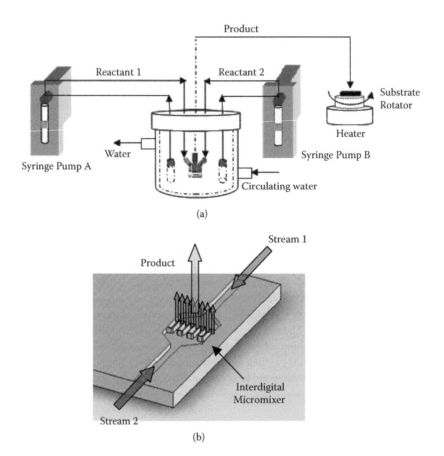

FIGURE 14.12 (a) A schematic diagram of the experimental setup. (b) A schematic diagram of the interdigital micromixer. (From S.-H. Liu and C.-H. Chang, *Chem. Eng. Technol.*, 30(3): 334–340 (2007). With permission.)

is a new approach that could be adopted for many chemical solution deposition processes. It could enable low-temperature deposition of many compound semiconductor thin films. We have successfully used MASD approaches for the deposition of CuS, CuSe [31], CuInS$_2$ [32], and CuInSe$_2$ [33]. Another variant of MAND is Microreactor-Assisted Nanoparticle Deposition (MANpD) involving the use of microreactor technology to implement real-time nucleation, growth, purification, and functionalization of NPs for deposition and assembly of NP films and structures. A variety of nanostructured thin films could be fabricated using MANpD. For example, nanoporous ZnO, flower-like ZnO, and ZnO nanorod arrays were fabricated by the MANpD technique and used as a channel layer for thin-film transistors [10], as a nanotextured surface for enhanced boiling [6], and as an antireflection coating layer for solar cells [34], respectively. In addition to inorganic materials, MAND is also suitable for the synthesis and deposition of organic nanomaterials. We have demonstrated the feasibility of microreactor-based continuous synthesis and deposition of dendrimers.

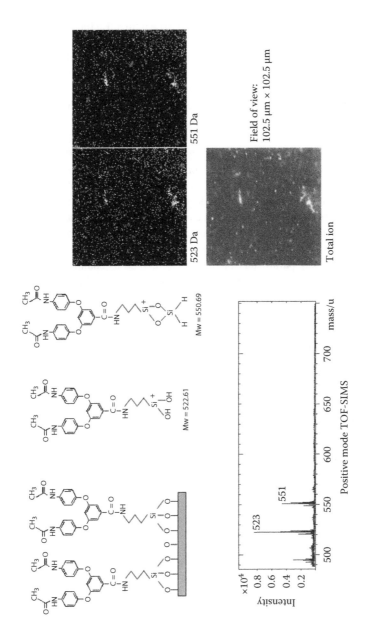

FIGURE 14.13 TOF-SIMS schematic diagrams of the dendron deposition process via a microreactor.

Rapid, continuous flow, high yield and selectivity, and most importantly, a facility for numbering up the process for industrial production scale, MAND opens the possibility to scale the production of nanomaterials. One approach to the high-production scaling of nanomaterial deposition is to use an equal-down/equal-up approach [35]. This approach starts from the process or product to be realized at the commercial scale, where the main requirements and parameters are identified for successful application in the marketplace. These key requirements (e.g., film properties, cycle times, etc.) are specified by industrial partners and used to specify design goals for laboratory-scale devices in an equal-down step. The design goals provide the demonstration requirements necessary to enable scale-up. The reaction conditions of the laboratory design are designed to be *equal* to the commercial-scale design with respect to the governing heat transfer, mass transfer, and reaction kinetics, leading to similarity with respect to key reactor geometries and materials, fluid dynamics, mixing methods, reaction engineering approaches, and thermal management strategies. A consequence of developing the laboratory-scale reactor and process chemistry, which controls the microscale heat transfer, mass transfer, and reaction kinetics, is the definition of the shape and structure of the active unit reactor cell that can be replicated to produce higher chemical production volumes. Unit cell results obtained from the laboratory-scale demonstration are then used for the detailed design of the commercial-scale reactor in an equal-up step. The key parameters of the unit cell, such as channel width, channel length, and modified residence time, are made to be the same in the commercial-scale reactor, with the only difference being the number of channels, the size of headers, and perhaps, the techniques used to fabricate the reactor.

ACKNOWLEDGMENTS

The authors acknowledge financial support from the W.M. Keck Foundation, National Science Foundation CAREER CTS-0348723, CBET-0654434, Department of Energy's Office of Energy Efficiency and Renewable Energy (EERE), Industrial Technology Program (ITP), Nanomanufacturing Activity through award NT08847 DOE ITP, instrumentation equipment grant from the Murdock Charitable Trust (2010004), and Oregon Nanoscience and Microtechnology Institute (ONAMI).

REFERENCES

1. A. Ennaoui, M. Weber, R. Scheer, and H.J. Lewerenz. Chemical-bath ZnO buffer layer for CuInS sub(2) thin-film solar cells. *Sol. Energ. Mater. Sol. Cell*, 54(1/4): 277–286 (1998).
2. C.H. Huang, S.S. Li, W.N. Shafarman, C.-.H Chang, E.S. Lambers, and L. Rieth. Study of Cd-free buffer layers using Inx(OH,S)y on CIGS solar cells. *Sol. Energ. Mater. Sol. Cells.*, 69(2): 131–137 (2001).
3. C.H.T. Tseng, B.K. Paul, C.-H. Chang, and M.H. Engelhard. Continuous precipitation of ceria nanoparticles from a continuous flow micromixer. *Int. J. Adv. Manuf. Technol.*, accepted (2012).
4. J.J. Hu and J.S. Zabinski. Nanotribology and lubrication mechanisms of inorganic fullerene-like MoS$_2$ nanoparticles investigated using lateral force microscopy (LFM). *Tribology Letters*, 18(2): 49–57 (2005).

5. S. Stewart, P. Albertus, V. Srinivasan, I. Plitz, N. Pereira, G. Amatucci, and J. Newman. Optimizing the performance of lithium titanate spinel paired with activated carbon or iron phosphate. *J. Electrochem. Soc.*, 155(3): A253–A261 (2008).

6. T.J. Hendricks, S. Krishnan, C. Choi, C.-H. Chang, and B.K. Paul. Enhancement of pool boiling heat transfer using nanostructured surfaces on aluminum and copper. *Int. J. Heat Mass Transfer*, 53(15–16): 3357–3365 (2010).

7. S. Krishnadasan, J. Tovilla, R. Vilar, A.J. deMello, and J.C. deMello. On-line analysis of CdSe nanoparticle formation in a continuous flow chip-based microreactor. *J. Mater. Chem.*, 14: 2655 (2004); E.M. Chan, R.A. Mathies, and A.P. Alivisato. Size-controlled growth of CdSe nanocrystals in microfluidic reactors. *Nano Letters*, 3: 199–201 (2003); H. Nakamura, Y. Yamaguchi, M. Miyazaki, H. Maeda, M. Uehara, and P. Mulvaney. Preparation of CdSe nanocrystals in a micro-flow-reactor. *Chem. Commun.*, 2844–2845 (2002); B.K.H. Yen, N.E. Stott, K.F. Jensen, and M.G. Bawendi, A continuous-flow microcapillary reactor for the preparation of a size series of CdSe nanocrystals. *Adv. Mater.*, 15: 1858 (2003).

8. S. Liu, C. Chang, B.K. Paul, and V.T. Remcho. Convergent synthesis of polyamido dendrimer using a continuous flow microreactor. *Chem. Eng. J.*, 135(1): S333–S337 (2008); S.-H. Liu, C.-H. Chang, B.K. Paul, V.T. Remcho, and B. Abhinkar. Convergent synthesis of dendrimers using continuous flow microreactors. Presented at AIChE Spring National Meeting, Orlando FL, April 24 (2006); C.-H. Chang, V.T. Remcho, B.K. Paul, et al. Progress towards chip-based high-throughput dendrimer synthesis. Presented at IMRET 8, Atlanta, GA (2005).

9. C.-H. Chang, B.K. Paul, V.T. Remcho, S. Atre, and J.E. Hutchinson. Synthesis and post-processing of nanomaterials using microreaction technology. *J. Nanoparticle Res.*, 10(6): 965–980 (2008).

10. S.-Y. Han, Y.-J. Chang, D.-H. Lee, S.-O. Ryu, T.J. Lee, and C-H. Chang. Chemical nanoparticle deposition of transparent ZnO thin films. *Electrochem. Solid-State Lett.*, 10(1): K1–K5 (2007).

11. G. Hodes. *Chemical Solution Deposition of Semiconductor Films*. Marcel Dekker, New York, 2003.

12. R. Ortega-Borges and D. Lincot. Mechanism of chemical bath deposition of cadmium-sulfide thin-films in the ammonia-thiourea system—In-situ kinetic study and modelization. *J. Electrochem. Soc.*, 140(12): 3464–3473 (1993).

13. H. Löwe, W. Ehrfeld, V. Hessel, T. Richter, and J. Schiewe. Micromixing technology. In *Proceedings of the 4th International Conference on Microreaction Technology (IMRET 4)*, 2000, pp. 31–47.

14. Y.-J. Chang, Y.-W. Su, D.-H. Lee, S.-O. Ryu, and C.-H. Chang. Investigate the reacting flux of chemical bath deposition by a continuous flow microreactor. *Electrochem. Solid-State Lett.*, 12(7): H244–H247 (2009).

15. P.-H. Mugdur, Y.-J. Chang, S.-Y. Han, A.A. Morrone, S.-O. Ryu, T.J. Lee, and C.-H. Chang. A comparison of chemical bath deposition of CdS from a batch reactor and a continuous flow microreactor. *J. Electrochem. Soc.*, 154(9): D482–D488 (2007).

16. Y.-J. Chang, P.-H. Mugdur, S.-Y. Han, A.A. Morrone, S.-O. Ryu, T.J. Lee, and C.-H. Chang. Nanocrystalline CdS MISFETs fabricated by a novel continuous flow microreactor. *Electrochem. Solid-State Lett.*, 9(5): G174–G177 (2006).

17. B. K. Paul, C.L. Hires, Y.-W. Su, C.-H. Chang, S. Ramprasad and D. Palo, A uniform residence time flow cell for the microreactor-assisted solution deposition of CdS on a FTO-glass substrate, *Crystal Growth Des.* (2012), in press.

18. K.M. McPeak and J.B. Baxter, ZnO Nanowires grown by chemical bath deposition in a continuous flow microreactor, *Crystal Growth Des,* 9: 4538–4545, (2009).

19. S.-Y. Han, Y.-J. Chang, D.-H. Lee, S.-O. Ryu, T.J. Lee, and C-H. Chang. Chemical nanoparticle deposition of transparent ZnO thin films. *Electrochem. Solid-State Lett.*, 10(1): K1–K5 (2007).

20. J.Y. Jung, N.-K. Park, S.-Y. Han, G.B. Han, T.J. Lee, S.O. Ryu, C.-H. Chang, The growth of the flower-like ZnO structure using a continuous flow microreactor, *Curr. Appl. Phys.*, 8(6): 720–724 (2008).

21. D.A. Tomalia, J.R. Dewald, M.R. Hall, S.J. Martin and P.B. Smith, *Preprints of the 1st SPSJ International Polymer Conference,* Society of Polymer Science Japan, Kyoto, 1984; p 65.

22. C. Wu, M.W. Breshbiel, R.W. Kozack, and O.A. Gansow. Metal-chelate-dendrimer-antibody constructs for use in radioimmunotherapy and imaging. *Bioorg. Med. Chem. Lett.*, 4(3): 449–454 (1994).

23. L.D. Margerum, B.K. Campion, M. Koo, N. Shargill, J. Lai, A. Marumoto, and P.C. Sontum. Gadolinium(III) DO3A macrocycles and polyethylene glycol coupled to dendrimers—Effect of molecular weight on physical and biological properties of macromolecular magnetic resonance imaging contrast agents. *J. Alloys Comp.*, 249(1–2): 185 (1997).

24. Y. Kim and S.C. Zimmerman. Applications of dendrimers in bio-organic chemistry. *Curr. Opin. Chem. Biol.*, 2(6): 733–742 (1998).

25. A.R. Menjoge, R.M. Kannan, D.A Tomalia, Dendrimer-based drug and imaging conjugates: design considerations for nanomedical applications, *Drug Discovery Today,* 15(5–6): 171–185 (2010).

26. E. Buhleier, W. Wehner, and F. Vögle. Cascade-chain-like and nonskid-chain-like syntheses of molecular cavity topologies. *Synthesis*, 2: 155 (1978).

27. D.A. Tomalia, H. Baker, J. Dewald, M. Hall, G. Kallos, S. Martin, J. Roeck, J. Ryder, and P. Smith. A new class of polymers—Starburst-dendritic macromolecules. *Polym. J.*, 17(1): 117 (1985).

28. J.M.J. Fréchet, Y. Jiang, C.J. Hawker, and A.E. Philippides. In *Proceedings of IUPAC International Symposium on Macromolecules,* Seoul, 1989, pp. 19–20.

29. S.-H. Liu, C.-H. Chang, B.K. Paul, and V.T. Remcho. Convergent synthesis of polyamido dendrimer using a continuous flow microreactor. *Chem. Eng. J.*, 135S: S333–S337 (2008).

30. S.-H. Liu and C.-H. Chang. High rate convergent synthesis and deposition of polyamide dendrimers. *Chem. Eng. Technol.*, 30(3): 334–340 (2007).

31. C.R. Kim, S.Y. Han, C.H. Chang, T.J. Lee, and S.O. Ryu. A study on copper selenide thin films for photovoltaics by a continuous flow microreactor. *Mol. Crystals Liquid Crystals*, 532: 455–463 (2010).

32. M.S. Park, S.-Y. Han, E.J. Bae, T.J. Lee, C.H. Chang, and S.O. Ryu. Synthesis and characterization of polycrystalline CuInS2 thin films for solar cell devices at low temperature processing conditions. *Curr. Appl. Phys.*, 10(3): S379–S382 (2010).

33. C.R. Kim, S.Y. Han, C.H. Chang, T.J. Lee, and S.O. Ryu. Synthesis and characterization of CuInSe$_2$ thin films for photovoltaic cells by a solution-based deposition method. *Curr. Appl. Phys.*, 10(3): S383–S386 (2010).

34. K. Mullins. Moth eyes inspire better optics. Northwest Science and Technology, Spring 2010.

35. F. Becker, J. Albrecht, G. Markowz, R. Schütte, S. Schirrmeister, K.J. Caspary, H. Döring, T. Schwartz, and E. Schüth. DEMiS: Results from the development and operation of a pilot-scale microreactor on the basis of laboratory measurements. IMRET 8, TK 131 f.2005.

Index

Printed and bound by CPI Group (UK) Ltd, Croydon, CR0 4YY

18/10/2024

01776261-0012